Contributions in Petroleum Geology & Engineering 3

Underground Storage of Natural Gas

This series includes:

Volume 1: Geologic Analysis of Naturally Fractured Reservoirs

Volume 2: Applied Open-Hole Log Analysis

Volume 3: Underground Storage of Natural Gas

Volume 4: Gas Production Engineering

Gulf Publishing Company
Book Division
Houston, London, Paris, Tokyo

Contributions in Petroleum Geology & Engineering 3

Underground Storage of Natural Gas

M. R. Tek

Contributions in Petroleum Geology & Engineering

Series Editor: George V. Chilingar, University of Southern California

Volume 3:
Underground Storage of Natural Gas

Library of Congress Cataloging-in-Publication Data

Tek, M. R. (Mehmet Rasin), 1927–
Underground storage of natural gas.

Includes bibliographies and index.
1. Gas, Natural—Underground storage. I. Title.
TP756.5.T437 1987 665.7′42 86-32017

ISBN 0-87201-913-6

CONTENTS

to Gretchen

PREFACE

Underground Storage of Natural Gas is the result of the author's work in the storage industry for more than 25 years. This interaction has been in the form of teaching, research, and consulting service.

The book is intended primarily for the practicing storage engineer. In addition, geologists, site/facility managers, and energy administrators, as well as educators and students, will find this a complete and indispensable reference.

An intensive summer course in underground storage, held annually in Ann Arbor, Michigan, and at times in Europe and Australia, had much to do with the assembly and specific sequence of the topics presented. Over many years, the research conducted at the University of Michigan on various aspects of underground storage provided part of the basis for the new data, procedures, and concepts presented here.

The book starts by focusing on properties of storage systems, then proceeds with the theory of their performance, leading to concepts and procedures for design. The day-to-day operation of storage fields is outlined in clear, practical terms, followed by chapters on what has been learned in research.

The chapters on movement of water, threshold pressure, and handling of liquids address topics of longstanding interest in storage technology. Other topics, such as numerical simulation of storage, are included as subjects to which undoubtedly much will be added in the coming years to augment our present understanding and capabilities.

Finally, two case histories in underground storage conclude the book as useful examples of the real storage world.

M. R. Tek

ACKNOWLEDGMENTS

Many individuals from industry and academia as well as friends and family not only made this book possible, but helped shape both its intent and content. Their support, encouragement, and personal interest are gratefully acknowledged. Former students, personal friends, company executives, engineers and associates, too numerous for a specific list, all contributed and all know who they are. I would, however, be remiss if I did not acknowledge specifically five individuals who influenced and inspired me to the greatest extent throughout the project. These are Karl Hachmuth, former executive engineer with Phillips Petroleum; Donald Van Curler, Architect; O. C. Davis, Chairman Midcontinent Corp.; and John B. Holstead, Partner of Vinson and Elkins, Attorneys in Houston, Texas—all four special friends. The fifth is my wife, Margaret Gretchen Tek, who provided invaluable help with the word processing and technical illustrations.

1

HISTORY AND TECHNOLOGY

Underground storage is the uniquely efficient process that matches the constant supply from long-distance pipelines to the variable demand of markets, which are subject to weather, for engineering and economic advantage.

The overall well-being of the entire Western World depends critically upon having sufficient energy resources, and the status of natural gas reserves and supply must be in the forefront of any long-range energy planning effort. Of over 80 quads consumed worldwide each year in energy, about 30% comes from natural gas, a figure only exceeded by oil (see Figure 1-1).[1]

With the recent declining outlook on oil and increasing prospects for pipeline-accessible gas in many areas of the world, development of new storage horizons near the markets—as well as effective use of existing storage resources—continues to be a priority, emphasizing the need for better understanding of storage performance.

In underground storage, natural gas is injected into subsurface reservoirs when market demand falls below the supply available from the pipeline. It is withdrawn from storage to supplement the steady supply from the pipeline when demand exceeds that supply. Because demand fluctuates with temperature in the residential/commercial markets of regions with intemperate climates, underground storage is not only desirable, but an economic necessity for efficient use of the pipelines. It permits high load factors in utilization of pipeline facilities and effective delivery during peak demand periods. Underground storage also indirectly helps conserve oil by utilizing associated solution gas which, otherwise, would have to be flared.

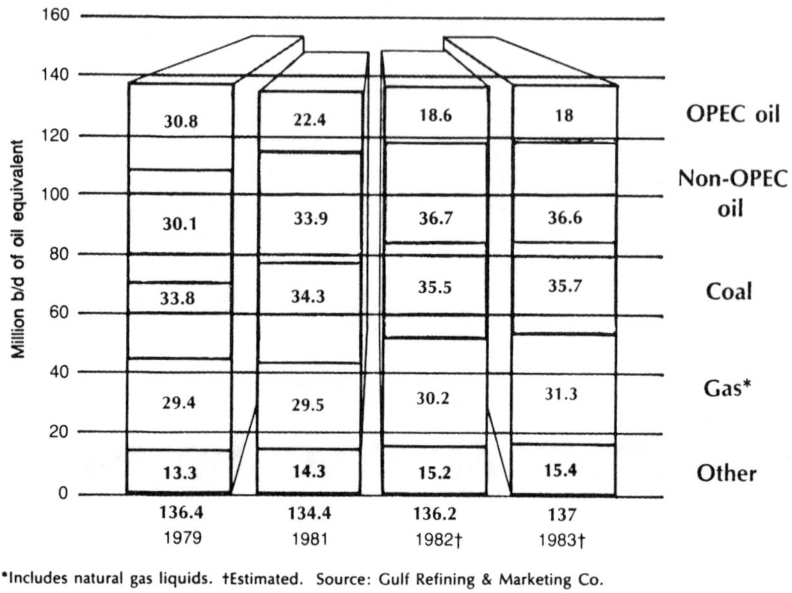

Figure 1-1. World primary energy consumption. Reprinted from International Petroleum Encyclopedia 1984, Pennwell Publishing Co. Data from Gulf Refining and Marketing Co.

The use of underground storage, which was first introduced in 1915, has experienced remarkable growth, resulting in about 8 trillion standard cubic feet of gas presently in storage. There now exist more than 400 pools in 26 states.

Table 1-1[2] shows the energy consumption and fuel shares from 1960 through 1978 in the free world. It is interesting to note that between 1960 and 1973, while the world consumption of energy doubled, the U.S. percentage share decreased from nearly 50% to 42%.

Table 1-2 shows the world natural gas production for 1974 through 1980. These figures are quite indicative of late trends which have resulted in the recent revival of interest in pipeline technology, underground storage, and conservation of methane, considered by many to be the premier fuel. This interest extends from Europe and the Middle East to China and the Far East.

ASPECTS OF ENERGY RELATED TO GAS STORAGE

Energy may be defined as *ability to do work*. Interchangeability between work in various forms and heat is an important subject in thermodynamics.

(text continued on page 7)

Table 1-1
World Energy Consumption and Fuel Shares: History, 1960, 1973, and 1978 (quadrillion Btu)

Region or Country	1960					1973					1978				
	Total Energy Consumed	Fuel Shares (Percent)				Total Energy Consumed	Fuel Shares (Percent)				Total Energy Consumed	Fuel Shares (Percent)			
		Coal	Oil	Gas	Other		Coal	Oil	Gas	Other		Coal	Oil	Gas	Other
United States[a]	44.2	23	45	28	4	75.1	18	47	30	5	78.8	18	49	25	8
Canada	3.9	14	49	10	27	8.1	8	46	20	26	9.0	6	42	22	30
Japan	3.9	48	36	1	15	15.3	14	79	2	5	14.9	13	73	5	9
Western Europe[b]	26.9	55	35	2	8	54.0	19	63	10	8	54.7	19	56	14	11
Finland/Norway/Sweden	1.8	13	49	0	38	3.9	5	58	0	37	4.1	5	50	2	43
United Kingdom/Ireland	8.0	70	29	0	1	10.4	35	51	11	3	10.1	34	44	17	5
Benelux/Denmark[c]	2.7	53	47	0	0	7.3	11	67	22	0	6.4	10	57	30	3
West Germany	6.2	73	25	0	2	11.5	30	58	10	2	12.0	26	54	16	4
France	3.7	51	35	3	11	8.2	15	70	7	8	8.2	15	61	11	13
Austria/Switzerland	0.9	24	34	6	36	2.0	8	58	7	27	2.1	5	53	11	31
Spain/Portugal	1.0	41	41	0	18	3.0	14	70	2	14	3.6	15	67	2	16
Italy	2.2	14	54	11	21	6.2	5	78	10	7	6.2	6	69	17	8
Greece/Turkey	0.4	41	55	0	4	1.4	21	76	0	3	1.9	30	64	0	6
Austria/New Zealand	1.4	54	43	0	3	2.9	35	50	6	9	3.5	40	42	10	8
Total OECD[b]	80.4	35	42	16	7	155.4	18	56	19	7	160.9	18	53	19	10
Total Non-OECD[b]	10.5	31	57	6	6	25.0	21	60	11	8	30.7	20	66	10	4
OPEC	1.7	3	75	21	1	5.0	1	66	30	3	6.2	0	71	24	5
Other	8.8	37	53	4	6	20.0	25	59	7	9	24.5	24	65	7	4
Total free world[b]	90.9	35	43	15	7	180.3	18	56	18	8	191.6	18	55	18	9

Source: From D.O.E.'s 1980 Annual Report to Congress, Volume 3. Courtesy API.
[a] Includes Puerto Rico, Virgin Islands, and purchases for the Strategic Petroleum Reserve.
[b] Numbers may not add to totals due to rounding
[c] Benelux countries are Belgium, the Netherlands, and Luxembourg.

Table 1-2
World Natural Gas Production—Twenty Leading Nations (billion cubic feet)

Nation	1974[r] Marketed Production[a] Total	1974[r] Marketed Production[a] Billion cu ft/day	1974[r] Percent of Free World Production	1974[r] Percent of Total World Production	Nation	1975[r] Marketed Production[a] Total	1975[r] Marketed Production[a] Billion cu ft/day	1975[r] Percent of Free World Production	1975[r] Percent of Total World Production
1. United States	21,600.5	59.18	61.6	45.8	1. United States	20,108.7	55.09	59.3	42.6
2. USSR	9,201.3	25.21	26.2	19.5	2. USSR	10,205.9	27.96	30.0	21.6
3. Canada	3,045.5	8.34	8.7	6.5	3. Netherlands	3,208.4	8.79	9.5	6.8
4. Netherlands	2,956.7	8.10	8.4	6.3	4. Canada	3,075.7	8.43	9.1	6.5
5. United Kingdom	1,230.0	3.37	3.5	2.6	5. People's Republic of China	1,400.0	3.84	4.1	3.0
6. People's Republic of China	1,200.0	3.29	3.4	2.5	6. United Kingdom	1,208.2	3.31	3.6	2.6
7. Romania	1,011.5	2.77	2.9	2.1	7. Romania	953.5	2.61	2.8	2.0
8. Iran	787.4	2.16	2.2	1.7	8. Iran	771.1	2.11	2.3	1.6
9. West Germany	713.2	2.00	2.0	1.5	9. West Germany	639.4	1.75	1.9	1.4
10. Mexico	560.9	1.54	1.6	1.2	10. Mexico	583.9	1.60	1.7	1.2
11. Italy	540.4	1.48	1.5	1.1	11. Italy	514.3	1.41	1.5	1.1
12. Venezuela	476.0	1.30	1.4	1.0	12. Venezuela	450.3	1.23	1.3	1.0
13. Libya	345.2	0.95	1.0	0.7	13. Libya	382.6	1.05	1.1	0.8
14. East Germany	273.1	0.75	0.8	0.6	14. East Germany	280.0	0.77	0.8	0.6
15. France	269.4	0.74	0.8	0.6	15. Argentina	271.6	0.74	0.8	0.6
16. Argentina	255.7	0.70	0.7	0.5	16. France	259.8	0.71	0.8	0.6
17. Saudi Arabia	219.0	0.60	0.6	0.5	17. Poland	210.6	0.58	0.6	0.4
18. Poland	202.7	0.56	0.6	0.4	18. Algeria	210.0	0.58	0.6	0.4
19. Algeria	198.5	0.54	0.6	0.4	19. Saudi Arabia	200.0	0.55	0.6	0.4
20. Kuwait	186.9	0.51	0.5	0.4	20. Brunei	186.5	0.51	0.6	0.4
Total free world	35,054.2	96.04	100.0	74.3	Total free world	33,931.1	93.00	100.0	71.9
Total world	47,171.5	129.24	—	100.0	Total world	47,207.3	129.34	—	100.0

Source: U.S. Bureau of Mines World Natural Gas, May 27, 1977. Courtesy API.

[r] Revised.

[a] Comprises all gas collected and utilized as fuel or as a chemical industry raw material, including gas used in oil and/or gas fields as a fuel by producers even though it is not actually sold.

Table 1-2
Continued

Nation	1976[r] Marketed Production[a,b] Total	1976[r] Marketed Production[a,b] Billion cu ft/day	1976[r] Percent of Free World Production	1976[r] Percent of Total World Production	Nation	1977[p] Marketed Production[a,b] Total	1977[p] Marketed Production[a,b] Billion cu ft/day	1977[p] Percent of Free World Production	1977[p] Percent of Total World Production
1. United States	19,952.4	54.64	55.4	39.6	1. United States	20,337.5	55.72	57.5	37.7
2. USSR	11,331.3	30.96	—	22.6	2. USSR	12,395.2	33.96	—	23.0
3. Netherlands	3,542.7	9.68	9.9	7.1	3. Canada	3,301.4	9.04	9.3	6.1
4. Canada	3,157.2	8.63	8.8	6.3	4. Netherlands	2,932.5	8.03	8.3	5.4
5. Iran	1,699.6	4.64	4.7	3.4	5. People's Republic of China	1,787.0	4.90	—	3.3
6. People's Republic of China	1,433.7	3.92	—	2.9	6. Iran	1,756.1	4.81	5.0	3.3
7. United Kingdom	1,202.3	3.28	3.4	2.4	7. United Kingdom	1,499.9	4.11	4.2	2.8
8. Romania	854.6	2.33	—	1.7	8. Romania	1,005.5	2.75	—	1.9
9. Mexico	790.8	2.16	2.2	1.6	9. West Germany	907.7	2.49	2.6	1.7
10. Nigeria	632.6	1.73	1.8	1.3	10. Mexico	745.3	2.04	2.1	1.4
11. Italy	562.1	1.54	1.6	1.1	11. Nigeria	717.4	1.97	2.0	1.3
12. West Germany	544.0	1.49	1.5	1.1	12. Libya	488.6	1.34	1.4	0.9
13. Libya	439.2	1.20	1.2	0.9	13. Italy	448.6	1.23	1.3	0.8
14. Venezuela	396.0	1.08	1.1	0.8	14. Venezuela	392.8	1.08	1.1	0.7
15. Argentina	280.2	0.77	0.8	0.6	15. Indonesia	351.9	0.96	1.0	0.7
16. Algeria	256.9	0.70	0.7	0.5	16. Algeria	284.5	0.78	0.8	0.5
17. France	250.7	0.68	0.7	0.5	17. France	256.8	0.70	0.7	0.5
18. Indonesia	248.6	0.68	0.7	0.5	18. Argentina	250.0	0.68	0.7	0.5
19. Australia	209.3	0.57	0.6	0.4	19. Australia	236.7	0.65	0.7	0.4
20. Chile	175.8	0.48	0.5	0.4	20. Chile	208.9	0.57	0.6	0.4
Total free world	36,006.5	98.50	100.0	71.5	Total free world	35,393.2	96.97	100.0	65.7
Total world	50,367.1	137.74	—	100.0	Total world	53,883.7	147.63	—	100.0

Sources: U.S. Energy Information Administration, United States only; other nations, *Oil and Gas Journal*. Courtesy API.

[r] Revised. [p] Preliminary.

[a] Comprises all gas collected and utilized as fuel or as a chemical industry raw material, including gas used in oil and/or gas fields as a fuel by producers even though it is not actually sold.

[b] United States is reported marketed production; all others may include some gross production.

(table continued on next page)

Table 1-2
Continued

Nation	1978 Marketed Production[a,b] Total	1978 Marketed Production[a,b] Billion cu ft/day	1978 Percent of Free World Production	1978 Percent of Total World Production	Nation	1979 Marketed Production[a,b] Total	1979 Marketed Production[a,b] Billion cu ft/day	1979 Percent of Free World Production	1979 Percent of Total World Production
1. United States	19,974.0	54.72	56.6	37.1	1. United States	20,471.3	56.09	53.8	35.5
2. USSR	13,131.6	35.98	—	24.4	2. USSR	14,367.1	39.36	—	24.9
3. Canada	3,133.1	8.58	8.9	5.8	3. Canada	3,646.5	9.99	9.6	6.3
4. People's Republic of China	2,325.2	6.37	—	4.3	4. People's Republic of China	2,832.7	7.76	—	4.9
5. Iran	1,746.9	4.79	5.0	3.2	5. Netherlands	2,717.8	7.45	7.1	4.7
6. Netherlands	1,624.5	4.45	4.6	3.0	6. United Kingdom	1,965.7	5.39	5.2	3.4
7. United Kingdom	1,262.4	3.46	3.6	2.3	7. Romania	1,210.5	3.32	—	2.1
8. Romania	1,211.6	3.32	—	2.3	8. Mexico	974.9	2.67	2.6	1.7
9. Mexico	887.5	2.43	2.5	1.7	9. Iran	913.3	2.50	2.4	1.6
10. West Germany	643.2	1.76	1.8	1.2	10. West Germany	826.3	2.26	2.2	1.4
11. Nigeria	586.6	1.61	1.7	1.1	11. Indonesia	816.3	2.24	2.1	1.4
12. Indonesia	565.8	1.55	1.6	1.1	12. Italy	533.6	1.46	1.4	0.9
13. Italy	540.8	1.48	1.5	1.0	13. Norway	414.9	1.14	1.1	0.7
14. Venezuela	401.4	1.10	1.1	0.7	14. Venezuela	406.0	1.11	1.1	0.7
15. Libya	364.0	1.00	1.0	0.7	15. Nigeria	365.6	1.00	1.0	0.6
16. Norway	351.8	0.96	1.0	0.7	16. Chile	306.4	.84	0.8	0.5
17. Algeria	328.7	0.90	0.9	0.6	17. Brunei-Malaysia	302.7	.83	0.8	0.5
18. Brunei-Malaysia	325.7	0.89	0.9	0.6	18. Algeria	301.5	.83	0.8	0.5
19. Chile	320.7	0.88	0.9	0.6	19. Australia	254.1	.70	0.7	0.4
20. Argentina	254.3	0.70	0.7	0.5	20. Argentina	241.7	.66	0.6	0.4
Total free world	35,272.8	96.64	100.0	65.4	Total free world	38,075.1	103.58	100.0	66.0
Total world	53,911.1	147.70	—	100.0	Total world	57,666.5	157.26	—	100.0

(table continued on next page)

Table 1-2
Continued

Nation	1980[p]			
	Marketed Production[a,b]		Percent of Free World Production	Percent of Total World Production
	Total	Billion cu ft/day		
1. United States	20,090.0	55.04	53.7	34.4
2. USSR	15,355.5	42.07	—	26.3
3. People's Republic of China	3,469.0	9.50	—	5.9
4. Netherlands	2,799.6	7.67	7.5	4.8
5. Canada	2,668.3	7.31	7.1	4.6
6. United Kingdom	1,500.0	4.11	4.0	2.6
7. Mexico	1,190.5	3.26	3.2	2.0
8. Romania	1,176.4	3.22	—	2.0
9. Indonesia	1,028.4	2.82	2.8	1.8
10. West Germany	738.6	2.02	2.0	1.3
11. Norway	705.2	1.93	1.9	1.2
12. Pakistan	600.0	1.64	1.6	1.0
13. Italy	524.6	1.44	1.4	0.9
14. Venezuela	518.0	1.42	1.4	0.9
15. Algeria	517.0	1.42	1.4	0.9
16. Saudi Arabia	310.2	0.85	0.8	0.5
17. Argentina	296.9	0.81	0.8	0.5
18. Iran	292.0	0.80	0.8	0.5
19. Kuwait	291.3	0.80	0.8	0.5
20. Brunei	282.0	0.77	0.8	0.5
Total free world	37,388.9	102.44	100.0	64.0
Total world	58,458.8	160.16	—	100.0

Source: U.S. Energy Information Administration, United States only; other nations, *Oil and Gas Journal*. Courtesy API.

[p] Preliminary.

[a] Comprises all gas collected and utilized as a fuel or as a chemical industry raw material, including gas used in oil and/or gas fields as a fuel by producers even though it is not actually sold.
[b] United States is reported marketed production; all others may include some gross production.

When different forms of energy are considered, it is recognized that while all are convertible to heat, some are more easily convertible to work than others. This convertibility to work is measured by engineers and thermodynamicists by the concepts of *availability* and *entropy*.

For the purpose of this review, without going into unnecessary thermodynamic detail, Figure 1-2 should help one to recognize and understand the various forms of energy that occur in flow systems. It is interesting to note that the various categorical divisions are all binary, or dual, splits.

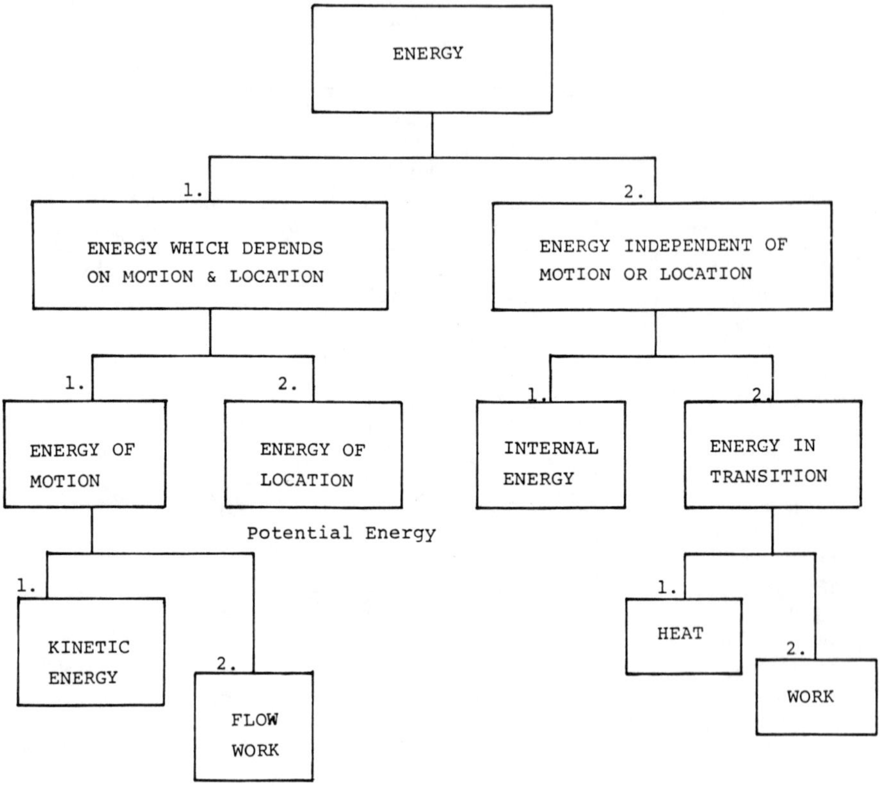

Figure 1-2. Various forms of energy in flow systems.

Energy manifests itself in two broad categories. These are:

1. Energy that depends upon motion or location
2. Energy that is independent of motion and location

Energy that depends upon motion occurs in one of two recognized forms:

1. *Kinetic energy*, usually computed as $V^2/2g_c$, in units of ft × lb_f/lb_m, where:

$$V = \text{velocity, ft/sec}$$

$$g_c = \text{conversion factor} = 32.17 \frac{lb_m}{lb_f} \times \text{ft/sec}^2$$

2. The flow-work for fluids in pipes or other conduits, usually computed as p/ρ, again in units of ft × lb_f/lb_m, where:

p = pressure, lb_f/ft^2 (psfa)
ρ = density of flowing system lb_m/ft^3

Energy that depends upon location is designated *potential energy*. Here, the word *potential* refers to the gravitational field of the earth. The potential energy is usually computed as gZ/g_c, ft × lb_f/lb_m, where:

g = acceleration due to gravity, 32.17 ft/sec^2
Z = elevation above an arbitrary datum, ft

The energy quantities that are independent of motion and location may also be classified into two categories. One is energy that is intrinsically possessed by matter. That is, energy related to molecular state of activity and dependent upon temperature, as well as pressure, surface, chemical potential, and other effects. It is called *internal energy*. It is usually designated by U, Btu/lb_m.*

The next category is energy that can only exist in transition. That, too, comes in two distinct forms:

1. *Shaft-work*, w ft × lb_f/lb_m†
2. *Heat*, q, which may be measured in Btu or, to be consistent with other units in Figure 1-1, in units of ft × lb_f/lb_m

In thermodynamics, when various systems are considered, energy as such is an *extensive* property; i.e., one that depends on the extent of the mass considered. When energy balances are performed on a steady-state flow system, on the other hand, the energy quantities are regarded on a per-pound-mass basis, thus rendering them *intensive*, or independent of the extent of mass involved.

Ecological Perspective

We depend on energy to live in our environment. The air we breathe, the water we drink, the food we cook, the house in which we sleep, and the car we drive all relate to energy. The exploration and exploitation of our resources to generate and use energy in various form(s) is the work of engineers. Accordingly, the engineering discipline is intricately enmeshed with environmental conditions, needs, and ecological resources.

---

* One Btu is equivalent to 778 ft × lb_f work units.

† The energy quantities in Figure 1-1 are based per unit pound mass of fluid involved, for convenience.

Because of their interrelation with the environment and with ecological concerns, the problems of engineering and technology attract a great deal of public attention. The possibility of irreversible damage to the earth's natural resources is raised by many who are concerned about the limited and delicate nature of our ecological resources. Ever since the crude oil crisis of 1972–73, triggered by the oil embargo on the Western World, our awareness of energy as related to engineering resources, on one hand, and to ecological values, on the other, has been a focal point of public attention throughout the world. Energy must be understood in all its complex aspects—particularly by engineers. It is only through greater understanding of energy: its supply, demand, and distribution, its global, political, and economic aspects, that one can hope to solve energy-related problems.

Engineering Methodology

In design or in performance analysis, the engineering methodology directed to energy always starts with implementation of three principles. These principles, held to be the enduring truths of engineering, may be summarized as:

1. The First Law of Thermodynamics (work into heat)
2. The Second Law (heat into work—subject to certain constraints)
3. The Principle of Conservation, which may be applied to mass, energy, momentum, species, etc.

In analyzing energy as it relates to engineering and economics, we often make use of The First Law expressed in the form of some overall efficiency. In terms of objective economics, that translates into so many dollars per unit product or effect. The cost of 1 Btu based on $28 to $35 per barrel of light Arabian crude oil from Saudi is a case in point.

It isn't very often, on the other hand, that we bother checking the design against The Second Law. The Second Law—often ignored—usually becomes controlling, particularly when economics become subjective due to institutional external factors. An excellent example of this is the desirability of liquefied natural gas (LNG) because it returns some of the Btu it has consumed in the form of Joule-Thomson cooling. During the early fifties, it was thought that liquefying the methane in Louisiana and transporting it by barge up the Mississippi River to Chicago would be economically feasible and profitable. During that project, at least part of the economic justification lay in the fact that revaporization into gaseous methane would generate a substantial amount of Joule-Thomson cooling, which could be used to refrigerate meat products as a side benefit. On the face of it, the value in dollars per Btu so generated was obvious. What was not so apparent was the effect of temperature level on the exchange between Btu and dollars. One Btu at room or normal temperature may be worth *x* dollars; but, how much would it be worth at $-258°F$? As one does not need that deep a refrigeration to freeze legs of lamb, it becomes quite apparent that LNG, as an alternative supply of energy, is more suited to steel making than food processing. As things turned out, the trade of LNG between Japan and the United States[3] proved that point during the late fifties.

EARTH'S ENERGY RESOURCES

The terms *energy shortage* and *energy crisis* are truly misnomers. To suggest that we are running out of energy resources is equally wrong. We really are not running out of energy sources. As the famous French scientist Lavoisier said, *"rien ne se crée, rien ne se perd"*—nothing is created, nothing is lost. Energy available in various forms around us simply changes from one kind to another and, unfortunately, always goes downhill to a lower and lower grade.

The engineering challenge is to be able to convert sufficient amounts of earth's energy into the form needed and dictated by market demand. Figure 1-3 is an organization of the energy resources of the earth. These resources may be broadly classified into two categories: the fossil fuels and the geophysical energy resources.

The fossil fuels are, in general, found in liquid, gas (or vapor), or solid phases. If they are liquid, they are called *crude oil.* If gaseous, they are called *natural gas.* When solid, they are called *coal.* The fossil fuels sometimes occur in a semi-solid state or an extremely viscous and heavy liquid form embedded in the porous matrix of sand and shales. These are called *oil shale* and *tar sands.*

The other energy resources shown in Figure 1-3, all derived from the sun, are generically referred to as *geophysical energy resources.* They include: direct solar energy, hydroelectric energy, and elements of minerals from which nuclear energy may be derived. Nuclear energy may be in the form of fission, which uses uranium ore or refined uranium-235 as its basic energy source. The uranium ore must be concentrated into fissile isotope through nuclear processing from 0.7% to 3% concentration. The fissile isotope is eventually spent and must be replaced. The uranium ore resources are finite, also. A subsequent process, in the limelight of controversy some time back, is the *fast breeder reactor* concept. This process would extend the useful reserves of uranium by breeding the fissile isotope in an efficient regeneration scheme. Unlike fission, which must originate from a fissionable mate-

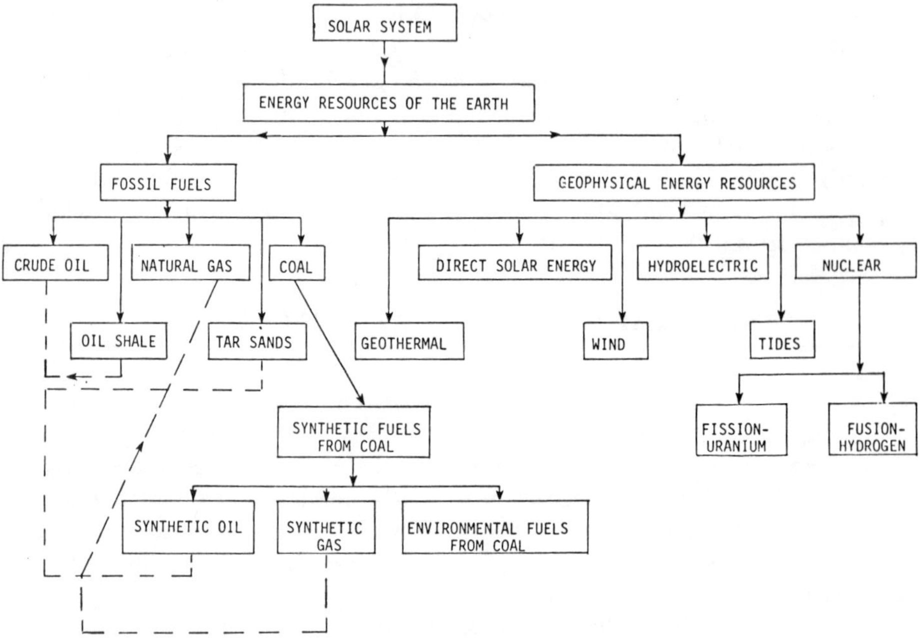

Figure 1-3. The energy resources of the earth.

rial, fusion from hydrogen has the potential to provide a clean and virtually unlimited supply of energy. According to those knowledgeable in the area,[4] the technology is such that several decades appear to be needed for the break-even lab scale experiments and the critical development of controlled energy from fusion. Beyond that, nobody knows how many decades must elapse before the electrical energy from fusion reaches the busbars. Besides solar, hydroelectric, and nuclear, the earth's resources also include: geothermal, wind, tide, and thermal energy from temperature gradients in the oceans.

Some Interesting Facts and Figures on Energy

1 Quad = 1Q = 10^{15} Btu/year
2Q = 1 MMBbl/DOE; MM = 10^6
MMB/DOE = Million barrels per day of oil equivalent

In the United States, the average household uses 3,000 KWh each year.

Table 1-3 shows a percent breakdown of the primary energy resources in the U.S. for the years 1970 and 1976.

Table 1-3
Primary Energy Resources in the United States—Distribution of Supply[5,6]

Source	1970	1976
Oil	43%	45%
Natural gas	33%	29%
Coal	23%	17%
All other sources (nuclear, geothermal, hydroelectric, solar, wind, tides, etc.)	1%	9%
Total	100%	100%

In 1978 the United States, with 6% of the world's population, consumed 40% of world energy; at the same time it produced about 30% of the world gross product.[7,8]

24% to 25% of the primary energy input in the U.S. went to generate electricity (1972 and 1976).

52% of oil consumption went to the transportation market (1976).[5]

62% of U.S. homes use natural gas for heating and cooking. Natural gas provides more than 30% of our primary energy.

United States Electrical Generating Capacity

1977	500,000 megawatts
Projection for year 2000	1,000,000 megawatts

HISTORICAL BACKGROUND ON FOSSIL FUELS

In historical and geological perspective, the chronicles of civilization show that man's evolution has been intimately related to his discovery and use of fossil fuels for energy. Fundamental relationships which underlie our activities—past, present, and future—relate both to materials in our environment and to energy continuously in flux between the solar system and our environment.

Fossil Fuels in Perspective of the Geological Time Scale

The earth is about 4.5 billion years old. The time scale for geology is quite staggering when viewed from the limited perceptions of our mortal existence. We rarely live to be 100 years old. It takes us about 5 years to attend college, perhaps 10 to 15 to complete our primary and secondary education. As professionals, we set goals and arrange our lives into 5, 10, at most 20 year plans. Our governments and leaders are evaluated and often changed every 4 years. On a slightly longer scale, our national defense, land utilization, and economic plans span periods of 20, 30, perhaps 50 years. We speak in terms of centuries but rarely in terms of millenia. In stark contrast, the planet we inhabit is 45 million millenia old.

Figure 1-4 depicts this time scale from the beginning of the earth until the present. That scale, shown at top, is divided into two significant periods. The Pre-Cambrian period spans from the beginning to 3,800 million years on the time scale. This period is distinct because Pre-Cambrian rocks, in general, do not have fossil fuels occurring in them. The Post-Cambrian period, about 700 million years long, is depicted on the next time scale just below the first. That is the period when the fossil fuels were formed, found, and utilized by man.

At the very end of the Post-Cambrian period, about the last one to two million years, man evolved and inhabited the earth. When this one to two million year scale is expanded in the bar below, one may observe three historical landmarks: First, the year 1200 A.D., marking the use of mined coal in old England. Next, 1858, when petroleum seepage was reportedly noted in Rumania. Third, and perhaps most significant, the first oil well drilled by

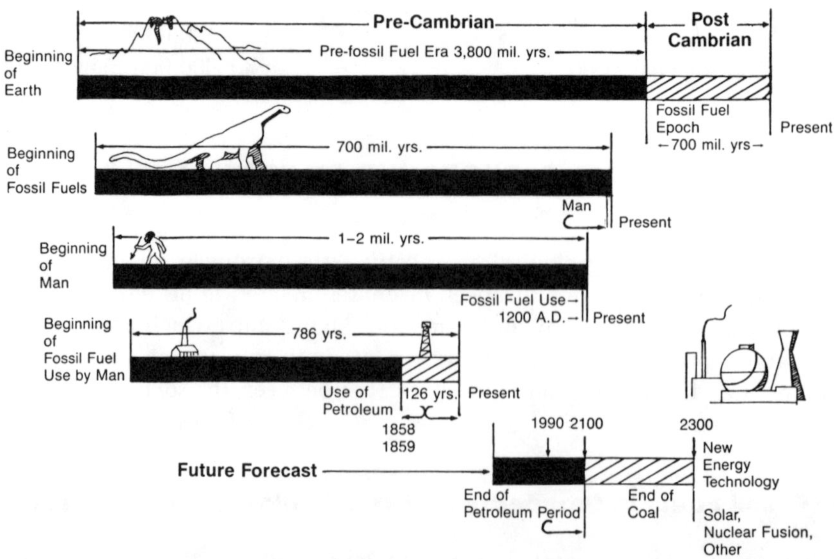

Figure 1-4. Historical perspective on fossil fuels.

man in 1859 in Titusville, Pennsylvania. If the period from 1200 to the present is viewed as the fossil fuel era, according to Figure 1-5 it will terminate within the next few centuries. In a comparative geological time scale, it is notable that *if oil* runs out before the year 2100, the era of crude oil will have been a mere 240 years! The corresponding period for natural gas will perhaps be a few centuries longer, and that of coal may be a half millenium or so.

It was M. King Hubbert who first pointed out that the era of fossil fuels (oil, gas, and coal) will be almost over when cumulative production of them reaches 85% of their initial total reserves (see Figure 1-5). It seems reasonable that once oil nearly runs out or becomes too expensive to explore and produce, natural gas will be exploited to the extent that it is economically feasible, and coal will follow as the last fossil fuel.

As the alternatives in energy shift from oil to gas to coal, hopefully this period of transition will be followed by the eras of nuclear fuels, solar energy, and, sometime in the future, fusion.

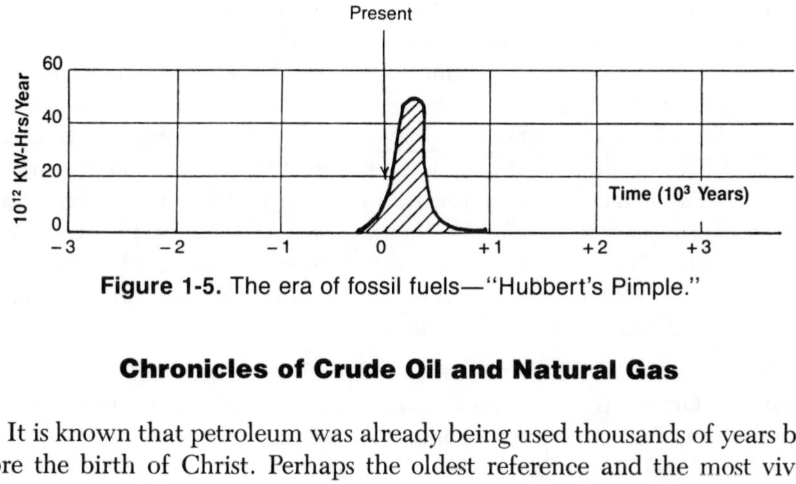

Figure 1-5. The era of fossil fuels—"Hubbert's Pimple."

Chronicles of Crude Oil and Natural Gas

It is known that petroleum was already being used thousands of years before the birth of Christ. Perhaps the oldest reference and the most vivid chronicle is the one given in the Old Testament. The book of Daniel, Chapter 3, speaks of the "eternal fires" as related to King Nebuchadnezzar. The eternal fires, some 40 cubits and nine tall, have since been reported as being periodically self-igniting natural gas seepage distributed along a circular pattern somewhere near the spot where the rivers Tigris and Euphrates meet. It was reported to the author while in Abadan, Iran a few years ago that the Eternal Fires are still there (near Ahwaz) but are only 10 feet tall now. Petroleum was also being used thousands of years before Christ by the Sumerians, who knew how to apply bitumen as insulating material for buildings.[7] The famous Greek historian Herodotus referred to oil seepage in

Persia. Alexander the Great is said to have discovered crude oil on the banks of the River Oxus.

History records the burning of pitch along with sulfur during the Battle of Palatea in Greece in the year 479 B.C. About two centuries before Christ, there is a record of a fire-spitting machine in the fleet of the Island of Rhodes. The fire was probably that of a petroleum product.

In U.S. folklore, when the drums along the Mohawk were heard, the flaming tips of Indians' arrows were most likely dipped in petroleum before being ignited. The use of oil was reported many times in history before 1859; however, its role changed drastically after 1859.

The Drake well in Titusville, Pennsylvania was the harbinger of a remarkable era of gasoline, the mass-produced automobile, the light bulb, and the industrial revolution led by the likes of Ford, Edison, and Steinmetz. Prior to the Drake well, salt-well drillers occasionally found "dirty black stuff"—oil—a fluid which was not particularly cherished. An enterprising American, S. M. Kier, used to buy the oil and sell it as a panacea for all kinds of ailments. In 1857, G. H. Bissell of New Haven, Connecticut thought of drilling especially to find oil, which he wanted to try as an alternative to the prevailing lamp oil of the time. The lamp oil, extracted from vegetation or whales, was getting expensive. Bissell formed a company which hired E. L. Drake, a former railway guard, and gave him the task of drilling for oil. Drake worked with a salt-well driller, A. W. Smith (called "Uncle Billy" in the profession). The two encountered enormous technical difficulties as well as financial adversities. However, they never even considered accepting defeat. Finally, on their historic try between July 1 and August 28, 1859, they drilled 60 ft in about 50 days and found what they were looking for: crude oil. At present it is not unusual to drill 1,500 ft/day or more through reasonably soft formations.

Even though the Drake well was not profitable and was referred to by many as "Drake's folly," oil fever broke out in Pennsylvania around the Titusville area. Refining, transportation, marketing, and research quickly followed. Colonel Drake died a virtually forgotten man, but within one century, hundreds of companies were formed to explore, produce, transport, refine, and sell products from petroleum. A decade after the Drake discovery, production moved along the Allegheny to its tributaries and to other states as well. Oil became the accepted illuminant and lubricant. Refining and transportation became realities. The first oil pipeline ever built was operated out of Pithole and constructed by Sam Van Sycle. The first significant economic contribution of the Drake well was the prevention of complete economic chaos and collapse after the Civil War by providing this country with a positive balance of payments.

Three years after "Drake's folly" one man had the vision of oil becoming an integral part of the American economy. His name was John D. Rockefeller. In 1926, along with Maurice B. Clark and Samuel Andrews, he formed a partnership to operate a small refinery outside Cleveland, Ohio. In about 10 years, this venture grew into Standard Oil Company.

No chronicle of the history of petroleum would be complete without reference to Henry L. Doherty, who pioneered the conservation ethic long before the turn of the century and nursed the natural gas industry during its infancy. It was through his persistent efforts that the first rejection of the "rule of capture" and repudiation of wasteful flaring of associated gas came about. Doherty also pioneered the large-diameter high-pressure natural gas pipeline and was involved in the first natural gas storage fields in the U.S.

The first gas wells drilled in the U.S. were by-products of the search for oil. The first offshore oil was the result of a directionally drilled hole in 1894 in Santa Barbara, California. The first offshore wells were from platforms erected in Caddo Lake, Louisiana and in Goose Creek, Texas. The first well drilled in high seas was in Cameron Parish, Louisiana in 1938. The search for natural gas, the exploration, drilling, and production expressly for natural gas, gained momentum after the advent of offshore exploration and benefited most from the recent advances in geophysics, particularly seismic work and data processing with electronic computing.

Another crucial development was the discovery of oil in the Middle East. As described in *Adventure in Oil* by Longhurst[9], the discovery of oil in Masjid-i-Suleyman, the first oil well in Persia, can strangely be traced to two brothers, Tom and John Morgan, in Rockhampton, Queensland, Australia. It seems these two brothers, while prospecting for the mythical lode of silver in 1899, found gold at a place called River Dee in Queensland. With the help of a young British attorney named Darcy, they formed a gold-mining company which helped all three become fabulously rich. Eventually, Darcy sold his shares to the Morgans and moved back to England. He was contacted by a French geologist who persuaded him to provide financial backing for a venture to look for oil in Persia, where seepage to the surface had been reported. The venture, which started near the turn of the century and apparently ran into many mishaps and difficulties, found no oil and just about ruined Darcy's entire fortune. Darcy, even though on the brink of bankruptcy, became so obsessed with the search that he persuaded a fellow London Club member to come to his aid. That friend, who worked in the Admiralty, had heard about the goings on in America in oil exploration. He thought oil could be the answer to efficiency in the British Fleet. He, in turn, persuaded the Admiralty to join the search with Darcy in Persia. The British Petroleum Corporation was formed.

In 1908, the venture resulted in the discovery of the first oil well in the Middle East, a long way from and many years after that memorable midsummer night when Tom Morgan found his gold nugget in the River Dee. The young Admiralty clerk who made the find possible was Winston Churchill.

EXPLORATION METHODS FOR NATURAL GAS

In the early days, exploration for oil and gas was mainly based on searching for oil seeps. The first wells were always drilled where surface evidence indicated the presence of oil. Today, equipped with modern tools, positive science, and accumulated correlated data from past drilling, the search takes a more systematic pattern.

The oldest scientific method of exploration, *surface geology*, attempts to find out what is below from surface observations. Aerial photographs, study of rock outcrops, MAD (*magnetic anomaly detection*) by airplanes, gravity surveys, and remote sensing are typical tools used in conjunction with surface geology.

Because surface geology can only provide partial answers at best, *subsurface geology* came into the exploration picture. One basic tool for acquiring subsurface data in a given area is *core drilling*. The diamond-bit core drill permits capturing of continuous rock specimens in sequence drilled from the surface. Exploratory holes drilled in the ground are usually surveyed by special electrical equipment for keys and markers correlating various strata. Such surveys, numerous in kind and detail, are called *electrical logs*. The most prominent of electrical logs is the *self-potential-resistivity* log. This log measures the self potential and resistivity between two points in the hole joined by an electrical path through the formation. As depicted in Figure 1-6, such a log permits one to differentiate between sands and shales occurring at various depths.

Figure 1-7 shows how electrical logs obtained from three different wells at different locations may be used to correlate several strata of water sand, oil sand, gas sand, and shale across a wide area. The log not only differentiates between the shales and sands, but when used along some cores, permits identification of the fluid saturating the pore space. It can, for example, readily distinguish between water sands and oil sands.

Other logs based on radioactivity, such as gamma-ray neutron and microlateral logs, permit location of gas sands, gas/water interface, and other similar features.

Cross-correlation of various strata according to their geologic age may also be obtained from examination of fossils and microfossils.

Figure 1-7 shows the cross-correlation between several electric logs used to reconstruct the subsurface zones extending across three well locations. Driller's logs, mineral content of cuttings, fossils, temperature surveys, and microfossils such as foraminifers are also helpful when correlating various subsurface sediments according to their sequence, depth, and age.

Bacteriology directed toward study of microorganisms that thrive on hydrocarbons, and geophysical methods such as magnetometer surveys or gravity surveys are also valuable tools in locating anomalies which may be traced to large oil or gas deposits. Several oil fields in the mid-continent United States have been discovered by MAD.

In geophysical prospecting, one of the more recent and successful methods for determining subsurface structural features is *seismic exploration*. In seismic work, a hole a few inches in diameter and several feet deep is drilled near the surface, packed with a pound or two of dynamite, and sealed off with mud. The charge is detonated, sending a point-source shock wave

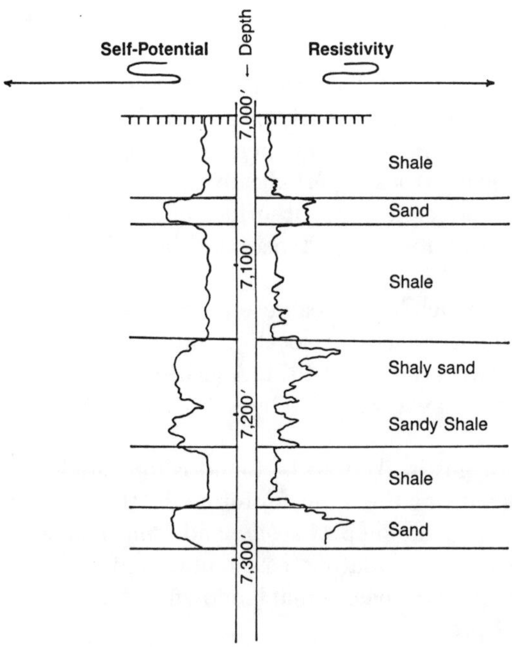

Figure 1-6. Typical self-potential log.

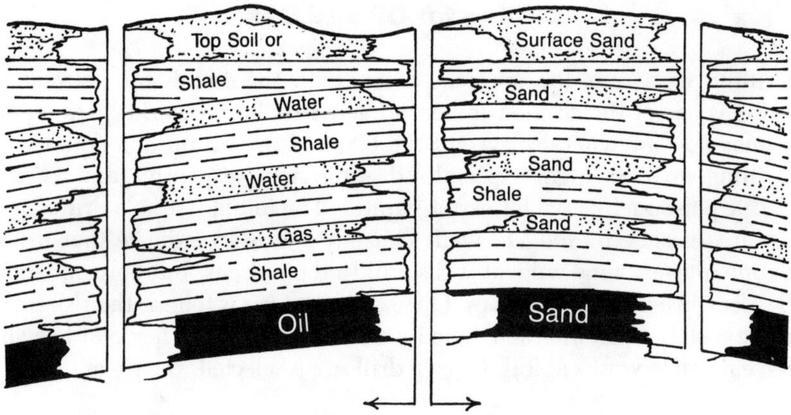

Figure 1-7. Cross-correlation of several electrical logs.

propagating into deep layers of sedimentary rocks. The elastic wave so generated is reflected off the interface of various strata, propagating with definite velocity and angle toward the surface where geophones pick it up and record it on a strip chart. The geophones are numerous and are located about 150 ft apart on a straight line 3 to 4 miles long. The data on reflected waves at various locations are analyzed with the aid of a computer program based on an information theory program which helps separate *random noise*, thereby leaving *coherent signal*. The coherent signal thus obtained is indicative of subsurface structure, revealing such anomalies as faults, anticlines, and traps.

The seismic work thus conducted by a full complement of crew and electronic equipment costs in the range of $3,000 to $4,000 per mile*. A cheaper but less reliable alternative to the reflection seismometer is *vibroseis*, where a large metallic plate under a heavy truck is periodically vibrated, sending similar continuous waves to be reflected from subsurface and recorded by geophones. The use of vibroseis costs about $2,000 per mile* but is limited to places accessible to trucks.

The reflection seismograph, which really came on strong during the 1950s, is credited with phenomenal success in locating Silurean-Niagaran reefs in the state of Michigan between 1964 and 1974. A success ratio as high as 90% of reefs found to wells drilled was achieved by AMOCO and Shell in Michigan in the mid-seventies.

Once a structure where oil or gas is likely to be found is indicated, the *leasing* becomes important in acquiring the rights for future development of a field. Leasing is carried out with the help of *scouts* and *land men* who make a career of locating prospects, negotiating lease rentals and bonuses, and determining royalties to be paid to the original landowner in the event of discovery, development, and production.

Drilling for Oil and Gas

Drilling represents the moment of truth in the search for petroleum. Drilling is one of the costliest industrial activities in the world. A single oil well may cost anywhere between $100,000 and several million dollars.

Drilling involves highly specialized equipment, expensive and knowledgeable personnel, and a lot of risk capital. The initial test hole drilled in a given area is called a *wildcat* well. Current* statistics indicate that in the U.S. one in every nine wildcat wells will find oil or gas, but only one in 50 will discover significant reserves. Once the oil or gas is found, the risk is less and the wells drilled are then called *development wells*. They cost less and involve much less risk capital. Once a drill site is selected, the rig and equip-

* 1974 figures[5]

ment must be brought in—by truck when possible. In remote and inaccessible areas, the rig is moved piece-by-piece by aircraft and helicopters. In offshore locations, barges are used before the drilling platform is erected.

A modern drilling rig stands 100 feet tall. It is assembled in two or three pieces. The power is provided by diesel, gasoline, or electric engines. Conventional oil drilling rigs are classified as *cable tool* and *rotary.*

Cable Tool Drilling

The cable tool, first used by the Chinese thousands of years ago (springpole drilling, 259 B.C.), is based upon the principle of impacting the earth repeatedly with a heavy guillotine-like tool hanging from the end of a steel cable. As the hole progresses due to the crushing action of the tool (called a *trepan*) the hole is filled with mud. The mud is bailed out, the bailer being a cylindrical bucket designed to remove the cuttings and wellbore fluids. In normal practice, the cable tool drill exposes various sediments to the atmosphere in open-hole fashion. That is why cable tool drilling is just like a continuous drill-stem test of all the formations penetrated. This method is usually not used in holes over 5,000 feet deep.

Rotary Drilling

The *rotary* is the standard drilling rig. It is fast, efficient, and economical and can reach depths up to about 30,000 feet. The basic elements of a rotary drill are shown in Figure 1-8. The platform through which the drill pipe is rotated is called the *kelly.* Usually, drillers' depths readings are given with reference to the *kelly bushing,* the square hole which fits onto and rotates the top portion of the drill pipe. The rig structure is about 90 feet tall and has a crown block at top called the *crow's nest,* a pulley that connects to the traveling block and to a block-and-tackle arrangement designed to lift the mud hose and swivel block. Immediately under the kelly bushing one can see the mud-pipe leading to the shale shaker table where samples from drill cuttings are collected, examined, and marked. The prime movers on the platform provide the energy for the draw works and mud pumps. The mud pit, the blowout preventers, the conductor pipe, and the surface casing can be seen under the kelly bushing in Figure 1-8.

The various drill pipes are mechanically joined across drill collars and the last pipe has the drill bit attached to it. The drill bit consists of three rotating cones studded with sharp teeth made of tough alloy steel. Sometimes the teeth are embedded with diamonds for cutting through particularly hard formations. The drill pipe assembly is called a *drill string.* It is made of steel of high tensile strength and is usually 4 to 5 inches in diameter. The total

Figure 1-8. Major components of a rotary drilling rig.

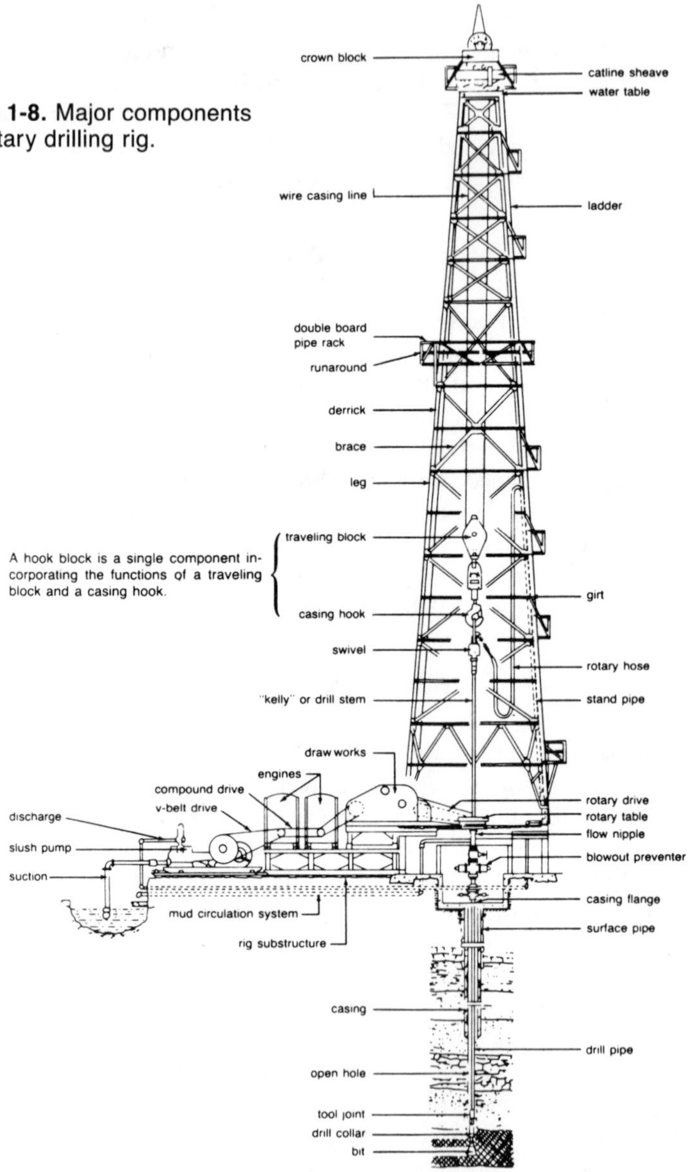

weight of all successive drill pipes extending 20,000 feet from the kelly would be of the order of 400,000 pounds.

The commencing of drilling operations is called *spudding in*, similar to the opening serve in a tennis game. The first conductor pipe is usually set at about 20 feet from the surface. As the drill pipe rotates, the weight resting on the bit crushing the earth, a special fluid called *drilling mud* is circulated through the drill string, out around the bit, and back up through the annular space created between the drill pipe and drilled hole. This drilled hole is isolated from the surrounding formations by the cementing of a large diameter steel pipe called *casing* to the sides of the hole. The purpose of the casing is twofold:

1. To prevent seepage into and contamination of water-bearing sands near the surface.
2. To anchor the drilled hole and to protect it structurally against blowouts, cave-ins, or high pressure effluents coming through.

As the hole progresses, additional casing of smaller and smaller diameter is cemented into the drilled hole before the bit reaches its objective depth, called *pay horizon*. When the big becomes dull and worn, a new one must be installed at the end of the drill string. One by one, all the drill pipes must be lifted out, unscrewed, removed, and stacked at the side of the rig. After the last one has had the drill bit replaced, the reconnecting and lowering process starts and the operation is reversed. This operation is called *making a trip*.

Drilling mud is a mixture of water, clay, and chemical additives such as barium sulfate, hydroxyethylcarboxymethyl-cellulose. The additives are usually selected specifically for control of viscosity and water loss. The barium sulfate is a weighing agent. In solution, it results in a density of fluid about twice as high as that of the water, permitting maintenance of high pressure near the bit. Drilling mud has four functions:

1. To clean and cool the bit
2. To carry cuttings to the surface
3. To seal off the side of the drilled hole
4. To control the pressure at the bottom, preventing cave-ins and blowouts

As the mud is circulated through the center of the drill pipe, out, and up around the annulus between the drill pipe and the previously drilled hole, if some unusually high permeability strata is encountered, water loss could conceivably occur from the drilling mud being circulated. Excessive water loss could result in a loss of continuity of the mud column and interruption of pressure maintenance across the bit. If the drillers came across a high pressure gas or condensate-bearing formation just then, the unfortunate coincidence could result in a blowout.

While cable tool drilling historically came into prominence with Colonel Drake's discovery well in Pennsylvania in 1859, the rotary drill became established when Captain Lucas drilled into the Spindletop reservoir in Beaumont, Texas in 1901. The event, in historical perspective, was the harbinger of the modern internal combustion engine and the industrial revolution.

The cost of drilling wells varies from $8.00 to $150.00 per foot drilled. During 1956, 58,000 wells were drilled in the United States. During the early eighties, the total number of wells reached 80,000 or more per year, then sharply dropped to less than 30,000 in the mid-eighties. The average cost was $90,000 per well. The overall cost of the well is closely related to the rate of drilling activity. The normal rate of penetration of the bit is a function of the weight it carries, pressure at the bit, and the hardness of the formation being drilled. It is usually 75 to 100 feet per hour during actual drilling. The rate is higher if low pressure is maintained at the bit. Sometimes a rate of 150 to 200 feet per hour is achieved in cutting through soft rocks. The basic cost of drilling per foot may be calculated from the following equation:

Cost per foot = (rig time-cost/hr) (rotating hrs + trip hrs) + (bit cost/bit footage)

The past trend in drilling research was toward developing the useful power at the bottom of the hole rather than conveying it down the drill pipe. Developments in this area included:

- Percussion drilling where hard steel microspheres constantly bombard the formation at the bit
- Turbo drilling where the circulating mud helps rotate the bit by going through the vanes at the bottom
- Use of laser energy to cut through the rock

The latest turbodrills develop 300–450 HP at the bit, rotating at speeds of 500 to 900 RPM with bit loadings at 20 to 35 tons.

Table 1-4 gives a comparison of various drilling power quantities developed at the bit.

Table 1-4
Comparison of Drilling Power Developed at the Bit[5]

Tool	Power available at 12″ diameter bit
12″ cable tool	26 hp
12″ rotary	70 hp
10″ turbo drill	190–285 hp (max. 300)
2–8″ turbo drills in tandem	210–285 hp (max. 385)

Core Drilling. Core drilling involves drilling through an area in the shape of concentric annuli recovering a cylindrical core of the material drilled at the middle. Core drilling uses diamond-studded core drill bits and is much more expensive. Besides conventional core bits, retrievable wire line core bits and side wall cutting tools are occasionally used.

The identification color photographs, and canning and sealing of cores are important for the tests to be conducted in laboratories. From these laboratory measurements one determines:

1. Porosity
2. Permeability
3. Oil-water-gas saturation
4. Gas-oil, gas-water, or water-oil contacts

Air-Gas Drilling. In order to achieve a high drilling rate, low pressure is required at the bit. This is sometimes realized by aerating the column or circulating drilling mud. The gas lifting the mud out of the hole by high-pressure air or gas has been tried with limited success in areas where no water intrudes into the hole. Engineering equations for circulation of aerated drilling mud were first developed by Poettmann and Bergman (1955).

Well Completion. Well completion involves the final finishing of drilled, cased, and cemented hole. Wells are completed by either having the casing set on top of the pay horizon (open hole) or being perforated by shaped charges after being cemented across the producing formation.

Well Stimulation. Well stimulation involves increasing the productivity of the well by fracturing and acidizing techniques. Initial cleaning of the wells involves mud-acid cleaning. The mud acid is usually 15% hydrochloric and hydrofluoric acid containing corrosion inhibitors.

Drill-Stem Testing. Drill-stem testing is the term used to designate a special well-flow test run during drilling in order to evaluate productive potential or a designated producing interval. It involves conduct and interpretation of both unsteady-state drawdown and buildup tests.

Directional Drilling. Directional drilling is a special technique which permits slanting the drilled hole through a desired number of degrees. It is very specialized, expensive, and sometimes hazardous. Some holes in California are known to be whipstocked almost 90 degrees. It permits reaching desired pay horizon locations which have inaccessible surface features, like reaching sides of salt domes likely to have oil reservoirs without having to drill through the massive salt body. On occasion, directional drilling permits salvaging a drilled hole by bypassing stuck drill pipe. Retrieval of stuck pipe in the drilled hole is done with specialized equipment called *fishing tools.*

References

1. Tek, M. R. and W. L. Shepherd. "Perspectives, Problems and Prospects on Energy in the Pacific," Master's Thesis, School of Chemical Engineering, The University of New South Wales, Kensington, N.S.W., Australia.
2. Ikoku, Chi U. *Natural Gas Engineering: A Systems Approach*, Tulsa, Oklahoma: Pennwell Publishing, 1980.
3. "LNG is Coming of Age," *Chemical Engineering*, McGraw Hill, Dec. 1970, p. 94.
4. Halpern, G. M. "Fusion Energy," *Kirk-Othmer Encyclopedia of Chemical Technology*, John Wiley & Sons, Vol. 11, 1980, pp. 590–603.
5. Tek, M. R. "Energy, Environment and Fossil Fuels," Course Notes, Chemical Engineering, University of New South Wales, 1976.
6. Tek, M. R. and W. L. Shepherd. "Perspectives, Problems and Prospects on Energy in the Pacific," Proceedings P.A.Ch.E.C., Denver, Colorado, 1977.
7. Hillard, J. H. "Gas, Natural," *Kirk-Othmer Encyclopedia of Chemical Technology*, John Wiley & Sons, Vol. 11, 1980, pp. 590–603.
8. *Gas Facts*, American Gas Association, Arlington, Virginia, 1976.
9. Longhurst, H. *Adventure in Oil, a Story of British Petroleum*, London: Sidgwick and Jackson, 1959.

2

PROPERTIES OF STORAGE RESERVOIRS, ROCKS, AND FLUIDS

NATURAL GAS RESERVOIRS SUITABLE FOR STORAGE

The process by which natural gas is formed in source beds, migrates through porous and permeable rocks, and is eventually entrapped is similar to that of crude oil. Theories related to differential migration and selective entrapment help discern the particular types of formations where oil or gas is likely to be discovered. Such formations include anticlinal geometry, structural highs, stratigraphic lenses, faults, and salt domes.

The gas reservoirs are either *dry* (in terms of crude oil, which may or may not be present in small quantities) or part of an oil-gas system forming a gas-cap. In oil fields where the gas has evolved from solution in crude oil, the gas is designated *associated gas*. Non-associated gas reservoirs are sometimes called *wet* in reference to their high specific gravity, resulting from their high content of heavier (normally liquid) hydrocarbons.

In considering depleted gas reservoirs as prospects for underground storage, one thinks of conventional dry gas reservoirs and conventional condensate reservoirs, in that order.

In addition to conventional gas, our natural gas resources include tight sands, shales, methane in coal, and natural gas in geopressured reservoirs. These unconventional resources are not likely to provide an opportunity for effective underground storage in the foreseeable future. The exception may be the high-pressure aquifers, where deep aquifer storage could possibly offer a chance to recover some of the native dissolved gas as well.

The study of conditions and processes prevailing in reservoirs is known as *reservoir engineering*. This discipline is similar for production of natural gas and crude oil. There are, however, some subtle differences with respect to production of gas as compared to storage. Historically, the necessity for precise underground data was first recognized in 1928. The first bottom hole pressure gauge was described by Sclater and Stephenson[1]. The need for accurate measurement of bottom hole pressure was emphasized by Millikan and Sidwell[2]. Early students of physical properties of rocks included Fancher, Lewis, and Barnes[3]; and Wycoff, Muskat, and Reed[4]. Measurement of permeability was first contributed by Wycoff and Botset[5]. Multiphase flow through reservoir rocks was first recognized by Leverett and Lewis[6]. Other significant developments which followed historically include bottom-hole sampling[7], measurement of connate water[8,9], and bottom hole temperatures and geothermal gradients[10].

Starting in the 1930s, material balance, reserve and recovery estimates, water encroachment, and unsteady-state flow methods were developed.

From the forties on through the fifties and sixties, optimal use of reservoir energy, multiphase flow, secondary and enhanced recovery, and highly sophisticated computing hardware and software contributed to the science of reservoir engineering, once defined as the "art of developing and producing oil and gas fields in such a manner as to obtain a high economic recovery."[11]

Properties of natural gas are important both in design and in performance-analysis work. Those properties which must frequently be used by engineers responsible for development and operation of underground storage fields will be discussed in this section.

Composition of Natural Gas

Natural gas is a complex mixture of essentially straight chain, paraffinic hydrocarbons and certain impurities. In natural gas, one normally finds methane, ethane, propane, butanes, and pentanes; and smaller amounts of hexanes, heptanes, octanes, and heavier hydrocarbons. In addition to the above main components, one may find occasional carbon dioxide, hydrogen sulfide, nitrogen, and almost always some water vapor. Occasional small quantities of aromatics and helium or argonne have also been known to occur in natural gas. Table 2-1 shows typical compositions of a native and a storage gas.

PROPERTIES OF NATURAL GAS

Practical engineering problems involving production or storage of natural gas often include calculations related to density, z-factor, pressure-volume relationships, specific transport properties, and hydrates. These properties will be presented and reviewed in the following pages.

Table 2-1
Typical Native and Storage Gas Compositions

Component	Native Gas mole %	Storage Gas mole %
Nitrogen	7.270	5.227
CO_2	1.070	1.394
H_2S	0.0	0.0
O_2	0.0	0.530
C_1	75.570	83.194
C_2	6.150	7.289
C_3	5.090	1.621
C_4	2.480	0.473
C_{5+}	2.370	0.230
	Σ = 100.000	Σ = 100.000

Density

Natural gas is a mixture of mostly methane, some ethane and propane, and minute amounts of butanes, pentanes, hexanes, and heavier. The calculation of density is based upon Boyle's law and rules related to mixture properties. According to Charles' and Boyle's Law, the density of an ideal gas can be derived as:

$$\rho = \frac{(MW)p}{RT} \tag{2-1}$$

where ρ = Density of the ideal gas, lb_m/ft^3
p = Pressure of the gas, lb_f/ft^2
MW = Molecular weight, lb_m/lb-mole
R = Gas constant, $1{,}544 \dfrac{ft \times lb_f}{\text{lb-mole °R abs.}}$
T = Absolute temperature °R abs. (= 460 + °F)

In underground storage and pipeline flow applications, quite often the gas is identified by its *specific gravity*. The specific gravity is defined as the ratio of the density of that particular gas to the density of the reference gas, the air, at a given temperature and pressure. Accordingly, it follows that the specific gravity is the ratio of the molecular weight of the gas to the molecular weight of the air.

$$(SG) = (MW)/29 \tag{2-2}$$

Equation 2-1 can, therefore, be written as:

$$\rho = (SG)29\frac{p}{RT} \tag{2-3}$$

The specific gravity as defined above is dimensionless and, as such, unitless.

For a real gas:

$$\rho = \frac{29(SG)\ p}{ZRT} \tag{2-4}$$

where z is the non-ideality factor, also called *compressibility* factor or sometimes *incompletely super compressibility* factor.

Compressibility Factor

For a pure component, z is a function of reduced temperature and pressure (see Figures 2-1, 2-2, and 2-3).

$$z = z(T_{r'}, P_r) \tag{2-5}$$

$$T_r = \frac{T}{T_c} \tag{2-6}$$

$$p_r = \frac{p}{p_c} \tag{2-7}$$

T_c and p_c represent the critical temperature and pressure for natural gas, as computed from the composition or estimated from the gravity correlation given in Figures 2-1 and 2-2.

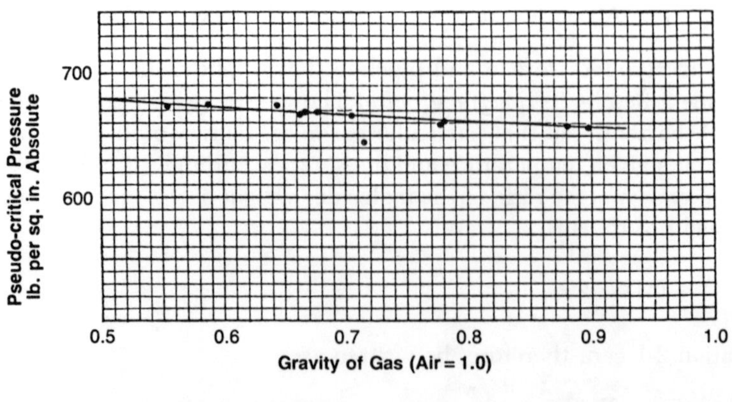

Figure 2-1. Approximate pseudo-critical pressure as a function of gas gravity.[12]

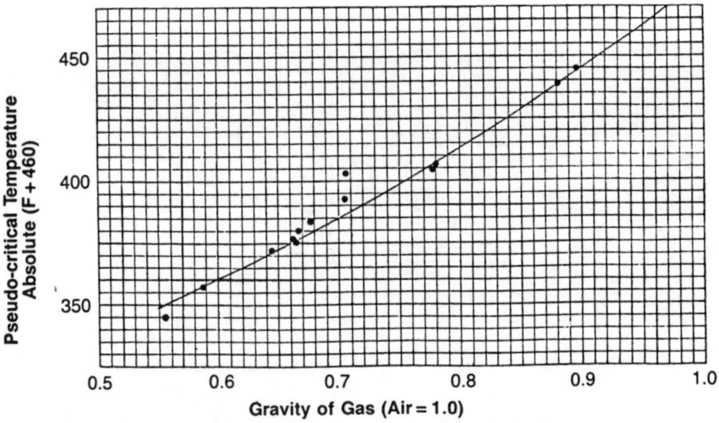

Figure 2-2. Approximate pseudo-critical temperature vs. gas gravity.[12]

HYDRATES OF NATURAL GAS

When free water is present, the natural gas-water mixtures are known to form solids called *hydrates* at high pressures and temperatures above the freezing point of water. Since water is almost invariably present in production of hydrocarbons, water-hydrocarbon systems are of interest all the way from production through gathering, processing, pipelining, and distribution of natural gas. The formation of gas hydrates, which resemble ice or wet snow, is undesirable because they plug up the pipe and often interrupt the flow.

Water Content of Natural Gas

The amount of water held in equilibrium in natural gas is correlated as a function of temperature and pressure and given in the *Engineering Data Book*[12] of the Natural Gas Processors Suppliers' Association and in *The Handbook of Natural Gas Engineering* by D. L. Katz.[13] The curves given by McCarthy, Boyd, and Reid in the first publication also provide corrections for salinity and gravity. These are shown in Figures 2-4 and 2-5. Knowing the temperature and pressure of the gas in the reservoir, by calculating the temperature and pressure at the wellhead, one can calculate how much free water will drop out per MMCF of total gas produced.

Figures 2-4 and 2-5 show the water-vapor content of natural gas in saturation from temperature and pressure of the gas.

Background on Gas Hydrates

Hydrates are solutions of gas in crystalline solids. They were first discovered by Faraday in 1823. Early experiments on hydrates were conducted by

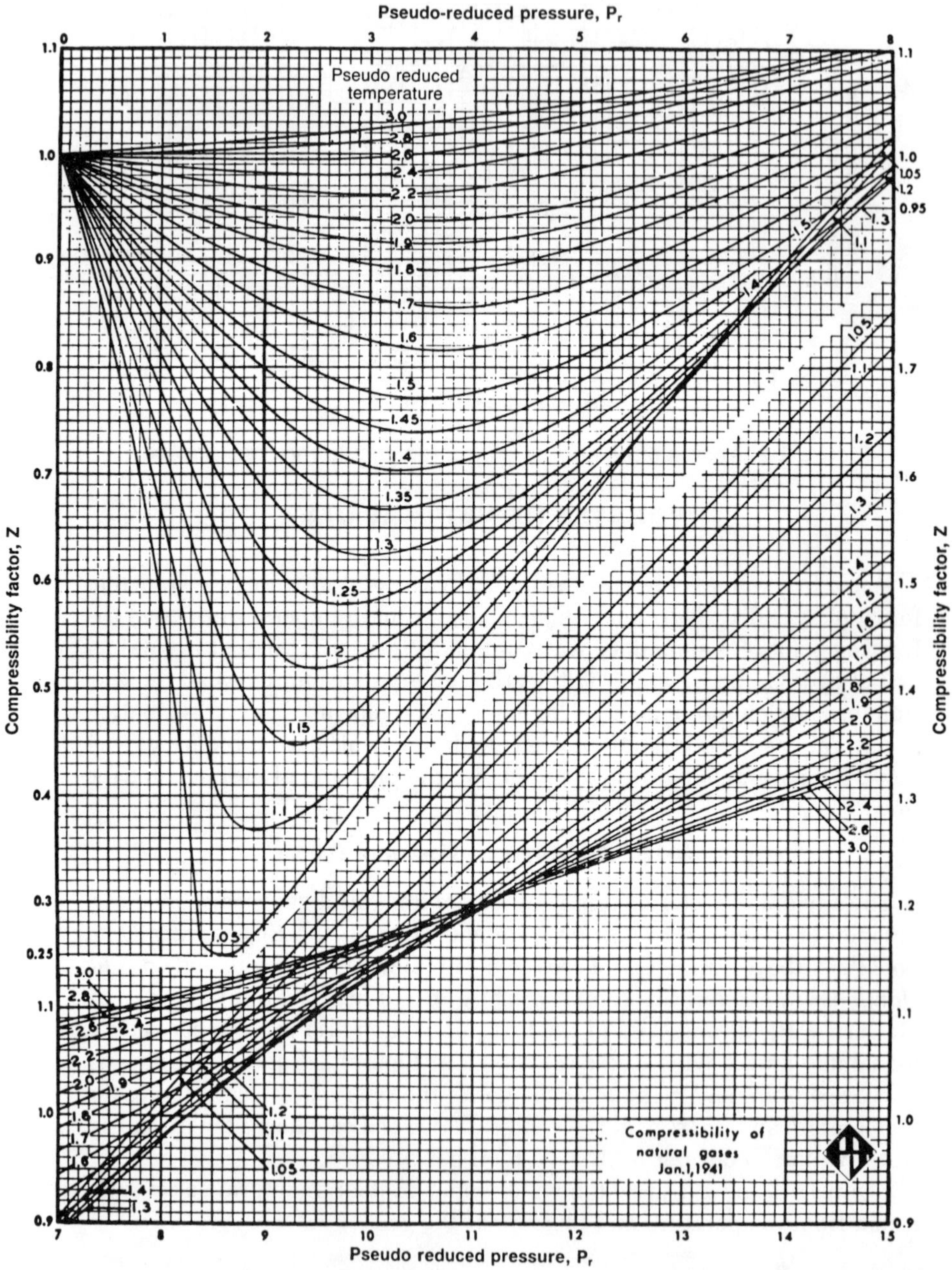

Figure 2-3. The compressibility factor for natural gases as a function of pseudo-reduced pressure and temperature.[12]

Villard de Forecrand using methane, ethane, and CO_2. The formation of gas hydrates in natural gas transmission lines was first reported by E. G. Hammerschmidt in 1934.[14] It is important to realize that most hydrocarbons in natural gas form hydrates when sufficient free-liquid water is present.

The general chemical formula assigned to hydrates is:

$$C_nH_{2n+2} \cdot m\ H_2O$$

For instance:

$$C_3H_8 + 17H_2O \rightarrow C_3H_8 \cdot 17H_2O$$

Hydrate Forming Conditions

There exist several practical correlations in literature directly giving the hydrate threshold loci on the temperature-pressure quadrant.[12,13] Figure 2-6 shows the loci of hydrate-forming conditions in the pressure-temperature quadrant for methane and other natural gas mixtures specified by gravity.

Direct Calculation of Hydrate Forming Conditions

If the composition of the gas is known as a table of y_i (mole fraction of compound i) vs. the particular component i and, say, the pressure known, the hydrate-forming temperature is calculated from the solid-vapor mixture phase equilibrium constants K_{sv_i} for each particular component. If a component is included in the composition but does not form hydrates, K_{sv} for that component is automatically taken as infinity to make the corresponding x_i (the mole fraction in the solid hydrate phase) go to zero. Accordingly, given y_i vs. i, and p, the following steps are pursued:

1. Assume T = T, a temperature
2. Look up K_{vs_i} from p and T and get x_i
3. Calculate $x_i = \dfrac{y_i}{K_{vs_i}}$; repeat for all components
4. Calculate $\sum_1^N x_i$
5. If $\sum_1^N x_i = 1$, stop, and not only is the assumed T the correct temperature, but x_i is the very composition of the hydrate.

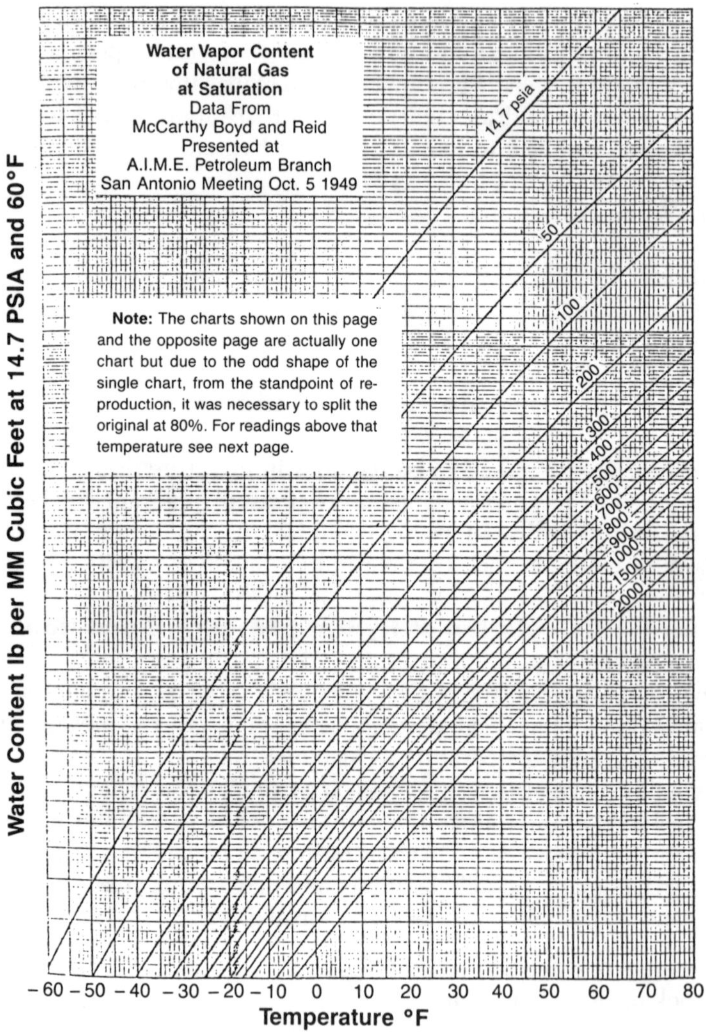

Figure 2-4. Water vapor content of natural gas at saturation.

6. If $\sum_{1}^{N} X_i \neq 1$, assume another T and repeat until $\sum_{1}^{N} X_i$ approaches 1.0.

Prevention of Hydrate Formation

Figure 2-7 is available for finding the quantity of methanol needed to provide a given lowering of the hydrate forming temperature for a given gas.

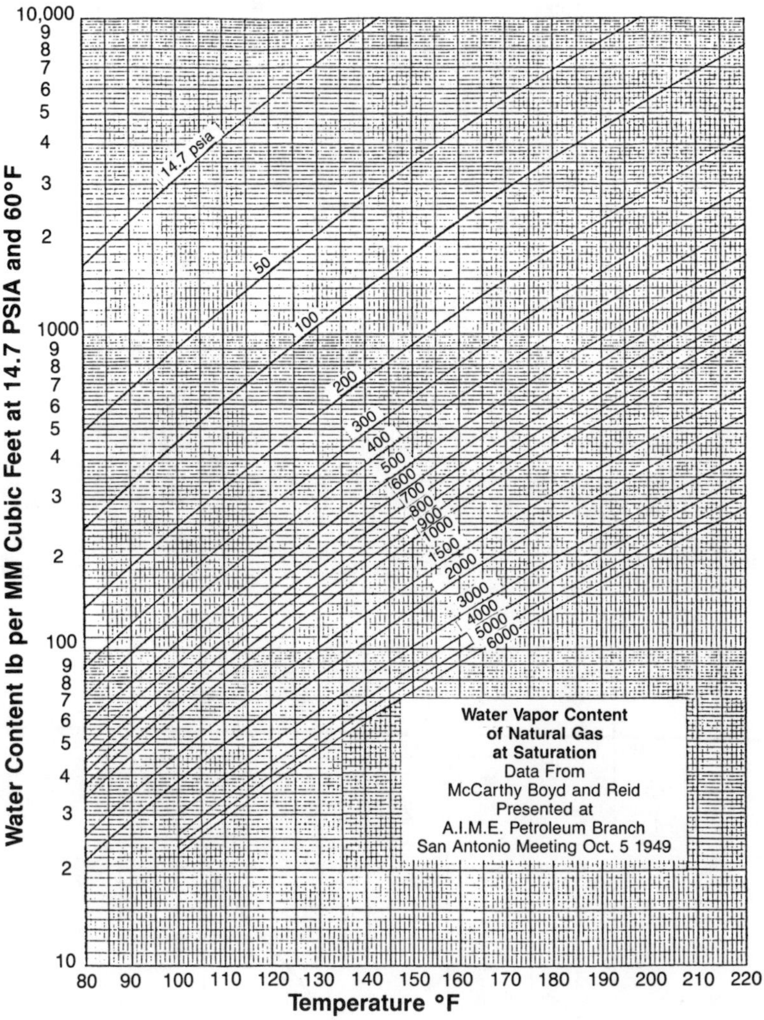

Figure 2-5. Water vapor content of natural gas at saturation.

Since methanol is an economical agent for this task, the procedure for finding the methanol requirement, developed for the Natural Gas Pipeline Company of America by Hammerschmidt[14] (an early pioneer in hydrate investigations), is given here. Figure 2-7 shows the lowering of the freezing point of the hydrate vs. methanol concentration in water.

Water forms in the liquid phase when a gas saturated within the reservoir is cooled, as when coming up the wellbore or flowing in the gathering lines. The alcohol addition must be sufficient to provide an alcohol concentration

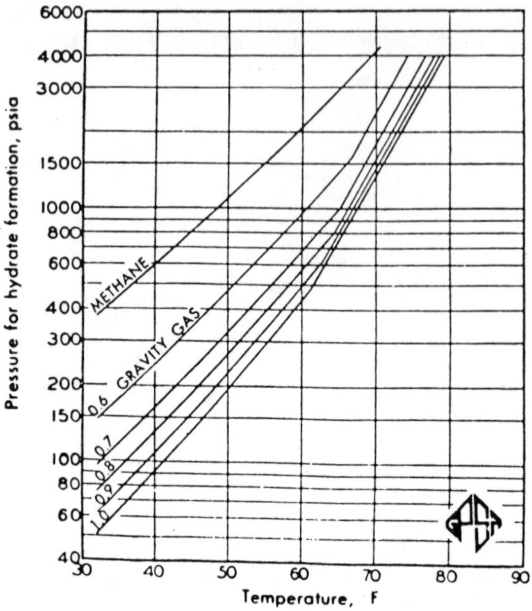

Figure 2-6. Pressure-temperature curves for predicting hydrate formation.[12]

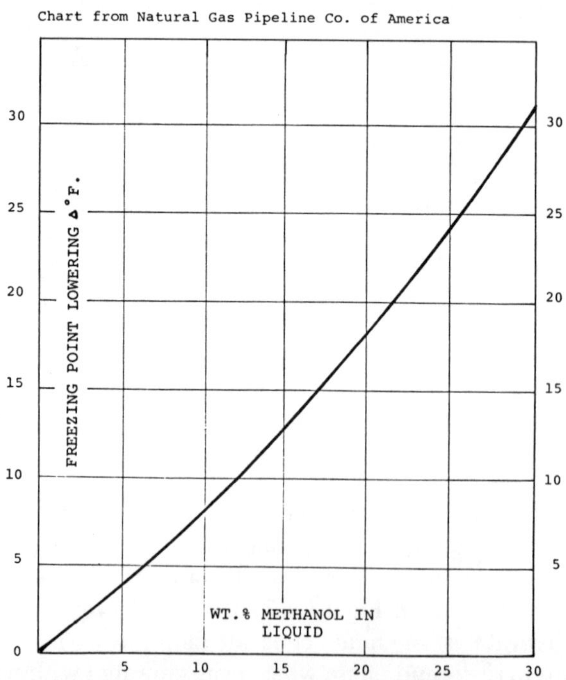

Figure 2-7. Gas hydrate freezing point depression vs. weight % methanol in liquid.

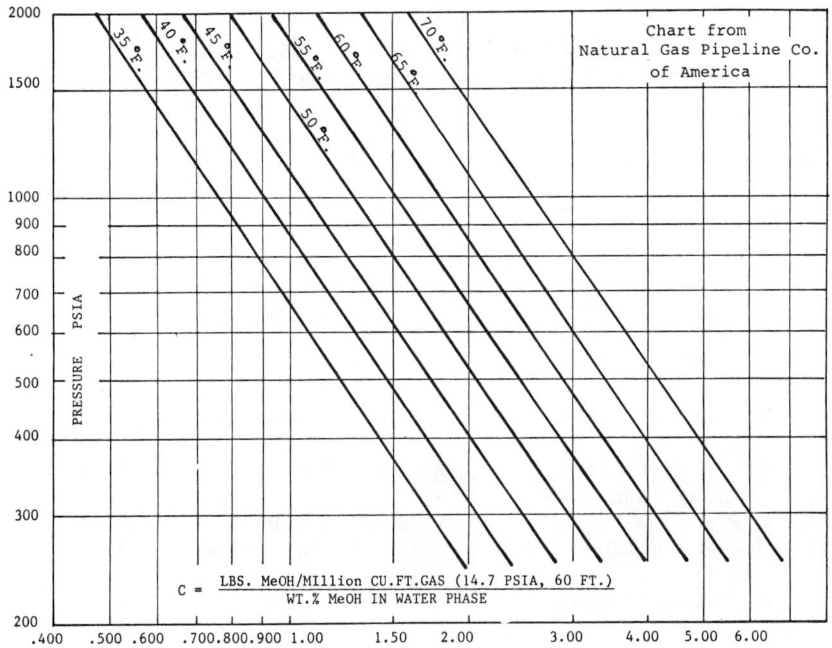

Figure 2-8. Ratio of methanol vapor to liquid composition vs. pressure.

in the water that will prevent hydrates from forming. To do this, there must be enough alcohol injected to provide for the alcohol which vaporizes into the gas phase, as well as that required to maintain a desired concentration in the liquid phase.

Figure 2-8 can be used to find the quantity of methanol in the vapor state. Since the partial pressure of methanol depends on the temperature of the aqueous solution and the methanol concentration, these variables are required to use the chart. The quantity of methanol, then, depends upon the total pressure of the gas.

Table 2-2 shows how one finds the methanol requirements for a given natural gas under specified conditions.

Example 2-1

Problem. Natural gas (0.6 gravity) is coming from the wellhead saturated with water at a temperature of 75°F. In transit to the dehydration plant, the gas will cool to 45°F. The pressure on the wellhead is 800 psia. How much methanol injection is required to prevent hydrate formation as the gas cools to 45°F?

Table 2-2
Work Sheet for Calculating Methanol Requirements to Prevent Hydrate Formation for a 0.60 Gravity Gas*

Step No.	Quantity	Units	Instructions	Example Data & Calculations
1	Line pressure	psia	Field data	700
2	Gas temperature at saturation**	°F	Field data	60
3	Temperature at which line must be protected from hydrates (minimum gas temperature)	°F	Field data	40
4	Water content of gas at saturation	lb/MMCF	Figure 5-8, p. XX	23
5	Water content of gas at minimum temperature	lb/MMCF	Figure 5-8, p. XX	11
6	Change in water content	lb/MMCF	4 minus 5	12
7	Hydrate temperature of gas	°F	Figure 5-40, p. XX Handbook	58.0
8	Freezing point lowering required	°F	7 minus 3	18.0
9	Weight % methanol in liquid	%	Figure 1	20
10	Vapor to liquid composition ratio	lb/MMCF/ wt. %	Using 1 & 3, find from Figure 2	1.17
11	Methanol concentration in gas	lb/MMCF	Multiply 9 by 10	23.4
12	Methanol to saturate liquid	lb/MMCF	Line 6 × Line 9/ 100 − Line 9	3.0
13	Methanol-injection rate	lb/MMCF	11 plus 12	26.4
14	Methanol-injection rate (pure methanol at 68°F)	gal/MMCF	Line 13/6.59 lb/gal	4.0

* Procedure developed at Natural Gas Pipeline Co. of America.
** If gas is undersaturated, use dew point temperature at pressure in number 1.

Solution. At 800 psia, the hydrate-forming temperature for 0.6 gravity gas is 58°F (from Figure 2-6). It will form hydrates long before it cools to 45°F unless methanol is injected.

The freezing-point lowering necessary is 58–45 = 13°F.

The necessary weight percent methanol required in the liquid phase is 15% (from Figure 2-7).

From Figure 2-8, at 800 psia and 45°F, C ≅ 1.32.

$$\frac{\text{lb of Me OH/MMCF gas}}{\text{wt \% Me OH in water phase}} = 1.32$$

methanol required = 15 × 1.32 = 19.80

Say 20 lb/MMCF

OTHER TRANSPORT PROPERTIES RELEVANT TO STORAGE

Viscosity of brines decreases with reservoir temperature and, in general, increases with pressure and salinity. Figure 2-9 shows the viscosity of water as a function of temperature with different salinity and pressure values.[15]

The viscosity of natural gas depends on the temperature, pressure, and composition. Figure 2-10, reproduced from The Natural Gas Supply Men's Association[12], shows the viscosity of hydrocarbon gases at 1 atmosphere as a function of temperature and molecular weight.

Figure 2-11 is a monograph permitting the calculation of hydrocarbon gas viscosity from specific gravity, temperature, and pressure[12]. Figure 2-12 gives the thermal conductivity of paraffinic hydrocarbons as a function of temperature[12]. Figure 2-13 gives the molecular weight of petroleum fractions as a function of boiling point and API gravity. It is used to determine pseudo-critical properties of remainder fractions such as C_{7+}.

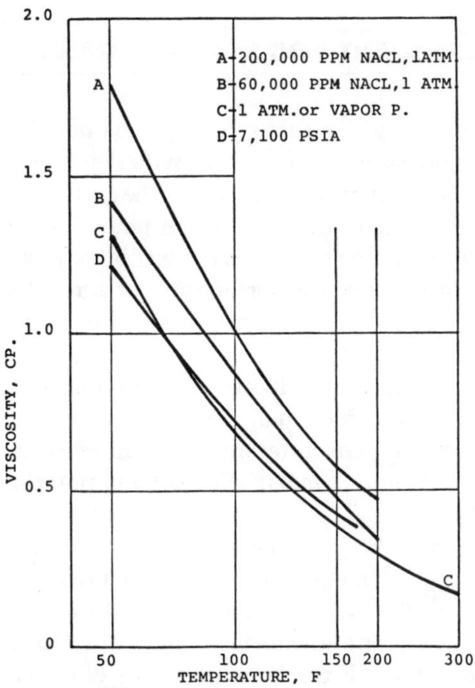

Figure 2-9. Viscosity of brines.[15]

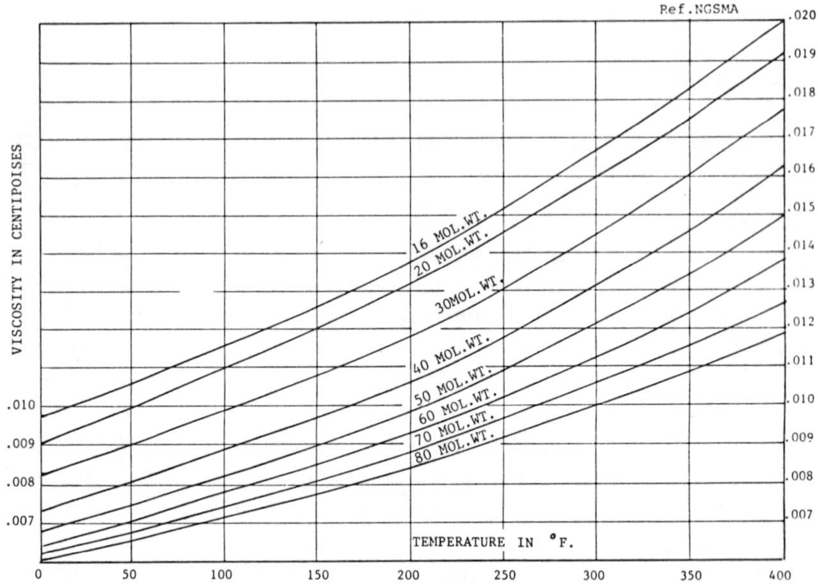

Figure 2-10. Viscosity of hydrocarbon gases.

JOULE-THOMSON COEFFICIENT FOR NATURAL GAS

As the storage gas flows through the wellbore during injection or withdrawal, both the temperature and pressure of the gas vary with depth. The change in the temperature is due to heat transfer between the gas in the wellbore and the surrounding environment, and changes in potential and kinetic energies of the gas. In addition, as the gas enters the wellbore, Joule-Thomson cooling due to pressure drop across the perforations or across the wellbore radius from porous rock into an open hole completion further affects the eventual temperature distribution in the wellbore. The change in pressure occurring in up or down flow in the wellbore is the result of static head changing with vertical position and frictional, i.e., viscous losses, as well as occasional two-phase flow effects. The latter usually manifest themselves in the form of *evaporative cooling* or *condensative heating* resulting from the exchange of latent heat.

A computer program for predicting the Joule-Thomson coefficient was developed, using the Benedict-Webb-Rubin equation of state, as a term paper at the University of Michigan. Its development is included in Appendix A. The results are presented in this chapter in the form of charts correlating the Joule-Thomson cooling coefficients in °F per psi with gravity of the gas as a function of different pressures at constant, given temperatures. These are shown in Figures 2-14 through 2-18.

The Joule-Thomson coefficient is a useful and practical property for estimating the isenthalpic temperature drop of the gas per unit pressure drop. It is defined as:

$$\mu_j = \left(\frac{\partial T}{\partial P}\right)_H \tag{2-8}$$

The Joule-Thomson coefficient characterizes the temperature behavior of gases in a throttling experiment; i.e., when no heat or work crosses the ther-

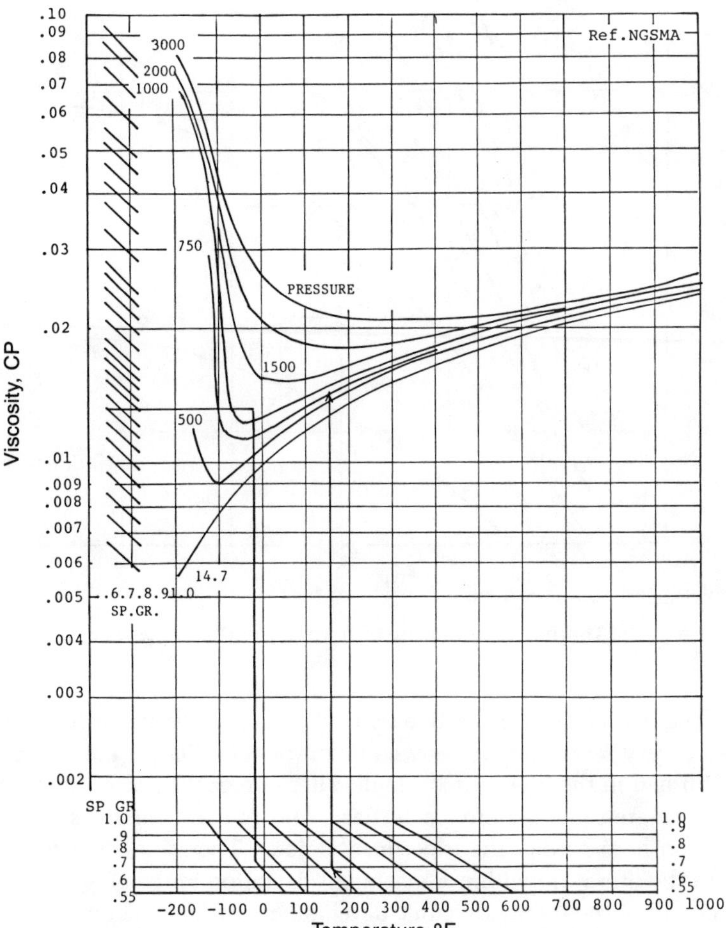

Figure 2-11. Hydrocarbon gas viscosity.

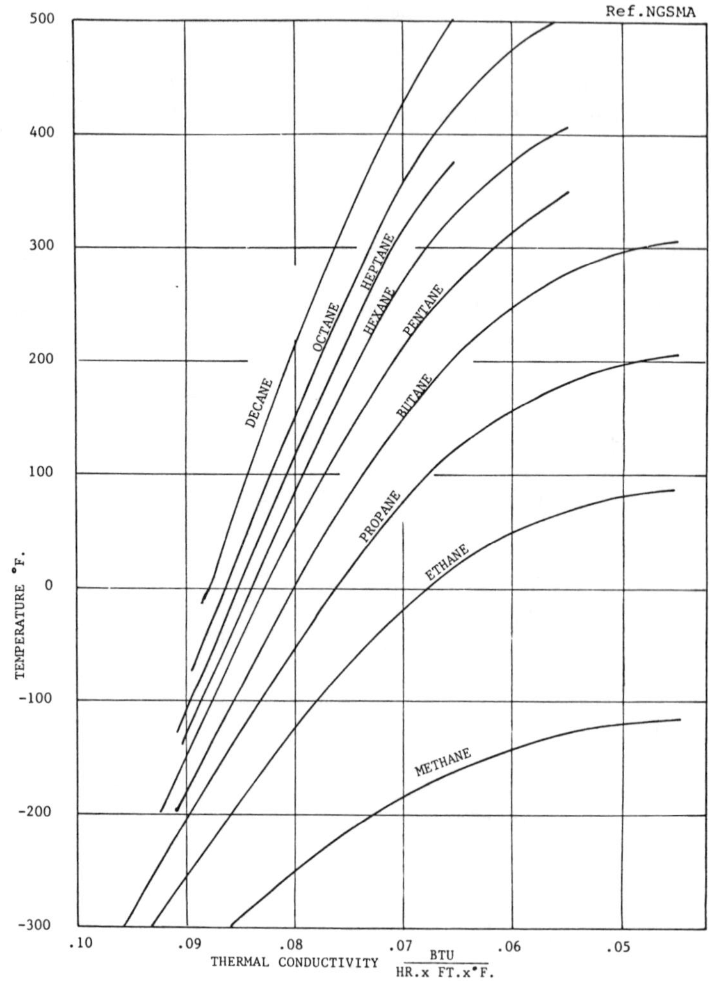

Figure 2-12. Thermal conductivity of normal paraffinic hydrocarbons.

modynamic boundary. It is positive for cooling and negative for heating. It provides a very severe test for any equation of state. There are not many data published in the literature on Joule-Thomson cooling of complex mixtures. There are some for binary, tertiary, and quinary mixtures.[16,17] The charts given here predict the experimental Joule-Thomson coefficients reported in the literature within a maximum deviation range of $\pm 4\%$.

The details of the thermodynamic development of Joule-Thomson coefficients from the Benedict-Webb-Rubin equation, a computer program developed for calculations, and comparison of results with literature data, as well as an example problem, are included in Appendix A.

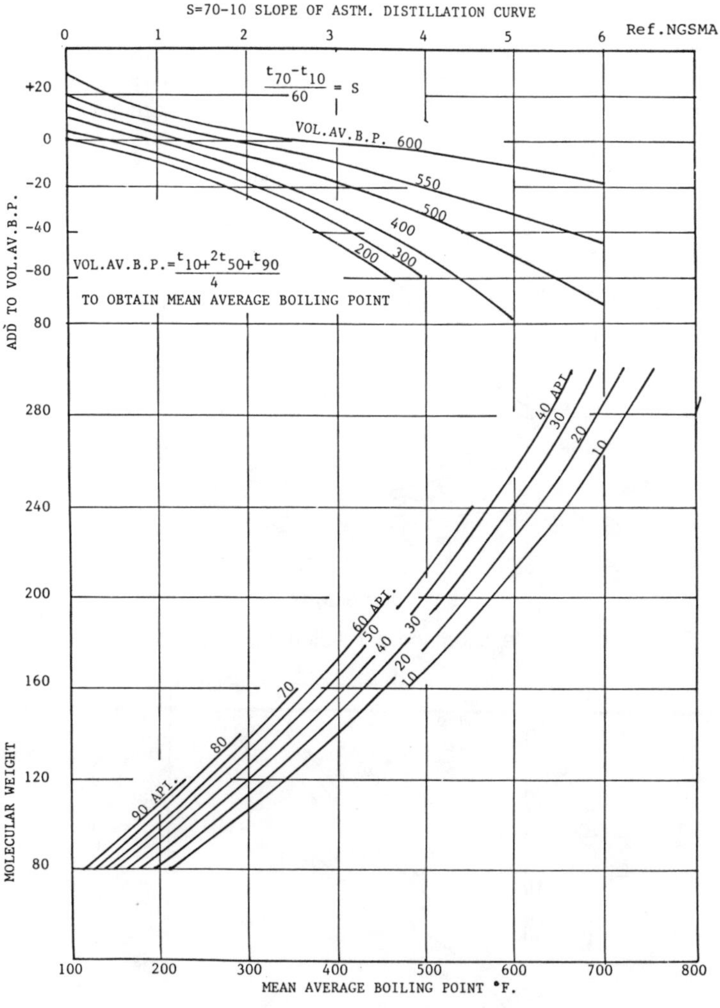

Figure 2-13. Molecular weight vs. boiling point and gravity for petroleum factors.

PROPERTIES OF RESERVOIR ROCKS: POROSITY AND PERMEABILITY

When natural gas is produced or stored in reservoir rock, the nature of the porous medium determines the quantity of the gas content of the rock and the ability of the fluids to move through the rock.

The fraction of the total bulk volume of the rock occupied by the pore space through which the gas is stored is called the *porosity* of the rock. Porosity is a dimensionless, fractional quantity, usually designated by the symbol "ϕ".

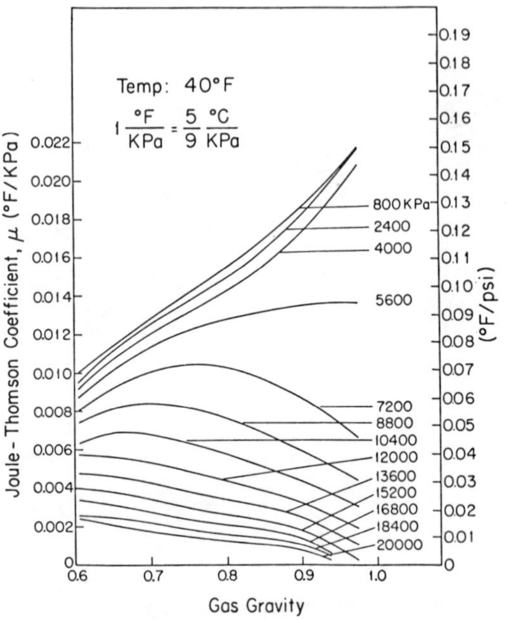

Figure 2-14. Joule Thomson coefficient for natural gas (T = 40°F).

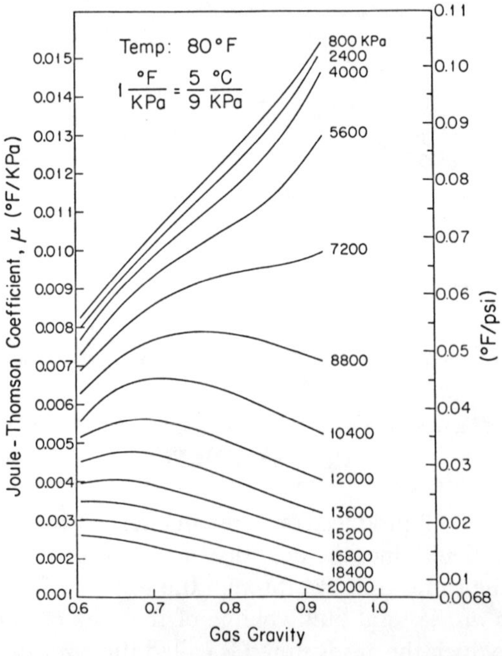

Figure 2-15. Joule Thomson coefficient for natural gas (T = 80°F).

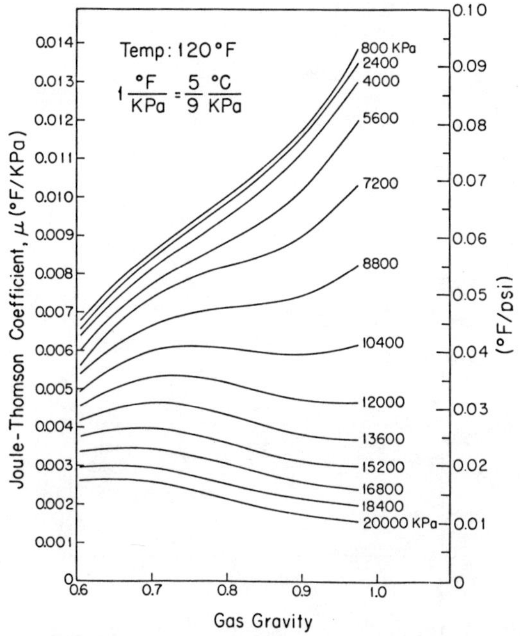

Figure 2-16. Joule Thomson coefficient for natural gas (T = 120°F).

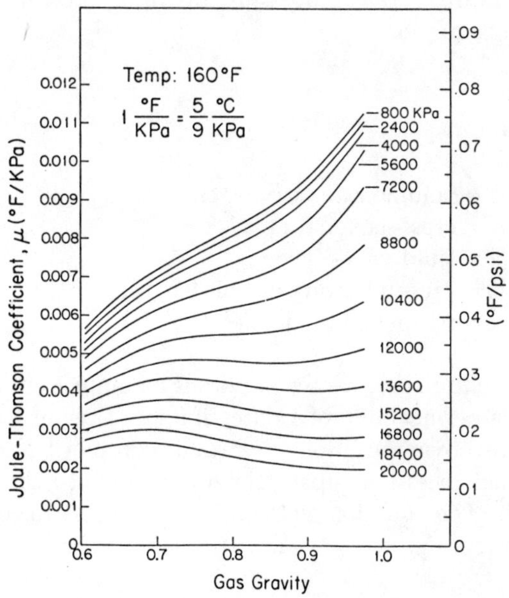

Figure 2-17. Joule Thomson coefficient for natural gas (T = 160°F).

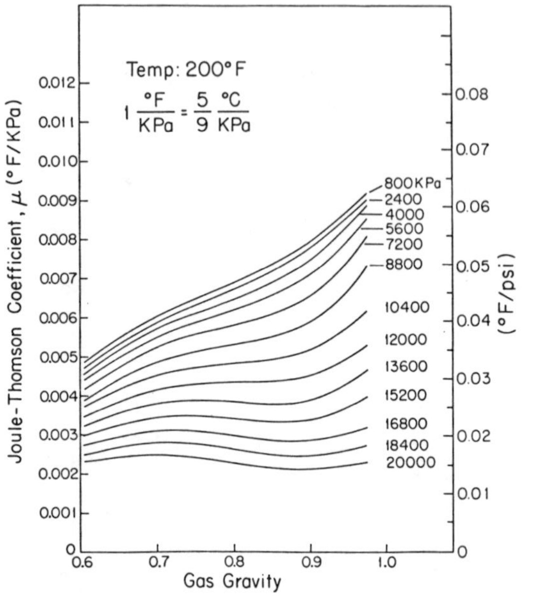

Figure 2-18. Joule Thomson coefficient vs. gas gravity.

The ability of the rock to conduct the flow of fluids through it is referred to as its *permeability*. Permeability is usually designated "k" and is defined by the following equation:

$$\frac{q}{A} = \frac{k}{\mu}\left(\frac{-dP}{dL}\right) \tag{2-9}$$

where
q = volume rate of flow, cm^3/sec
A = cross-sectional area, cm^2
μ = fluid viscosity, cp
$-dP/dL$ = pressure drop per unit length, atm/cm
k = permeability, darcies

According to Equation 2-9, a cube of porous medium 1 cm on edge ($A = 1$ cm^2) will have a permeability of 1 darcy if a rate flow of 1 cm^3 per second can be maintained by application of a pressure drop of 1 atmosphere across the front and back faces 1 cm apart, flowing a fluid of 1 centipoise viscosity (water at 68°F). The equation giving the permeability in reservoir or field units is as follows:

$$q = 0.00634\frac{kA}{\mu}\left(\frac{-dP}{dL}\right) \tag{2-10}$$

where $q = ft^3/day$
k = millidarcies
$A = ft^2$
P = psia
L = ft
μ = cp
$(-dP/dL)$ = psi/ft

Hydrologists define permeability somewhat differently. Since they are principally interested in aquifers and water flow, their coefficient of permeability includes the viscosity of water. In their equations, they have:

$$q = k_p \frac{AH}{L} \tag{2-11}$$

where k_p = coefficient of permeability
q = rate of discharge of an aquifer, gal/day at 60°F
A = cross-sectional area, ft^2
H = difference of hydraulic head, of water
L = length of the aquifer, ft

Measurement of Permeability

The permeability is measured by passing a fluid of known viscosity through a porous core of measured dimensions and measuring the flowrate when a known pressure drop is applied across the core. If a core plug of cross-sectional area A cm^2, L cm long is used with a fluid of viscosity μ centipoise, and if a rate-of-flow of q cm^3 per second is measured when a pressure drop of $P_1 - P_2$ atmospheres is maintained across the ends, then Equation 2-9 may be written to give the permeability K as:

$$k = q \frac{\mu L}{A(P_1 - P_2)} \tag{2-12}$$

using the same units as in Equation 2-9. For the flow of gas through a core, the rate-of-flow q varies with the pressure. The permeability with respect to gas flow, then, may be calculated with q evaluated at the average pressure, or by integrating Equation 2-9 and treating q as a function of pressure P. In this case, the equation giving the permeability of the core is given as follows:

$$k = \frac{2{,}000\ QL\mu P_b}{A(P_1^2 - P_2^2)} \tag{2-13}$$

where k = permeability, millidarcies
Q = gas flowrate measured at base pressure P_b atmospheres, cm^3/sec
L = length of core, cm
μ = gas viscosity, cp
P_1 = upstream pressure, atm
P_2 = downstream pressure, atm

Example 2-2

Problem. A core specimen is 3/4″ in diameter and 1 1/2″ long. In a laboratory test, an oil with a viscosity of 0.73 cp passes through the core at a rate of 97 cm^3 per hour for a pressure drop of 8.1 psi. What is the permeability of the core?

Solution.

$$k = \frac{q\mu L}{A(P_1 - P_2)}$$

$$A = \frac{\pi D^2}{4} = \frac{\pi \times (0.75 \times 2.54)^2}{4} = 2.85 \text{ cm}^2$$

$$k = \frac{97 \times 0.73 \times 14.7}{3{,}600 \times 2.85 \times 8.1} \cong 0.01 \text{ darcies}$$

Measurement of Porosity

Quite often, the same sample from reservoir rock is used to measure the porosity and the permeability. The specimen is usually extracted with an organic solvent in a special extractor and dried in an oven. The measurement principally revolves around the determination of the volume of rock solids and that of the bulk volume. Once the pores are completely filled by a liquid of known density, then the porosity may be calculated as follows:

$$\text{Porosity } \% = \frac{(\text{mass of liquid in pores})\ 100}{(\text{density of liquid})(\text{volume of core})}$$

Nature of Flow Through Porous Media

Just as observed in flow through pipes, the flow through a porous medium may be *laminar* or *turbulent*. If the rate of flow is proportional to the pressure drop, the flow is said to be *laminar*. At high rates of flow, when the pressure drop exceeds that value predicted by a laminar law, the flow is said to be *turbulent*. The foundation for the equations of laminar flow through porous media is Darcy's law, which is expressed mathematically in Equation 2-9. In flow of a fluid through a porous medium, the term *turbulent flow* is used to designate the flow regime where the pressure drop is more than proportional to the flowrate.

The basic relation given by Darcy has been extended to cover the flow of any fluid by several investigators who also defined its limitations and its range of applicability. From two relatively simple concepts, Darcy's law and the principle of conservation of mass, equations can be derived to describe isothermal laminar flow of incompressible, slightly compressible, or substantially compressible fluid flow in one, two, or three dimensions.

The phenomenon of turbulence has been observed for fluids flowing through beds with pores large enough to permit observations of turbulent eddies similar to those observed in pipes. The use of Darcy's law is restricted to conditions of laminar or streamline flow through the pore space. In laminar flow, the individual fluid particles flow along trajectories fixed and parallel to the boundaries of pores. In such a flow regime, inertial forces are negligible compared to viscous forces. If the flowrate is increased beyond a critical value at sharp corners of semipores, constrictions, and points where the direction of flow changes, eddy currents are set up. These eddies disrupt the smooth streamlines and consume energy in excess of the streamline flow. At still higher rates of flow, turbulent effects dominate the flow mechanism and viscous forces become negligibly small compared to inertial forces.

A reliable criterion for determining the rate at which the flow through the porous medium becomes turbulent is, as in pipe flow, the Reynolds number. When the Reynolds number becomes equal to or greater than 1.0, the observed data on flowrate versus pressure drop begin to deviate gradually from Darcy's law. The Reynolds number (R_e) is given by:

$$R_e = \frac{dV\rho}{\mu} \tag{2-14}$$

where ρ = fluid density, lb_m/ft^3
d = effective particle diameter, ft
μ = fluid viscosity, $lb_m/ft \times sec$
V = fluid velocity, ft/sec

If a friction factor is defined for the flow of fluids through porous media by:

$$f = \frac{(-\Delta P)dg_c}{\rho \; 2LV^2} \tag{2-15}$$

Then a dimensionless plot of f vs. R_e clearly indicates the regions where Darcy's law is applicable and the region where it gradually loses its applicability.

Generalized Darcy Equation

A generalized equation has been developed and published by Tek[16] to fully represent the behavior observed in friction factor vs. Reynolds number plots:

$$f = \frac{d^2}{2kR_e}\left(1 + l_f\frac{R_e}{\phi}\right) \tag{2-16}$$

where d = mean particle diameter, ft
k = the permeability, ft^2
R_e = the Reynolds number
ϕ = the porosity
l_f = a dimensionless factor, called the *lithology factor,* representing the particular porous medium

Equations 2-14, 2-15 and 2-16 may be combined to give:

$$-\frac{\Delta P}{L} = \left(\frac{\mu}{k}V + l_f\frac{d\rho}{\phi k}V^2\right)\frac{1}{g_c} \tag{2-17}$$

It may be noticed from Equation 2-17 that, at low rates of flow, the term in V^2 becomes negligible in comparison to the linear term, and Darcy's law will prevail. At high rates of flow, on the other hand, under fully turbulent conditions the term in V may be neglected, leaving $-\Delta P$ proportional to V^2 or f = constant, i.e., independent of the Reynolds number. It can be shown that Equation 2-16, developed by Tek from published data, can also be derived from the Forechheimer equation.[19]

$$-\frac{dP}{dL} = \frac{\mu}{k}V + \beta\,\rho\,V^2 \tag{2-18}$$

The turbulence factor β is given in reference 13, as correlated against the porosity and permeability values.

Example 2-3

Problem. Bell-jar B has a volume V_B between the inlet and exit valves A and B, as shown in Figure 2-19.

Configuration 1: The bell-jar is empty, and it is at a pressure P_1 while valves A and B are closed. Additional volume ΔV is originally at atmospheric pressure. It is connected to a U-tube manometer.

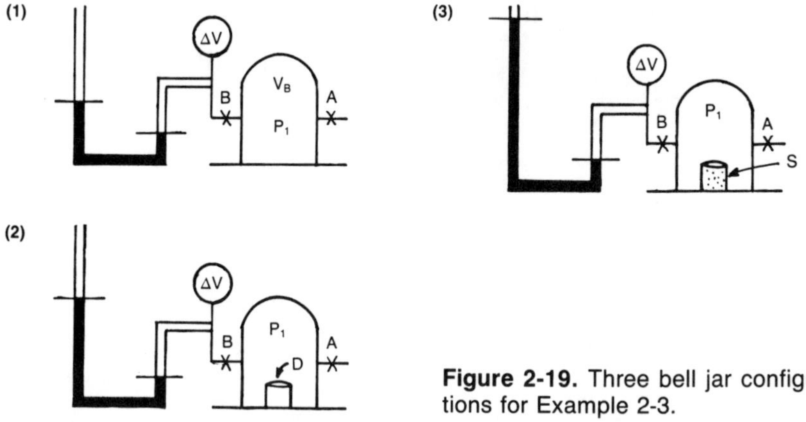

Figure 2-19. Three bell jar configurations for Example 2-3.

With A closed, B is opened, permitting air in V_B to expand into ΔV. When pressures are equalized, the manometer reads P_b in the chamber.

Configuration 2: A nonporous dummy of bulk volume V_D is placed under the bell-jar and the experiment (Configuration 1) is repeated. Now, pressure equalizes to P_d.

Configuration 3: A porous core of the same bulk volume as V_D is placed under the bell-jar and the experiment (Configuration 1) is again repeated as above. Again starting with P_1, this time pressures equalize to P_c.

What is the porosity of the core ϕ in terms of P_b, P_d, and P_c?

Solution.

$$P_1 V_B = P_b(V_B + \Delta V) \tag{1}$$

$$P_1(V_B - V_D) = P_b(V_B + \Delta V - V_D) \tag{2}$$

$$P_1(V_B - V_S) = P_c(V_B + \Delta V - V_s) \tag{3}$$

From Equation 1, $V_B + \Delta V = \dfrac{P_1 V_B}{P_b}$. Substitute in Equations 2 and 3.

$$P_1(V_B - V_D) = P_d\left(\frac{P_1 V_B}{P_b} - V_D\right) \tag{4}$$

$$P_1(V_B - V_S) = P_c\left(\frac{P_1 V_B}{P_b} - V_s\right) \tag{5}$$

Find V_D from Equation 4:

$$V_D(P_d - P_1) = P_d\frac{P_1V_B}{P_b} - P_1V_B$$

$$V_D = P_1V_B\left(\frac{P_d}{P_b} - 1\right)\frac{1}{(P_d - P_1)}$$

$$V_D = P_1V_B\left(\frac{P_d - P_b}{P_b(P_d - P_1)}\right)$$

$$V_s(P_c - P_1) = P_cP_1\frac{V_B}{P_b} - V_BP_1$$

$$V_s = V_BP_1\frac{(P_c - P_b)\ 1}{P_b\ (P_c - P_1)}$$

$$V_D = \frac{P_1V_B}{P_b}\frac{(P_d - P_b)}{(P_d - P_1)}$$

By symmetry:

$$V_s = \frac{P_1V_B}{P_b}\frac{(P_c - P_b)}{(P_c - P_1)}$$

$$\frac{V_s}{V_D} = \frac{(P_c - P_b)}{(P_c - P_1)}\frac{(P_d - P_1)}{(P_d - P_b)}$$

and:

$$\phi = (V_D - V_s)/V_D$$

$$= 1 - \frac{V_s}{V_D}$$

$$\phi = 1 - \frac{(P_c - P_b)}{(P_d - P_b)}\frac{(P_d - P_1)}{(P_c - P_1)}$$

If $(P_d - P_1) \cong (P_c - P_1)$, then:

$$\phi = \frac{(P_d - P_c)}{(P_d - P_b)}$$

Example 2-4

Problem. Figure 2-20 represents a gas reservoir discovered on a sandstone formation structurally adjacent to an aquifer in the underburden, comprised of blanket sand formation containing salt water.

The initial discovery pressure is 3,200 psia at bottom hole.

The reservoir temperature is 213°F.

The specific gravity of the gas is 0.7.

The average porosity is 0.172.

The connate water saturation from logs is determined to be 0.25 of pore space.

The seismic maps indicate that the structure of the gas sand S has a bulk volume of 415×10^6 ft^3.

1. From the data given above, calculate the initial gas-in-place reserves in standard cubic feet (SCF) at a base pressure of 14.7 psia and base temperature of 60°F.

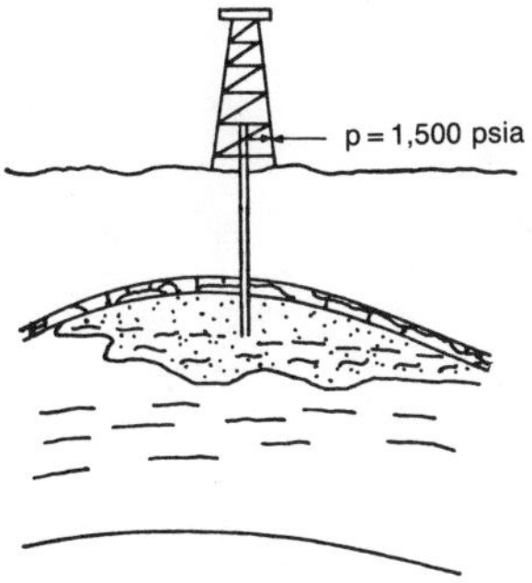

Figure 2-20. Cross section of gas reservoir for Example 2-4.

2. At the wellhead gathering system where the pressure is expected to be equal to 1,500 psia, what would be the hydrate-forming temperature for the gas?

3. After the reservoir has produced 935 million SCF of gas, the reservoir pressure is observed to decline to 2,925 psia. Assuming that the pore volume vacated by production of gas is fully occupied by influx of water, calculate how many cubic feet of water have invaded the reservoir between the discovery and the date of cumulative production of 935 MMCF.

4. If there was no water influx, what would the pressure have been after the cumulative production of 935 MMCF?

5. If the effective compressibility of native water and sandstone is constant and equal to 7×10^{-6} vol/vol × psi, estimate the approximate volume of the water in the aquifer surrounding the gas reservoir. (You may assume that during the production of gas the average pressure on the aquifer has declined from the original 3,200 psia to 3,000 psia.)

Solution.

1. Initial Reserves:

   Number of moles = n

   $$PV = znRT$$

   $$n = \frac{PV}{ZRT}$$

   $$T_c = 390\ °R$$

   $$P_c = 668\ \text{psia}$$

$$P_r = \frac{3{,}200}{668} = 4.79$$

$$T_r = \frac{460 + 213}{390} = \frac{673}{390} = 1.73$$

$Z = 0.882$

Net Pore Volume:

$$\begin{aligned} V &= V_B\phi(1 - Sw) \\ &= 415 \times 10^6 \times 0.172 \times (1 - 0.25) \\ &= 53.535\ 10^6\ ft^3 \end{aligned}$$

Original Moles:

$$n_o = \frac{53.535 \times 10^6 \times 3{,}200 \times 144}{.882 \times 1{,}544 \times 673}$$

$$= 26{,}916{,}546.17 \text{ moles}$$

$$Q_o = 26{,}916{,}546.17 \times 379.29$$

$$\begin{aligned} &= 1.02 \times 10^{10} \text{ SCF} \\ &= 10.2 \text{ MMMCF} \end{aligned}$$

2. At 1,500 psia, 0.7 gravity
Hydrate threshold $T_L = 68°F$

3. $n_p = \frac{935 \times 10^6}{379.29} = 2.47 \times 10^6$ cumulative moles

Remaining moles $= 26{,}916{,}546.17 - 2.47 \times 10^6$
$n = 2.45 \times 10^7$

New pore volume $= V_1$
$P_1 = Z_1 n_1 RT$

$$V_1 = Z_1 \frac{n_1 RT}{P_1}$$

$$P_{r1} = \frac{2{,}925}{668} = 4.38$$

$T_{r1} = 1.73$

$Z_1 = 0.872$

$$V_1 = \frac{0.872 \times 2.45 \times 10^7 \times 1{,}522 \times 673}{2{,}925 \times 144} = 5.27 \times 10^7\ ft^3$$

Water influx is the difference between original and later pore volumes.

$$V = 53.535 \times 10^6 - 5.27 \times 10^7$$
$$= 8.35 \times 10^5 \text{ ft}^3$$

4. If there was no water influx, the resulting pressure due to 925 MMCF depletion would have been:

$$P_1' = \frac{Z'_1 n_1 RT}{V_o}; \quad \text{assume } Z'_1 = Z_1 = 0.872, \quad n_1 = 26.916 \times 10^6 - 2.47 \times 10^6 = 2.45 \times 10^7 \text{ moles}$$

$$P_1' = \frac{0.872 \times 2.45 \times 10^7 \times 1{,}544 \times 673}{53.535 \times 10^6 \times 144} \cong 2{,}874 \text{ psia}$$

Check back Z_1: $Pr_1 = \dfrac{2{,}880}{668} = 4.3;\ T_r = 1.73;\ Z_1 = .870$ OK

5. $c = 7 \times 10^{-6}$ vol/vol × psi

$$c = -(1/V)(\Delta V/\Delta P)$$

$$v = +\Delta V/c(-\Delta P)$$

$$= \frac{8.35 \times 10^5}{7 \times 10^{-6}(3{,}200 - 3{,}000)}$$

$$= \frac{8.35 \times 10^5}{7 \times 10^{-6} \times 200} = 5.96 \times 10^8 \text{ ft}^3$$

Example 2-5

Problem. You want to generate 1 megawatt electrical power using a fossil fueled boiler/generator combination on natural gas feedstock.

1. With 30% overall efficiency and atmospheric air to support the combustion, determine the daily steady-state gas flow required.

2. What would be the cost of generating 1 kW-hr if the cost of natural gas is $4.00/MCF?

3. If, through compressed air injection, air is available at 50 psia, and the stoichiometric ratio and efficiency are maintained at the same levels, how much power can be generated?

   1 kW-hr = 3,412 Btu

   Heating value of natural gas = 972 Btu/SCF

Solution.

1. MW = 10^6 Watts
   = 1,000 kW

   Total energy generated in 1 day = 24,000 kW-hr

   1 kW-hr = 3,412 Btu

   Natural gas needed per day:

$$Q = \frac{24{,}000 \times 3{,}412}{972 \times .30} = 280{,}905.35 \text{ SCF}$$

$$Q = 280.905 \text{ MCF/D or about } 11{,}840 \text{ lb}_{m/D}$$

2. Cost at $4.00 MCF:

$$C = 280.905 \times 4 = \$1{,}123.62/\text{Day}$$

$$\frac{1{,}123.62}{1{,}000 \times 24} = \frac{\$1.123}{24}/\text{kW-hr}$$

$$= 4.68¢/\text{kW-hr}$$

3. If gas is pure methane:

   If the stoichiometric ratio is maintained, having air at 50 psia would make 50/14.7 = 3.4 times the number of moles of air available to combine burning 3.4 times as much methane and, as such generating 3.4 times the power; i.e., 3.4 megawatts.

Example 2-6

Problem. The composition from a newly discovered gas field is given as follows:

Component	Mole Fraction
C_1	0.8612
C_2	0.0591
C_3	0.0358
C_4	0.0172
C_5	0.0050
CO_2	0.0010
N_2	0.0207
	1.000

1. Calculate the specific gravity of gas.
2. Calculate the z-factor from composition.
3. Calculate the z-factor from specific gravity.
4. At discovery pressure 4,000 psia (bottom hole) and temperature T = 89°F reservoir, what would be the viscosity of gas?

Solution.

Comp.	(1) Mole Frac.	(2) (M.W.) (lb_m/lb-mole)	(3) T_c, °R	(4) P_c, psia	(1) × (2) Mass per mole
C_1	0.8612	16.04	344.2	673.1	13.814
C_2	0.0591	30.07	550.3	709.8	1.777
C_3	0.0858	44.09	666.3	617.4	1.578
n-C_4	0.0172	58.12	765.6	550.7	1.000
n-C_5	0.0050	72.15	845.5	489.5	.361
n-CO_2	0.0010	44.01	548.0	107.1	.044
N_2	0.0207	28.01	227.6	493.0	.580
					Σ = 19.154 lb_m/lb-mole

1. Calculate S.G.: S.G. = mole weight of gas/molecular weight of air
   Molecular weight of gas = Σ(1) × (2) = 19.154 lb/lb-mole
   Specific gravity of gas = 19.154/29 = 0.66

2. z – factor from composition:

$$z = f\left(\frac{T}{{}_pT_c}, \frac{P}{{}_pP_c}\right)$$

where T = 89°F or 549 °R

and P = 4,000 psia.

$$ {}_pT_c = \Sigma(1) \times (3) = (296.43 + 32.522 + 23.85 + 13.17 + 4.23 + .548 + 4.7113)\ °R = 375.46\ °R $$

$$ {}_pP_c = \Sigma(1) \times (4) = (579.69 + 41.95 + 22.10 + 9.47 + 2.447 + 1.071 + 10.21) = 666.92 \text{ psia} $$

${}_pT_r$ = 549/375.46 = 1.4622, pseudo-reduced temperature

${}_pP_r$ = 4000/666.92 = 5.997, pseudo-reduced pressure

From Figure 2-3, z(1.4622,5.997) = 0.85

3. Z-factor from specific gravity:

   Using the Z-factor chart for S.G. = 0.65, T_r = 375°R, P_r = 670 psia

   $Z(P_r = 5.97, T_r = 1.46) = 0.85$

4. From Figure 2-10, the viscosity of hydrocarbon gas having a molecular weight of 19.15 lb_m/lb-mole is 0.011 cp at 89°F and 1 atm.

   The pressure correction factor, from the *Handbook of Natural Gas Engineering*,[13] is 2.5 at 4,000 psia. So:

   $M = 2.5 \times 0.011 = 0.028$ cp

Alternate method. Using Figure 2-11 for 0.66 gravity, 89°F, and 4,000 psia, by extrapolation:

$M = 0.029$ cp

References

1. Sclater, K. C. and R. B. Stephenson. "Measurements of Original Pressure, Temperature and Gas Oil Ratio in Oil Sands," *Trans. AIME* (1928–1929) 82, 119.
2. Millikan, C. V. and C. V. Sidwell. "Bottom-Hole Pressures in Oil Wells," *Trans. AIME* (1931) 92, 194.
3. Fancher, G. H., J. A. Lewis, and K. B. Barnes. "Some Physical Characteristics of Oil Sands," The Pennsylvania State College, Bull. 12 (1933) p. 65.
4. Wyckoff, R. D., H. Botset, M. Muscat, and D. W. Reed. "Measurement of Permeability of Porous Media," *Bull. AAPG* (1934) 18, No. 12, p. 161.
5. Wyckoff, R. D., and H. Botset. "The Flow of Gas Liquid Mixtures through Unconsolidated Sands," *Physics* (1936) 7, p. 325.
6. Leverett, M. C. and W. B. Lewis. "Steady Flow of Oil-Gas-Water Mixtures through Unconsolidated Sands," *Trans. AIME* (1941) 142, p. 107.
7. Schilthuis, R. J. "Technique of Securing and Examining Subsurface Samples of Oil and Gas," *Drilling and Production Practice*, API (1936) pp. 171–180.
8. Pyle, H. C. and P. H. Jones. "Quantitative Determination of Connate Water Content of Oil Sands," *Drilling and Production Practice*, API (1936) pp. 171–180.
9. Schilthuis, R. J. "Connate Water in Oil and Gas Sands," *Trans. AIME* (1938) 127, 199–214.
10. Millikan, C. W. "Temperature Surveys in Oil Wells," *Trans. AIME* (1941) 142, 15.

11. Moore, T. V. "Reservoir Engineering Begins Second 25 Years," *Oil and Gas Journal* (1955) 54, No. 29, p. 148.
12. *Engineering Data Book*, Natural Gas Processors Suppliers Association, 1966.
13. Katz, D. L., et al. *Handbook of Natural Gas Engineering*, McGraw-Hill, 1959.
14. Hammerschmidt, E. G. "Formation of Gas Hydrates in Natural Gas Transmission Lines," *Industrial Engineering Chemistry* (1934) 26, 851.
15. Craft, B. C. and M. F. Hawkins, Jr. *Applied Petroleum Reservoir Engineering*, Prentice Hall, 1959.
16. Pasumarti, B. P. and M. R. Tek. "Predicting Joule Thomson Coefficients for Natural Gas," University of Michigan, August 1982.
17. Alhert, L. C. and L. A. Wenzel. *AIChE Journal* (1969) 15, 256.
18. Tek, M. R. "Development of a Generalized Darcy Equation," *Journal of Pet. Tech.* (1957) IX, No. 6, pp. 45–47, *Trans. AIME* (1957) 210, p. 45.
19. Forchheimer, P. "Wasserbewegun durch Boden," deitz, *Ver Deutsch, Ing.* (1901) 45, 1731.

See Appendix K for additional references on high velocity gas flow.

3

THEORY OF FLOW THROUGH STORAGE RESERVOIRS

INTRODUCTION

Production and pressure are the two related quantities that are measured and utilized in reservoir engineering perhaps more than any other parameters. When one is dealing with an oil or gas reservoir, the production is measured in stock-tank barrels of oil, or standard cubic feet (SCF) of gas produced. Often, the production is referred to bottom hole conditions where *in situ* reservoir barrels or bottom hole cubic feet of gas flow quantities are involved. Whether one is dealing with oil or gas reservoirs, the removal of hydrocarbons from the formation usually results in a pressure decline or gradient that is carefully measured. If the reservoir has been shut-in for a sufficiently long time to allow the pressure distribution to be equalized, one refers to *static pressure*. For condition of flow the pressures encountered are said to be *flowing pressures*.

Since the pressure, as measured, reflects the condition prevailing in the reservoir, the inventory in hydrocarbons, the degree of depletion, and other relevant reservoir information, it is considered to be the main dependent variable. It is a function of independent variables that relate to time and space. The latter are sometimes referred to as *time-space coordinates*.

If the pressure, P is known as a certain function P(x,y,z,t) of time-space coordinates, then one can compute the flowrate by first differentiating the pressures and obtaining the gradients prevailing at any time. From the gradients, the use of Darcy's law permits calculation of local superficial velocities over appropriate cross-sectional areas to give the rates of production in bottom hole conditions. The production at the bottom hole condition can readily be translated to the corresponding surface condition in units of stock tank barrels, SCF, or MCF, using some thermodynamic information related to solubility-shrinkage of oil, and pressure, temperature, and volume behavior of the gas. In implementing the procedure described above, one deals with several rock and fluid properties, called the reservoir or fluid parameters. The permeability k, the porosity ϕ, the compressibility c, and the formation volume factor B are typical reservoir or fluid parameters.

If the reservoir parameters are constant, the reservoir is said to be *homogeneous*. If some of the parameters have a directional dependency (for instance, vertical permeability being different from horizontal), then the reservoir is said to be *anisotropic*. Quite often the fluid properties vary with thermodynamic coordinates; i.e., pressure, temperature, and composition. If the reservoir events do change with time, the process is said to be *unsteady-state* or *transient*. When at least one variable changes with time, the process is *unsteady-state*. If, on the other hand, none of the variables change with time, the process is called *steady-state*. Since practically all production or injection operations involve decrease or increase of hydrocarbons in the reservoir, unsteady-state practically always prevails. Under certain conditions, however, either the rates or the pressures become constant or nearly constant, and the reservoir engineer finds it useful to look at the process as an approximate or "pseudo" steady-state. The equations describing the flow through porous media become simpler and more straightforward if the process is steady-state.

Sometimes, engineers also find it practical to view a process changing continuously with respect to time as a "succession of steady-states."

FUNDAMENTALS OF FLOW THROUGH POROUS MEDIA

The mathematical description of fluid flow through a porous medium almost always involves consideration of pressure, the dependent variable, as a function of time-space coordinates:

$$P = P(x,y,z,t) \quad (3\text{-}1)$$

In addition to the pressure, the complete description of the process of flow through porous media also requires determination of the velocities and the density of the fluid at any point (x,y,z) at any time t.

Accordingly:

$$U_x = U_x(x,y,z,t) \quad (3\text{-}2)$$

$$U_y = U_y(x,y,z,t) \quad (3\text{-}3)$$

$$U_z = U_z(x,y,z,t) \quad (3\text{-}4)$$

$$\rho = (x,y,z,t) \quad (3\text{-}5)$$

where p = pressure in porous media (units will be discussed later)
U_x = superficial velocity in x direction
U_y = superficial velocity in y direction
U_z = superficial velocity in z direction
ρ = density of the fluid
x,y,z = the three cartesian coordinates
t = the time variable

These equations indicate that we have five unknowns, i.e., five dependent variables. These are P, U_x, U_y, U_z, and ρ. Opposite these five unknowns we have, in general, four independent variables. These are x, y, z, t. Since we have five unknowns, in order to solve the problem completely we must have five equations. These equations are obtained by invoking three principles of general physics. These are:

1. Continuity equation—the law of conservation of mass
2. Darcy's law, or another suitable phenomenological rate expression
3. A thermodynamic equation of state.

Continuity Equation

Continuity equation, otherwise known as the principle of conservation of mass, is one of the main pillars of experimental and theoretical physics. It simply states that in flow through porous media, with respect to a considered system (an arbitrary region), the flux of mass input due to flow, minus the flux of mass output due to flow, plus the flux of mass introduced from some external source must be equal to the increase of mass content for the region. Consider, as shown in Figure 3-1, a three-dimensional rectangular element of sides Δx, Δy, Δz at time $t = t$ in a porous, permeable environment where the flow through porous media is taking place.

The element E shown is a parallelepiped adjacent to x, y, z and having its sides Δx, Δy, Δz. The net-mass flux-rate inbound in x direction is equal to $\rho U_x \, \Delta y \, \Delta z$.

The net-mass flux outbound in x direction is $\Delta y \, \Delta z \, [\rho U_x + \Delta(\rho U_x)]$. The net-mass flux-rate in the x direction is then $-\Delta y \, \Delta z \, \Delta(\rho U_x)$. Similarly, the net-mass flux-rate in the y and z directions are, respectively, $-\Delta x \, \Delta z \, \Delta(\rho U_y)$ and $-\Delta x \, \Delta y \, \Delta(\rho U_z)$.

Now, because the net total mass influx-rate in all three directions must be equal to net accumulation in a time increment Δt, one obtains:

$$-\Delta t \, [\Delta y \, \Delta z \, \Delta(\rho U_x)] - \Delta t \, [\Delta x \, \Delta z \, \Delta(\rho U_y)] - \Delta t \, [\Delta x \, \Delta y \, \Delta(\rho U_z)] = \Delta x \, \Delta y \, \Delta z \, \phi[\rho_{t+\Delta t} - \rho_t] \quad (3\text{-}6)$$

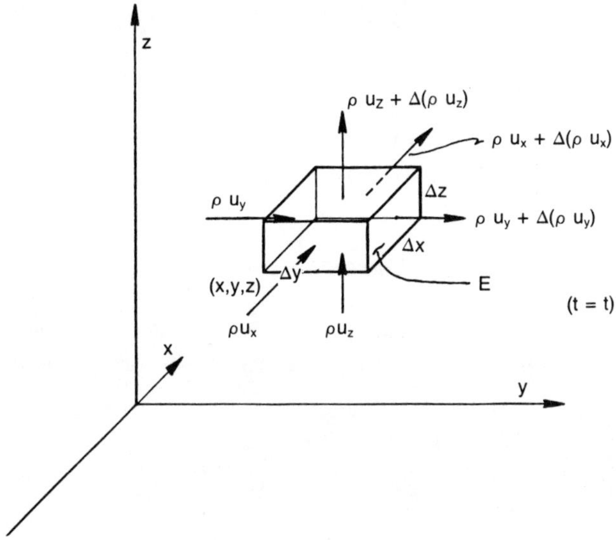

Figure 3-1. Element of volume in flow through porous media.

Equation 3-6 may be rewritten as:

$$-\frac{\Delta(\rho U_x)}{\Delta x} - \frac{\Delta(\rho U_y)}{\Delta y} - \frac{\Delta(\rho U_z)}{\Delta z} = \frac{\phi[\rho_{\Delta t+t} - \rho_t]}{\Delta t}$$

or proceeding to the limit as

$\Delta x \rightarrow 0$

$\Delta y \rightarrow 0$

$\Delta z \rightarrow 0$ and $\Delta t \rightarrow 0$

$$\frac{\partial}{\partial x}(\rho U_x) + \frac{\partial}{\partial y}(\rho U_y) + \frac{\partial}{\partial z}(\rho U_z) = -\phi\frac{\partial \rho}{\partial t} \quad (3\text{-}7)$$

Equation 3-7 is the continuity equation, which may also be written in short form as:

$$\text{div}(\rho\vec{U}) = -\phi\frac{\partial \rho}{\partial t}$$

where $\text{div}(\vec{\ }) = \frac{\partial}{\partial x}(\vec{\ })_x + \frac{\partial}{\partial y}(\vec{\ })_y + \frac{\partial}{\partial z}(\vec{\ })_z$ (3-8)

when the argument $(\vec{\ })$ is a vector quantity.

The continuity equation may also be derived in radial coordinates as:

$$\frac{1}{r}\frac{\partial}{\partial r}(r\rho\ U_r) = -\phi\frac{\partial \rho}{\partial t} \tag{3-9}$$

In the equation above, the porosity ϕ is assumed to be constant.

Darcy's Law

Darcy's law[1] is a phenomenological rate expression between a driving force, a potential gradient, and a resistance to the rate. It is usually expressed as:

$$\vec{U} = -\frac{k\rho}{\mu}\nabla\Phi \tag{3-10}$$

In Equation 3-10, U is the velocity vector, the magnitude of which is equal to volumetric rate of flow per unit cross sectional area, Φ the driving potential, $\nabla\Phi$ the gradient of the driving potential in the direction of flow, μ the viscosity of the fluid, k the permeability of the porous medium, and ρ the density of the fluid. The minus sign is to indicate positive flow in the direction of decreasing potential.

In Equation 3-10, the driving potential for flow through porous media has the dimension of L^2/T^2. It was M. King Hubbert[2] who showed that:

$$\Phi = \int_{p_o}^{p}\frac{dp}{\rho} + gz \tag{3-11}$$

where z was height above datum, p_o pressure at datum, and g, the acceleration of gravity.

If the vector Equation 3-10 is projected along three directions x, y, z, one obtains:

$$U_x = -\frac{\rho k_x}{\mu}\frac{\partial\Phi}{\partial x} \tag{3-12}$$

$$U_y = -\frac{\rho k_y}{\mu}\frac{\partial\Phi}{\partial y} \tag{3-13}$$

$$U_z = -\frac{\rho k_z}{\mu}\frac{\partial\Phi}{\partial z} \tag{3-14}$$

In Equations 3-12 through 3-14, the permeability is recognized to be, in general, different in three different directions of flow.

In substituting the value of Φ from Equation 3-11:

$$U_x = -\frac{k_x}{\mu}\frac{\partial p}{\partial x} \tag{3-15}$$

$$U_y = -\frac{k_y}{\mu}\frac{\partial p}{\partial y} \tag{3-16}$$

$$U_z = -\frac{k_z}{\mu}\left(\frac{\partial p}{\partial z} + g\right) \tag{3-17}$$

For radial flow in the horizontal direction, Darcy's law becomes:

$$U_r = -\frac{k_r}{\mu}\frac{\partial p}{\partial r} \tag{3-18}$$

In Equations 3-17 and 3-18, the units consistent with A.P.I.'s definition of permeability standards are:

U = cm/sec
k = Darcy
μ = centipoise
p = atmosphere
x,y,z,r = cm
ρ = atm × sec^2/cm^2

These units are called hybrid Darcy, cgs units.

Practical Oil or Gas Field Units in Permeability

In practical oil or gas field units, the equation giving the permeability may be written as:

$$q = 0.00634\,\frac{kA}{\mu}\left(\frac{-dp}{dL}\right) \tag{3-19}$$

where q = cu ft/day volumetric flow rate at the point in porous media where the pressure gradient is dP/dL
A = ft^2
μ = centipoise
k = millidarcies
dp/dL = psi/ft

Tensor Concept of Permeability

Let us examine Equation 3-10:

$$u = -\frac{k\rho}{\mu}\nabla\Phi \tag{3-20}$$

where $\Phi = \int_{p_o}^{p} \frac{dp}{\rho} + gz$ (3-21)

In combining Equations 3-20 and 3-21:

$$u = -\frac{k}{\mu}\rho\nabla\int_{p_o}^{p}\left(\frac{dp}{\rho} + gz\right) \tag{3-22}$$

Now, projecting along three directions and involving the Leibnitz rule for differentiation under the integral sign, for the general case of anisotropy in permeabilities:

$$u_x = -\frac{k_x}{\mu}\frac{\partial P}{\partial x} \tag{3-23}$$

$$u_y = -\frac{k_y}{\mu}\frac{\partial P}{\partial y} \tag{3-24}$$

$$u_z = -\frac{k_z}{\mu}\left(\frac{\partial P}{\partial z} + \rho g\right)^{*} \tag{3-25}$$

The relationships between the three velocity vector components in cartesian coordinates, pressure gradient vector components, and the gravitational vector may also be written in the form:

$$\vec{u} = -\frac{k}{\mu}(\nabla p + \nabla\rho g) \tag{3-26}$$

In Equation 3-26 we have proportionality between $\vec{u}$ vector $= \begin{bmatrix} u_x \\ u_y \\ u_z \end{bmatrix}$

and the sum of pressure gradient vector $\vec{\nabla}p = \begin{bmatrix} \partial p/\partial x \\ \partial p/\partial y \\ \partial p/\partial z \end{bmatrix}$

and the gravitational weight vector $\vec{\rho}g = \begin{bmatrix} 0 \\ 0 \\ -pg \end{bmatrix}$

* The equation is in absolute units.

When specific orientation in porous media is considered in addition to magnitude and direction, it becomes clear that K must be defined in all three main cartesian directions. It is therefore a tensor quantity, defined as:

$$k = \begin{Bmatrix} k_x & 0 & 0 \\ 0 & k_y & 0 \\ 0 & 0 & k_z \end{Bmatrix}$$

In general, in three-dimensional space a scalar is a tensor of zero order defined only by $3^0 = 1$ quantity, its magnitude. By comparison, a vector is a tensor of 1st order defined by 3^1 = three terms, i.e., having a direction defined by three unit vectors over and above its magnitude. And finally, a tensor of second order requires $3^2 = 9$ quantities to describe it, because it must have not only magnitude but directions for orientation as well.

In engineering mechanics, other typical examples of tensor quantities involve proportionalities between stress and deformation vectors.

Deviations from Darcy's Law

Klinkenberg Effect. When the fluid flowing through the porous media is a gas, it was observed by Klinkenberg[3] that at pressures sufficiently low, when the mean free path of the molecules becomes comparable to geometric dimensions of the interpore space, then the molecules slip through, giving high apparent and pressure-dependent permeability. It was shown that such permeabilities correlate linearly with $1/p$, reaching a limiting value of k_L at $1/p = 0$, i.e., $P = \infty$. The value of k_L is the permeability to single-phase liquid, readily measureable using oil, brine, or pure distilled water.

Accordingly:

$$k_g = k_L(1 + b/P) \tag{3-27}$$

where k_g = permeability to gas
k_L = permeability to gas at infinite pressure (which should be equal to permeability to liquid)
p = pressure
b = a constant depending upon the particular gas porous medium

High Flowrates—Inertial and Turbulence Effects. It has been known for a long time that Darcy's law becomes inadequate in describing high velocity flow through porous media. Fancher, et al.[4] measured pressure drop and flowrate for a large number of consolidated and unconsolidated sands. When dedimensionalized and plotted on a log-log scale, the friction factor corresponding to the pressure drop was correlated vs. a Reynolds number describing the flow through porous interstices. The correlation showed typical − 1 slope at low Reynolds numbers with gradual deviations to a nearly horizontal trend at high Reynolds numbers. These data and behavior were generalized in a dimensionless equation by Tek[5] utilizing a lithology factor, a dimensionless permeability number, a friction factor, and Reynolds number:

$$f = \frac{d^2}{2k}\frac{1}{R_e}\left[1 + \frac{\alpha}{\phi} R_e\right] \tag{3-28}$$

where d = average grain diameter
k = permeability (d^2/k = dimensionless permeability number)
$R_e = \frac{\rho v d}{\mu}$, Reynolds number
α = dimensionless lithology factor
ϕ = porosity

In the past, various investigators attributed deviations from Darcy's law to turbulence (Fancher and Lewis, 1933; Elenbaas and Katz, 1948; Cornell and Katz, 1953).* Others observed that in the rigid constraints imposed upon pore structure, the geometry did not permit truly random turbulence to develop in the sense it does in pipes (Tek, 1957; Bakhmeteff and Feodoroff, 1937). Some treated turbulence in the sense of a factor influencing permeability measurement (Carlson and Eastman, 1940). Others correlated the deviations from Darcy's law on the basis of a dedimensionalized friction factor/Reynolds number relationship (Brownell and Katz, 1947; Brownell and Dombrowski, 1950).

Johnson and Taliaferro (1938) correlated data taken at the Bureau of Mines using an nth power rate term in the pressure-drop equation. Green and Duwez studied high velocity gas flow through sintered metals. They used the classic Forchheimer equation with squared rate correction to the pressure-drop term:

$$-\frac{dp}{dL} = \mu\frac{u}{k} + \beta\rho u^2 \tag{3-29}$$

where β has subsequently been called the turbulence factor and correlated by Katz and Cornell, Katz and Janicek, and more recently recorrelated by Katz and Firoozabadi. In some cases a third, velocity cubed term was suggested in addition to the quadratic term in Equation 3-29.

In the area of immediate deviation from Darcy's law and before the onset of random turbulence, lengthening of flow lines and inertial effects were suggested as the cause for deviation from Darcy's regime (Hubert, 1950;

* These references are listed in Appendix K.

Humpert, 1953). An excellent review of inertial effects occurring in high velocity gas flow was also given by J. C. Slattery during 1963.

The more recent and generally accepted consensus on deviations from Darcy's law is that, as the velocity is increased, initial deviation is due to inertial effects, followed by even higher velocities with turbulent effects (Wright, 1968).

Another form of Equation 3-29, which incorporates laminar-inertial-turbulence effects in the sequence of Darcy-Transition-Non-Darcy regime, is given in the form of (LIT) flow effects correction factor by Govier (1961) and Wattenbarger and Ramey[6].

$$u = -\delta \frac{k}{\mu} \frac{dp}{dx} \tag{3-30}$$

where $\delta = 1/\left(1 + \beta \frac{\rho k u}{\mu}\right)$

$\delta = 1$ gives Darcy's law

For an anisotropic porous medium, δ is different in x, y, z directions, where it too can be considered a tensor quantity modifying the permeability tensor.

$$u = -\frac{1}{\mu} k\delta \nabla p \tag{3-31}$$

where $\delta = \begin{bmatrix} \delta_x & 0 & 0 \\ 0 & \delta_y & 0 \\ 0 & 0 & \delta_z \end{bmatrix}$

Equation of State

For an ideal gas, the pressure-volume-temperature relationship is given as:

$$pV = \frac{m}{(MW)} RT \tag{3-32}$$

where
p = pressure
V = volume
m = mass
MW = molecular weight
R = gas constant
T = absolute temperature

Since $\rho = m/V$, i.e., mass per unit volume, the density of an ideal gas may thus be expressed as:

$$\rho = \frac{p(MW)}{RT} \tag{3-33}$$

As the gas flows through the porous rock, the flow is isothermal, so:

$$\frac{\partial \rho}{\partial t} = \frac{(MW)}{RT}\frac{\partial p}{\partial t} \tag{3-34}$$

For a real gas, on the other hand:

$$\rho = \frac{(MW)}{ZRT}\,p \tag{3-35}$$

Partial Differential Equation for Liquid Flow

For isothermal flow of fluids of small and constant compressibility, by combining Darcy's law, continuity, and an expression of density as a function of compressibility, one obtains:

$$\left(k_x\frac{\partial^2 p}{\partial x^2} + k_y\frac{\partial^2 p}{\partial y^2} + k_z\frac{\partial^2 p}{\partial z^2}\right) + c\left[k_x\left(\frac{\partial p}{\partial x}\right)^2 + k_y\left(\frac{\partial p}{\partial y}\right)^2 + k_z\left(\frac{\partial p}{\partial z}\right)^2\right] + \left(\frac{\partial p}{\partial x}\frac{\partial k_x}{\partial x} + \frac{\partial p}{\partial y}\frac{\partial k_y}{\partial y} + \frac{\partial p}{\partial z}\frac{\partial k_z}{\partial z}\right) = \phi\mu c\frac{\partial p}{\partial t} + \mu\frac{\partial \phi}{\partial t} \tag{3-36}$$

In Equation 3-36, the effect of gravity is neglected and the viscosity is assumed constant. Provided that c is small, k is constant and isotropic, and porosity is constant; and, if the pressure gradients squared terms are neglected, Equation 3-36 reduces to:

$$\frac{\partial^2 p}{\partial x^2} + \frac{\partial^2 p}{\partial y^2} + \frac{\partial^2 p}{\partial z^2} = \frac{\phi\mu c}{k}\frac{\partial p}{\partial t} \tag{3-37}$$

For radial flow, in the general case, we have:

$$\frac{1}{r}\frac{\partial}{\partial r}\left(r\frac{\partial p}{\partial r}\right) + \frac{1}{k_r}\frac{\partial k_r}{\partial r}\frac{\partial p}{\partial r} + c\left(\frac{\partial p}{\partial r}\right)^2 = \frac{\phi\mu c}{k_r}\frac{\partial p}{\partial t} + \frac{\mu}{k_r}\frac{\partial \phi}{\partial t} \tag{3-38}$$

Again, provided that c is small, gravity is neglected, viscosity is constant, k is constant, and $(\partial p/\partial r)^2$ is negligibly small, Equation 3-38 becomes

$$\frac{\partial^2 p}{\partial r^2} + \frac{1}{r}\frac{\partial p}{\partial r} = \frac{\phi \mu c}{k}\frac{\partial p}{\partial t} \tag{3-39}$$

Equation 3-39 describes radial flow of a slightly compressible liquid through porous media in the unsteady-state. It is used in calculating the movement of water adjacent to storage reservoirs. It must be remembered that it is only through the assumptions of constant rock properties, small and constant compressibility, and negligibly small pressure gradients that the linear partial differential equation (3-39) is obtained. Otherwise, it would be nonlinear on several counts. Both Equations 3-37 and 3-39 are recognized as parabolic linear partial differential equations, often encountered in rate processes and called the *diffusivity equation*. The lumped parameter $k/\phi\mu c$ is called the *diffusivity*.

Equations 3-37 and 3-39 may be represented in short by:

$$\nabla^2 p = \frac{\phi \mu c}{k}\frac{\partial p}{\partial t} \tag{3-40}$$

The compressibility c is the total, saturation-weighted compressibility of the entire reservoir system. The diffusivity equation is also encountered in heat conduction, molecular diffusion, and electric potential distribution.

Steady-state, point source, and more general unsteady-state solutions to Equation 3-40 will be discussed later.

Partial Differential Equation for Gas Flow

The steps in deriving the partial differential equations describing the flow of natural gas in reservoirs are the same as those used for liquid flow. In this instance, however, one must use for the relationship between the pressure and density either the ideal gas law or the one that includes the gas compressibility factor.

Ideal Gas. When Darcy's law and the continuity equation are combined with:

$$\rho = \frac{M}{RT}p \tag{3-41}$$

for constant gas viscosity, constant rock properties, and neglecting gravity and squared pressure gradient terms, one obtains:

$$\frac{\partial^2 p^2}{\partial x^2} + \frac{\partial^2 p^2}{\partial y^2} + \frac{\partial^2 p^2}{\partial z^2} = \frac{2\phi\mu}{k}\frac{\partial p}{\partial t} \tag{3-42}$$

In the case of radial flow:

$$\frac{\partial^2 p^2}{\partial r^2} + \frac{1}{r}\frac{\partial p^2}{\partial r} = \frac{2\phi\mu}{k}\frac{\partial p}{\partial t} \tag{3-43}$$

or

$$\nabla^2 p^2 = \frac{\phi\mu}{kp}\frac{\partial p^2}{\partial t} \tag{3-44}$$

Equation 3-44 is not general, as it only applies to ideal gas, and only where one may justify neglecting the squared gradient terms. It is also non-linear.

Equations for Real Gas Behavior. When the non-ideality of natural gas is accounted for in using the z-factors, the effect of pressure on gas viscosity must be included as well. This is accomplished through the use of *pseudo-pressure*, which will be discussed shortly in a separate section.

Historical Perspective of Gas Flow Equations

Problems related to deliverability and flow of natural gas in reservoirs have been of interest for almost half a century. All through the development of petroleum reserves, the understanding and technology in gas production lagged behind the advent of oil production and developed at a slower pace. There were two reasons for this:

1. Development of resources and markets for natural gas remained years behind oil exploration and production.
2. Analytical difficulties for realistic description of gas flow through porous media persisted because of the nature of math involved.

During the early years, Muskat[7] suggested a simplified approach using constant low compressibility for natural gas. This was tantamount to adopting oil flow equations for natural gas. It proved reasonably adequate for high-pressure gas wells subject to low drawdowns. It was not applicable to producing fields, which encounter high pressure gradients due to particular high rates or heterogeneities.

The partial differential equation describing the unsteady-state gas flow was first developed for an ideal gas in the Darcy flow regime (Equation 3-44). It was nonlinear on several counts. In its derivation, the high-order gradient squared terms had been neglected. During the early years of developing gas technology, there had been numerous attempts to linearize the partial differential equation. Most of these resulted in the use of p^2 terms rather than p, in extending reservoir engineering procedures developed for oil wells to gas wells. An excellent overview that puts the literature in perspective is given by Al-Hussainy and Ramey[8]. During the intermediate years of computer science and numerical analysis applications (late fifties and early sixties), several researchers sought numerical solutions to nonlinear gas-flow problems using finite difference approximations[9,10,11,12].

Finally, in 1966, the most general descriptive approach to gas flow requiring the least amount of idealization and assumptions was developed by Al-Hussainy and Ramey[8] in the form of real gas potential, also called the *pseudo-pressure*. When the real gas potential, rather than p^2 term is used, they showed that the equation describing the gas flow indeed reduced to the familiar parabolic diffusion equation, which could be treated as quasi-linear under certain conditions.

Differential Equations Describing Real Gas Flow

For the most general case—laminar, inertial, turbulent flow of real gas with pressure- and temperature-dependent viscosity, compressibility, and Z-factor—one of the constitutive equations (modified Darcy or Forchheimer) is combined with continuity and the expression of the equation of state to generate the partial differential equation describing the flow:

$$\nabla \left(\frac{p}{\mu z} \cdot k\delta \nabla p \right) = \frac{\partial}{\partial t} \left(\phi \frac{p}{z} \right) \tag{3-45}$$

where p = pressure
ϕ = porosity
k = permeability tensor
z = compressibility factor
t = time
∇ = operator (vector over a scalar, scalar (gradient) over a vector)
δ = LIT correction factor and it has the notation of a tensor

$$\delta_{ii} = \frac{1}{1 + \beta \rho k_{ii} u_{i/\mu}} \tag{3-46}$$

where (in appropriate units) β = turbulence factor (from Forchheimer eq.)[13]
ρ = density of gas

u_i = velocity vector (i direction)
μ = viscosity of the gas

When the real gas potential is defined as:

$$m(p) = 2\int_{o}^{p} \frac{p}{z\mu}\, dp \tag{3-47}$$

for Darcy's flow, a change of variable from p to m(p) reduces Equation 3-45 to:

$$\nabla^2 m(p) = \frac{\phi\mu c}{k}\frac{\partial}{\partial t} m(p) \tag{3-48}$$

For the more general case of non-Darcy flow in a radial direction, using the δ_r LIT modifier, one readily obtains:

$$\frac{1}{r}\frac{\partial}{r}\left[r\ \delta_r \frac{\partial m(p)}{\partial r}\right] = \frac{\phi\mu c}{k}\frac{\partial m(p)}{\partial t} \tag{3-49}$$

where c, the compressibility of the real gas, is defined as:

$$c = \frac{1}{p} - \frac{1}{z}\left(\frac{\partial z}{\partial p}\right)_T \tag{3-50}$$

Detailed derivations for Equations 3-48 and 3-50 are included in Appendix B.

Steady-State Solutions

The steady-state solutions result from integrating the differential equations with proper boundary conditions, with the time-dependent term set equal to zero. Steady state solutions may be due to constant pressure or no-flow boundary. In the case of no-flow boundary they are called semi-steady state.

Steady-state solutions provide useful flow equations for design and long-time operation of storage fields.

Liquid Flow. The steady-state solution for single-phase radial flow of liquids through porous media is given for the *constant pressure boundary* condition in field units:

$$Q = 0.00709\,\frac{kh}{\mu}\frac{(p_1 - p_2)}{\ln r_1/r_2} \tag{3-51}$$

where Q = flowrate into the wellbore, Bbls/day
k = permeability, millidarcies
h = formation thickness, ft
μ = viscosity of liquid, centipoise
r = radius, ft (r_1 at outer boundary, r_2 at wellbore)
p = pressure, psia (p_1 at outer boundary, p_2 at wellbore)

For semi-steady-state with constant liquid flowrate subject to no flow boundary condition, one obtains:

$$\bar{p} - p_{wf} = \frac{q\mu}{2\pi kh}\left[\ln r_{eD} - \frac{3}{4}\right] \tag{3-52}$$

Equation 3-52 is in darcy units

$r_{eD} = r_e/r_w$

where r_e = exterior radius (of no flow boundary), cm
r_w = interior radius (of wellbore), cm
$\bar{p}$ = average pressure in the drainage matrix, atm
p_{wf} = semi steady state flowing pressure in the wellbore, atm
q = production into the wellbore, cc/sec
μ = viscosity, cp
k = permeability, darcies
h = formation thickness, cm

It must be noted that, when the flow is steady-state with *no flow* boundary condition, the rate of decline of wellbore pressure with time becomes constant and equal to:

$$\frac{\partial p_{wf}}{\partial t} = -\frac{q}{ch\pi\ r_e^2\ \phi} \tag{3-53}$$

Rate of pressure decline becomes inversely proportional to fluid-filled pore volume.

The steady-state liquid flow equations given above are helpful in certain calculations related to aquifer storage fields.

Gas Flow. The steady-state gas flow equations with constant-pressure boundary condition will be presented below for both laminar (Darcy) and laminar-inertial-turbulent (general) cases.

First, the approximate solution for constant-pressure (both internal and external) boundaries and average μ and z-factor throughout the drainage matrix will be given. Next, the more complete solution for the z-factors varying throughout and, finally, the steady-state solutions both from the integration of Forchheimer's equation and those corresponding to the m(p) equation will be discussed.

Steady-State Radial Darcy Flow of Gas (Constant Pressure Boundaries)

Figure 3-2 represents the radial-cylindrical flow model upon which the flow equation will be based. Upon integrating the partial differential equation of the gas flow through a porous reservoir, the following steady-state, radial viscous flow equation may be derived:

$$Q = \pm \frac{0.0007030\ hk\ (P_1^2 - P_2^2)}{zT\mu \ln r_1/r_2} \tag{3-54}$$

where h = sand thickness, ft
k = permeability, millidarcies
P = pressure, psia
z = compressibility factor (average over the drainage area)
T = absolute flowing temperature, °R
μ = viscosity, centipoises (average)
Q = steady state gas flow rate measured at 14.7 psia and 60 F, Mcf/day
r = radius, feet

The above set of units is called the *field units for gas flow*. For gas flowing from a reservoir to a well, point 1 is out in the reservoir boundary and point 2 is at the wellbore. When gas is being injected into a well, then point 1 is at the wellbore and point 2 is at the outer boundary. The sign + or − denotes withdrawal or injection, respectively.

For the case of constant pressure boundaries but variable z throughout, an approximate steady-state solution can be derived as follows:

$$Q = \frac{2\pi\ kh}{\mu} \frac{1}{\ln r_b/r_w} \frac{T_f}{T_{sc}} \frac{1}{p_{sc}} \int_{p_w}^{p_f} \frac{p}{z}\, dp \tag{3-55}$$

For a gas of known composition or gravity, using the z-factor charts, one can plot p/z vs. p then graphically integrate.

In Equation 3-55 the units are Darcy units, i.e.:

Q in std cc/sec

k in darcies

h in cm

T_f in °R abs
T_{sc} in °R abs
p_{sc} in atm
p in atm
μ in cp

p_f is pressure at the formation boundary. p_w is pressure at the wellbore. Radius of external boundary is r_b. Radius of the wellbore is r_w. The p/z integrals have been tabulated and are available in the literature.

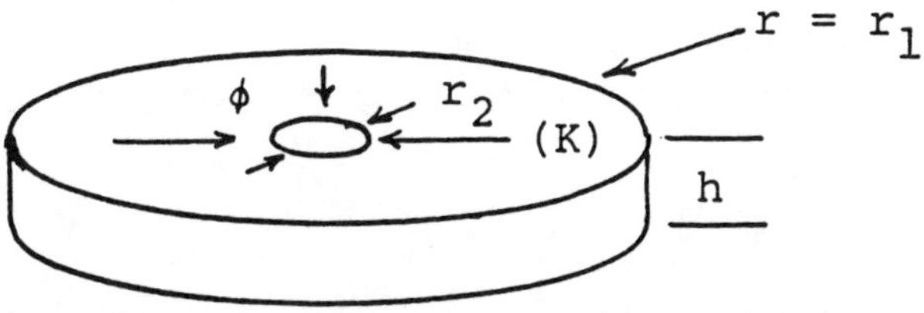

Figure 3-2. Model for horizontal radial flow.

Steady-State Radial Turbulent Flow of Gas (Constant Pressure Boundaries)

Equation 3-54 is valid when Darcy's law is valid or the flow is in the laminar range. When the inertial effects become appreciable, one must use in the steady state the radial turbulent flow equation, which is given as follows:

$$p_1^2 - p_2^2 = \frac{1.424\ zTQ\ln(r_1/r_2)}{hk} + \frac{3.161 \times 10^{-12}\ GQ^2zT(1/r_2 - 1/r_1)}{h^2} \tag{3-56}$$

Equation 3-56 may be derived by direct integration of the Forchheimer equation. The derivation is given in Appendix B.

Equation 3-56 is in field units, i.e.:

p_1 = pressure at $r = r_1$ reservoir boundary, psia
p_2 = pressure at $r = r_2$ wellbore radius, psia
Q = std rate of production, (SCF/D)
μ = average viscosity, cp
z = average z-factor, dimensionless
r_1 = radius, ft
r_2 = radius, ft

h = formation thickness, ft
k = permeability, millidarcies
β = turbulence factor from Forchheimer eq., ft^{-1}
G = specific gravity in gas = (MW)/29, dimensionless.

Steady-State Solution in Terms of Real Gas Potential

For constant terminal rate, Equation 3-48 is solved analytically subject to appropriate initial and boundary conditions for constant μc product.

For pseudo-steady-state, and a finite reservoir having r_w and r_e as wellbore- and exterior-boundary radii, the equation for constant terminal rate is given as[8]:

$$m[p(r_e,t)] - m[p(r_w,t)] = 5.0304 \times 10^7 \frac{p_{sc}\,T}{T_{sc}\,kh} q_{sc} \left[\ln r_e/r_w - 3/4 + s + Dq_{sc}\right] \quad (3\text{-}57)$$

In Equation 3-57, s is the dimensionless coefficient describing the skin effect occurring near the wellbore. It can be estimated from the far and near zone permeabilities by the equation given by Hawkins[14]:

$$s = \left(\frac{k}{k_s} - 1\right) \ln \frac{r_s}{r_w} \quad (3\text{-}58)$$

where
k = permeability of far zone, millidarcies
k_s = permeability of near zone, millidarcies
r_s = radius of skin zone, ft
r_w = radius of wellbore, ft
$m[p(r,t)]$ = real gas potential, $psia^2/cp$
p_{sc} = standard pressure = 14.7 psia
T_{sc} = standard temperature = 520 °R abs
T = formation temperature °R abs
q_{sc} = constant terminal production rate MMCF/D
h = formation thickness, ft

The turbulence coefficient D may be evaluated from

$$D(\mu_w, r_w) = \frac{7.675 \times 10^{-13} G\beta k}{h\,\mu_w} \left(\frac{1}{r_w} - \frac{1}{r_d}\right) \quad (3\text{-}59)$$

D has the dimension of $1/q_{sc}$.

Equation 3-59 is in field units, where k is in millidarcies, h in feet, μ_w in cp and β in ft^{-1}; r_w and r_d are in feet. The drainage radius r_d can be calculated from $t_D = 1/4$.*

$$t_D = \frac{2.637 \times 10^{-4} kt}{(\mu c)_i \, r^2} \tag{3-60}$$

The product (μc), which varies with pressure, is recommended to be evaluated at $p = p_i$ corresponding to the initial condition in an equilibrated storage reservoir. t = time in hours.

Unsteady-State Solutions

A vast amount of literature is available on unsteady-state solutions to the linear diffusivity equation in terms of pressure, real gas potential (pseudo-pressure) in gas flow. We shall review and highlight liquid unsteady-state solutions first. In underground storage, these apply to water movement, pump testing, design and development calculations applied to aquifer storage.

Unsteady State Solution for Radial Flow of Liquids (Small and Constant Compressibility)

Basically, the differential equation describing the unsteady-state flow of liquids in porous media is nonlinear and not susceptible to direct analytical solution. However, there are certain simplifying assumptions that remove the nonlinearities and permit analytical solutions, but have also proven practical in the field. These solutions[15] will be highlighted in this section.

There are three basic solutions as related to the geometric shapes and conditions prevailing at time-space boundaries. These are for:

1. Infinite reservoir, radial flow
2. Finite reservoir, radial flow, no flow outer boundary
3. Finite reservoir, constant pressure, outer boundary

In all three cases above, the following assumptions are necessary before a solution can be obtained:

1. Well penetrating the entire thickness of formation
2. Homogeneity k = constant, ϕ = constant

---

* Katz, et al. suggested $t_D = 1/1.498$ for radius of drainage.
Tek, Grove, and Poettmann recommend 1/4.29.
Van Poolen as well as Matthews and Russel recommends 1/4.

3. Isotropy $k_x = k_y = k_z = k_r = k$
4. Constant h, formation thickness
5. Small and constant c, effective compressibility
6. Constant μ, viscosity
7. Small pressure gradients (negligible squared gradients)
8. Negligible gravity forces

Infinite Reservoir—Line Source (Liquid Flow)

$$p(r,t) = p_i - \frac{q\mu}{4\pi\ kh}\left[\frac{1}{2}Ei\left(\frac{\phi\mu\ cr^2}{4kt}\right)\right] \tag{3-61}$$

$$Ei(x) = \int_x^{\infty} \frac{e^{-u}}{u}\,du \tag{3-62}$$

$$\text{for } x < \frac{1}{100},\ Ei(x) = \ln\left(\frac{1}{x}\right) - 0.5772$$

Equation 3-61 may be approximated by:

$$p_{wf} = p_i - \frac{q\mu}{4\pi kh}\left[\ln\frac{kt}{\phi\mu c{r_w}^2} + 0.80907\right] \tag{3-63}$$

Equation 3-63 is in Darcy units. A step-by-step derivation is included in Appendix C.

Finite Reservoir Radial Flow—No Flow Boundary (Liquid Flow)

$$p_{wf} = p_i - \frac{q\mu}{2\pi kh}\left[\frac{2t_{Dw}}{{r_{eD}}^2} + \ln r_{eD} - \frac{3}{4} + 2\sum_{n=1}^{\infty}\frac{e^{-n^2 t_{Dw}}J_1^2(\alpha_n\ r_{eD})}{n^2[J_1^2(\alpha_n\ r_{eD}) - J_1^2(\alpha_n)]}\right] \tag{3-64}$$

In Equation 3-64

$$r_D = \frac{r}{r_w},\ r_{eD} = \frac{r_e}{r_w},\ t_{Dw} = \frac{kt}{\phi\mu c{r_w}^2}$$

also:

α_n are the roots of:

$$J_1(\alpha_n r_{eD})Y_1(\alpha_n) - J_1(\alpha_n)Y_1(\alpha_n r_{eD}) = 0$$

where J_1 and Y_1 denote Bessel functions of first order first kind. Units in Equation 3-64 are Darcy units.

For sufficiently large t_{Dw} Equation 3-64 reduces to:

$$p_{wf} = p_i - \frac{q\mu}{2\pi kh}\left[\frac{2t_{Dw}}{r_{eD}^2} + \ln r_{eD} - \frac{3}{4}\right] \tag{3-65}$$

Finite Reservoir Radial Flow—Constant-Pressure Outer Boundary (Liquid Flow)

$$p_{wf} = p_i - \frac{q\mu}{2\pi kh}\left[\ln r_{eD} - 2\sum_{n=1}^{\infty}\frac{e^{-\beta_n^2 t_{Dw}}J_o^2(\beta_n r_{eD})}{n^2[J_1^2(\beta_n) - J_o^2(\beta_n r_{eD})]}\right] \tag{3-66}$$

where $r_{eD} = \dfrac{r_e}{r_w}$

$$t_{Dw} = \frac{kt}{\phi\mu c r_w^2}$$

β_n is a root of

$$J_1(\beta_n)Y_o(\beta_n r_{eD}) - Y_1(\beta_n)J_o(\beta_n r_{eD}) = 0$$

J's and Y's are Bessel functions, first kind first and zeroth order. Units are Darcy units.

Unsteady-State Solution for Radial Gas Flow

The approximate analytical solution for the unsteady-state radial flow of real gas is obtained by treating the product (μc) as constant. The solution is obtained first for constant terminal rate, for Darcy flow with no-skin effect. The effect of extra pressure drop for skin and turbulence are superposed to the basic solution by inspectional analysis in proper dimensions.

$$m[p(r_w,o)] - m[p(r_w,t)] = 5.0304 \times 10^7 \frac{(p_{sc}\,T)}{(T_{sc}\,kh)} q_{sc}[1/2 \ln t_D + 0.405 + s + Dq_{sc}] \tag{3-67}$$

Equation 3-67 is in field units where q_{sc} is the constant rate of production std MMCF/D. The *s* represents the dimensionless skin-coefficient. It is usually calculated from pressure build-up data. It may also be estimated from an equation given by Hawkins[14] as related to far-zone and near-zone permeabilities:

$$s = \left(\frac{k}{k_s} - 1\right) \ln \frac{r_s}{r_w} \qquad (3\text{-}68)$$

k and k_s are, respectively, far- and near-zone permeabilities, r_s and r_w are near-zone (or skin) and wellbore radii.

In Equation 3-67, m(p) represents the real gas potential in $psia^2/cp$
t is time in hrs
p_{sc} and T_{sc} are the standard pressure and temperature at which q_{sc}, the constant terminal rate, is measured in MMCF/D
T is the formation temperature, °R abs
k is the permeability in millidarcies
h is the formation thickness in feet

$$t_D = \frac{2.637 \times 10^{-4}\ kt}{\phi(\mu c)_i\ r_w^2} \qquad (3\text{-}69)$$

The product $(\mu c)_i$ is the viscosity × compressibility product at initial conditions.

μ is in cp, c is in vol/vol per psi
ϕ is fractional porosity

If r is substituted for r_w in Equations 3-67 and 3-69, one then obtains the real gas potential distribution in time and space throughout the drainage matrix.

Even though, in actuality, during unsteady-state flow the value of (μc) continues to change in time, comparison between the above approximate solution (treating (μc) as constant at their initial value) with more exact numerical solutions obtained by finite difference (FDM) or finite element (FEM) techniques indicates reasonable agreement between approximate analytical solutions and more nearly exact numerical results.*

The turbulence coefficient D has the dimension of $1/q_{sc}$ or (T/L^3). It can be calculated approximately from:

$$D = \frac{7.675 \times 10^{-3} G\beta k}{h\mu_w}\left[\frac{1}{r_w} - \frac{1}{r_d}\right] \qquad (3\text{-}70)$$

* In a recent paper, Tek, Lee, and Logan have shown that the above is true for only slight or moderate turbulence intensity.[18]

While D actually changes from point to point and keeps increasing due to decreasing pressures even during the steady state, that effect is shown to be small.[16]

In Equation 3-70:

G = specific gravity of the gas, dimensionless
β = turbulence factor, ft^{-1}
k = permeability, millidarcies
r_d = drainage radius, ft
r_w = wellbore radius, ft

Note that soon after the flow is established at a constant terminal rate, the value of r_d becomes large enough to have its inverse become negligible as compared to $1/r_w$.

General Unsteady-State Equations for Production Pressure Behavior of Gas Wells (For Variable Terminal Rate)

$$m(p_w)_i - m(p_w)_N = 5.0304 \times 10^7 \frac{p_{sc}\,T}{T_{sc}\,kh} \left[q_o p_D(N\Delta t_D) + \sum_{i=1}^{N} (q_i - q_{i-1}) p_D\,[N - (i - 1)]\Delta t_D + D_{qN}\right] \quad (3\text{-}71)$$

where: $m(p) = 2\int_o^p \frac{p}{z\mu}\,dp$ the real gas potential (pseudo-pressure), $psia^2/cp$

p_{sc} = standard pressure, psia
T_{sc} = standard temperature, °R abs
T = reservoir temperature, °R abs
k = permeability, millidarcies
q_o = initial rate at time $t = 0$, MMCF/D
t_D = dimensionless time, $= 2.637 \times 10^{-4} kt/\phi\mu c r_w^2$
t = actual time, hours
ϕ = porosity, fractional quantity
μ = gas viscosity, cp (at average or initial reservoir conditions)
r_w = wellbore radius, true or effective, feet
c = effective compressibility (counter for time steps), vol/vol × psi
N = an integer number
i = an integer number
p_D = dimensionless pressure-drop function

Dimensionless Pressure-Drop Function—Radial Flow, Unsteady-State

$$p_D = 1/2 \ln t_D + 0.4045 + S \tag{3-72}$$

Combined Radial and Linear Flow (Unsteady-State with Vertical Fracture)

$$p_D = \sqrt{\pi t_{Dxf}}\ \mathrm{erf}\left(\frac{1}{2\sqrt{t_{Dxf}}}\right) - \mathrm{Ei}\left(-\frac{1}{4t_{Dxf}}\right) \tag{3-73}$$

$$\text{where } t_{Dxf} = 2.637\ 10^{-4}\ \frac{kt}{\phi(\mu c)_i x_f^2} \tag{3-74}$$

x_f = half-fracture length, ft

Completely Heterogeneous Reservoirs (Unsteady-State with Non-Ideal Geometry)

$p_D = f(t_D)$ As computed using Equation 3-71 from field data.

Simplified Version of Equation 3-71 for Constant Terminal Rate or Constant Terminal Pressure (Darcy Flow)

$$m(p_w)_i - m(p_w)_N = 5.0304 \times 10^7 \frac{p_{sc}}{T_{sc}} \frac{T}{kh} q_o p_D(N\Delta t_D) \tag{3-75}$$

a. For constant terminal rate of q_o MMCF/D Equation 3-75 gives:

$$m(p_w)_N = f(t)$$

Knowing m_N one can calculate or look up p_N

b. For constant terminal pressure, maintain $m_i - m_N$ constant, calculate q_o from Equation 3-75 with N = 1, then use Equation 3-71 for each Δt.

To calculate q_1:

$$m_i - m_1 = 5.0304 \times 10^7 \frac{p_{sc}}{T_{sc}} \frac{T}{kh} [q_o p_D(1\Delta t_D) + (q_1 - q_o) p_D(1\Delta t_D)]$$

Then set N = 2 to calculate q_2:

$$m_i - m_2 = 5.0304 \times 10^7 \frac{p_{sc}}{T_{sc}} \frac{T}{kh}$$

$$[q_o p_D(2\Delta t_D) + (q_1 - q_o) p_D(2\Delta t_D) + (q_2 - q_1) p_D(1\Delta t_D)] \quad (3\text{-}76)$$

and so on . . .

Application of Superposition Equations with the Real Gas Potential Problem

For an aquifer storage reservoir, Field A, the following data are available:

p_{sc} = pressure base, 14.7 psia
T_{sc} = temperature base, 520°R abs
T = reservoir temperature, 553°R abs
kh = 3,512 millidarcy-feet
p_o = 1,700 psig wellhead

depth to mid perforation = 4,000 feet
specific gravity of gas = 0.59

1. Approximately calculate the plot m(p) versus P for the range 2,000 psia – 600 psia
2. Using ϕ = 0.12
   μ = 0.0137 cp at 2,000 psia
   h = 100 feet
   r_w = 1/4 feet

Calculate Q versus time for t = 1, 2, 3, 4, 5 days—for a constant terminal pressure case with 100 psi ΔP at bottom hole.

Wellbore Storage Effects

During pressure build-up tests, it is well known that while the well is shut-in at the surface, the flow will continue for some time into the wellbore. The quantity that flows into the wellbore after the well has been shut-in at the surface is referred to as *after production*. In pressure build-up plots, it is identified occasionally as the early S-shaped trend.

When a storage well, initially shut-in for some time, is turned on to flow, the unloading of the wellbore dominates the early flow conditions during the draw down. This early flow, similar to after production during early build-up, will influence and complicate reservoir performance unless it is properly accounted for.

For a quantitative evaluation of wellbore storage effect, a dimensionless constant was introduced by Everdingen[17]:

$$\bar{c} = \frac{V_w c_{ws}}{2\pi h r_w^2 \phi\, c} \tag{3-77}$$

where V_w = volume of the wellbore, cu ft
c_{ws} = average wellbore fluid compressibility, vol/vol × psi
h = reservoir formation thickness, ft
r_w = reservoir boundary radius, ft (inner boundary)
ϕ = porosity, fraction
c = compressibility of fluid in the drainage matrix, vol/vol × psi

During after-production or early flow, as the contents of both the reservoir and the wellbore decline in pressure, $\bar{c}$ will change, reflecting the wellbore storage effect. In numerical calculations, the boundary conditions for early flow times must be modified to account for the change in $\bar{c}$.

Example 3-1

Problem. A newly discovered natural gas well is completed on a horizontal disk shaped formation of very large radius. The permeability of the formation is believed to be constant and equal to 100 millidarcies. The net thickness of the producing formation is 95 feet.

The discovery pressure was 6,000 psia bottom hole, the porosity 12%, and the formation temperature is 175°F.

During an initial productivity test, the well is allowed to flow for 90 days at the constant rate of 12 MMCF/D.

1. Calculate, approximately, the pressure at the wellbore as of the end of the three month flow period by point-source approximation.
2. What would be the pressure at the end of 3 months using the constant terminal rate equation with the real gas potential?
   The specific gravity of the gas G = 0.714
   Viscosity = 0.0135 cp
   Radius of the wellbore r_w = 1/4 ft
   Skin effect s = 0

Solution.

1. By point source approximation:

$$p_w = p_o + \frac{162.6(\pm q)\mu}{kh}\left[\log_{10}\frac{\phi\mu c r_w{}^2}{0.00633\ kt} - 0.3513\right]$$

With approximately 6,000 psia bottom hole pressure, q in reservoir Bbls/day must be calculated first:

$$q = 12 \times 10^6 \times \frac{14.7 \times z}{6{,}000} \times \frac{635}{520} \times \frac{1}{5.615};\ z = 1.05$$

q = 6,719.61 Bbls/day; use + for production:

$$p_{w\ 90\ days} = 6{,}000 + \frac{162.6 \times 6{,}719.61 \times 0.0135}{100 \times 95}$$

$$\left[\log_{10} \frac{0.12 \times 0.0135 \times 1}{6{,}000 \times 16 \times 0.00633 \times 90} - 0.3513\right]$$

$$= 6{,}000 + 1.55\,[\,-\,7.53 - .3513]$$

$$= 5{,}987.79 \text{ psia}$$

2. By real gas potential:

Approximate equation for real gas potential:

$$m(p) \cong 2\,\frac{1}{z\mu}\,\frac{p^2}{2}$$

$$P_c = 670 \text{ psia}$$

$$T_c = 390°\text{R abs}$$

$$P_r = \frac{6{,}000}{670} = 8.96$$

$$z = 1.05$$

$$m(p) = \frac{1}{1.05 \times 0.0135}\,p^2$$

$$= 70.55\,p^2$$

$$m(p)_i - m(p)_N = 5.0304 \times 10^7 \times \frac{14.7 \times 635}{520 \times 100 \times 95}\,q_o p_D(N\Delta t_D)$$

For $\Delta t = 1$ day:

$$\Delta t_D = \frac{0.006336 \times 1 \times 100}{0.12 \times 0.0135 \times 1/6{,}000 \times 1/16}$$

$$= 37{,}546{,}667.67$$

$$p_D = 1/2 \ln t_D + 0.4045 + S$$

$$m(p)_N = m(p)_i - 5.0304 \times 10^7 \times \frac{14.7}{520} \frac{635}{100 \times 95} q_o p_D(90\Delta t_D)$$

$$= 70.55\ \overline{6{,}000^2} - 5.0304\ 10^7 \frac{14.7}{520} \frac{635}{100 \times 95} q_o p_D(90\Delta t_D)$$

$$= 2.5398 \times 10^9 - 12{,}974{,}704.21$$

$$m(p)_N = 2.256825 \times 10^9$$

$$p_N = \left(\frac{2.256825 \times 10^9}{70.55}\right)^{1/2}$$

$$= 5{,}984.65 \text{ psia}$$

Example 3-2

Problem. Starting with Darcy's law written in oil field units, derive a steady-state radial flow equation.

Solution.

$u = \pm \frac{k}{\mu} \frac{dp}{dr}$, where μ is superficial velocity

$q = + \frac{kA}{\mu} \frac{dp}{dr}$, with q production rate called positive

$+ \frac{dp}{dr} = \frac{q\mu}{kA}$, for Darcy flow purely laminar

$\frac{dp}{dr} = \frac{\mu}{kA} q + \beta\rho \frac{q^2}{A^2}$, when non-Darcy flow turbulence effects are included

Now, rewrite equation in field units:

$$q = 0.00633 \frac{kA}{\mu} \frac{dp}{dr} \tag{1}$$

where k is in millidarcies
A in ft^2
μ in cp
p in psia
r in ft
q in cu ft/day

$q = q(r)$ and $p = p(r)$ and, of course, $n = n(r)$ moles/day

$$qp = znRT$$

$$q_{sc}\, p_{sc} = 1nRT_{sc} \qquad \text{(sc for standard condition)}$$

$$\frac{q}{q_{sc}}\frac{p}{p_{sc}} = z\frac{T}{T_{sc}}$$

$$q = q_{sc}\frac{p_{sc}}{p}z\frac{T}{T_{sc}} \qquad \text{(substitute in Equation (1) above; assume } z = z_{av}\text{)}$$

$$q_{sc}\frac{p_{sc}}{p}z_{av}\frac{T}{T_{sc}} = 0.00633\ k\frac{A}{\mu}\frac{dp}{dr}$$

for horizontal, radial, centripetal flow

$$q_{sc}\frac{p_{sc}}{p}z_{av}\frac{T}{T_{sc}} = 0.00633\ k\frac{2\pi rh}{\mu}\frac{dp}{dr} \tag{3}$$

$$q_{sc}\frac{p_{sc}}{p}z_{av}\frac{T}{T_{sc}} = 0.03973\frac{kh\ r}{\mu}\frac{dp}{dr} \tag{4}$$

In the equation above, q_{sc} is in SCF/D.
If Q is in MCF/D, $q_{sc} = Q \times 1{,}000$.

$$1{,}000\ Q \times \frac{p_{sc}}{p}z_{av}\frac{T}{T_{sc}} = 0.03973\frac{kh}{\mu}r\frac{dp}{dr}$$

$$Q\ p_{sc}\ z_{av}\frac{T}{T_{sc}}\frac{1{,}000}{0.03973}\int_{r_1}^{r_2}\frac{dr}{r} = \frac{kh}{\mu}\int_{p_1}^{p_2}p\ dp$$

$$Q\ p_{sc}\ z_{av}\frac{T}{T_{sc}}\ 25{,}170\ \ln\frac{r_2}{r_1} = \frac{kh}{\mu}(p_2^2 - p_1^2) \tag{5}$$

In Equation (5), Q is in MCF/D
$p_{1,2}$ is in psia
T, T_{sc} is in °R abs
z is dimensionless
k is in millidarcies
h is in ft
using T_{sc} = 520°R abs
p_{sc} = 14.73 psia

$$Q = \frac{1.986 \times 10^{-5} \times 520}{14.7}\ \frac{kh(p_2^2 - p_1^2)}{\mu T z_{av} \ln (r_2/r_1)}$$

$$Q = 0.0007025 \frac{kh(p_2^2 - p_1^2)}{\mu z_{av} T \ln (r_2/r_1)}$$

Example 3-3

Problem. A newly discovered oil well in an infinitely large reservoir is turned on for production at a constant rate of 200 Bbls/day in bottom hole wellbore conditions.

1. Derive the relationship giving the time-rate of decline of pressure in the wellbore as a function of time.
2. Numerically, what would be the pressure-drop-rate in psi/day at t = 0, t = 1 day, t = 10 days, t = 100 days?

$k = 350$ mdcy
$\mu = 1$ cp
$h = 80$ ft
$\phi = 0.15$

Solution.

$$p_w = p_0 + \frac{(\pm q_0)162.6}{kh}\left[\log \frac{\phi\mu c r_w^2}{0.00633\ kt} - 0.3513\right]$$

Time-rate of decline of pressure at the wellbore:

$$\frac{dp_w}{dt} = \frac{(\pm q_0)162.6}{kh} \times \frac{0.00633\ kt}{\phi\mu c r_w^2} \times \frac{\phi\mu c r_w^2}{0.00633\ k} \times \left(-\frac{1}{t^2}\right)$$

$\left(\frac{-dp_w}{dt}\right) = +\frac{(\pm q_0)162.6}{2.303\ kht}$; use $+q_0$ for production decline of pressure at:

1. t = 0 days ∞ psi/day, vertical decline on p = p(t) curve
2. t = 1 day $\quad q_0 = 200$ Bbls/day
   $\mu = 1$ cp, k = 350 mdcy, h = 80 ft

$$-\frac{dp_w}{dt} = -\frac{200 \times 162.6}{350 \times 80 \times 1 \times 2.303} = 0.5 \text{ psi/day} \qquad \text{(1 day)}$$

$$-\frac{dp}{dt} = 0.05 \text{ psi/day} \qquad \text{(10 days)}$$

$$-\frac{dp}{dt} = 0.005 \text{ psi/day} \qquad \text{(100 days)}$$

Example 3-4

Problem. Natural gas is being injected into a well at a constant terminal pressure of 2,000 psia (bottom hole). At time t = 0 and previously, the pressure at the wellbore was 1,500 psia. Please calculate and plot the rate of injection as a function of time. (It is suggested that you calculate q_0 at 0.01 day then q_1 at 1 day, q_2 at 2 days. Assume Darcy flow.)

p_{sc} = 14.7 psia
T_{sc} = 520°R abs
kh = 1,800 millidarcy-feet
k = 30 millidarcies

The porosity ϕ = 0.12
μ = 0.018 cp

The pseudo-criticals for the gas: p_c = 670 psia
T_c = 380°R abs

Formation Temperature T = 89°F
r_w = 1/4 ft.

The reservoir is horizontal and circular.

Skin coefficient S = 0

Solution. Find q_0 first:

$$m_0 - m_N = 5.304 \times 10^7 \frac{p_{sc}}{T_{sc}} \frac{T}{kh} 2\, q_0\, p_D(0.01\Delta t_D)$$

$$m = \int_0^p \frac{2p}{z\mu} dp; \text{ assume } (z\mu) \cong \text{constant}$$

$$\mu = 0.018$$

$$p_r = \frac{1{,}750}{670} = 2.61 \qquad \text{(using average pressure)}$$

$$T_r = \frac{549}{380} = 1.44$$

$$z = 0.76$$

$$m = \frac{1 \times 2\, p^2}{0.76 \times 0.018 \times 2} = 73.10\, p^2$$

$$m_0 - m_N = \frac{5.304 \times 10^7 \times 14.7 \times 549}{520 \times 1{,}800} q_0\, p_D(0.01\Delta t_D)$$

$$= 457{,}317.0\, q_0\, p_D(0.01\Delta t_D)$$

take 1 Δt = 1 day

$$t_D = \frac{0.00633 \times 30 \times 1}{0.12 \times 0.018 \times (1/1750) \times (1/16)} = 2{,}464{,}000$$

$$0.01\, t_D = 24{,}640$$

$$p_D(24{,}640) = 1/2 \ln 24{,}640 + 0.4045 + s = 5.46$$

$$q_0 = \frac{(m_0 = m_N)}{457{,}317 \times 5.46};\ q_0 = \frac{73.10\,(\overline{1500}^2 - \overline{2000}^2)}{457{,}317.0 \times 5.46} = -51.23 \text{ MMCF/D}$$

Next set N = 1 and calculate q_1 for t = 1 day

$$m_o - m_n = 5.304 \times 10^7 \frac{p_{sc}}{T_{sc}} \frac{T}{kh} [q_0\, p_D\, (1\Delta t_D) + (q_1 - q_0)\, p_D\, (1\Delta t_D)$$

$$= 457{,}317.0\,[-51.23\, p_D\, (2{,}464{,}000) + (q_1 - q_0)\, p_D\, (2{,}464{,}000)]$$

$$p_D\, (2{,}464{,}000) = 7.76$$

$$m_o - m_n = 457{,}317.0 \times 7.76\, q_1$$

$$q_1 = -36.05 \text{ MMCF/D}$$

Next set N = 2 $\quad p_D\, (2\, t_D) = 8.11$

$$m_0 - m_2 = 457{,}317\, [q_0 \times 8.11 + (q_1 - q_0)\, 8.11 + (q_2 - q_1)\, 7.76]$$

$$= 457{,}317\, [q_1 \times 8.11 + (q_2 - q_1)\, 7.76]$$

$$= 457{,}317\, [-36.05 \times 8.11 + (q_2 + 36.05)\, 7.76]$$

$$m_0 - m_2 = 457{,}317\, [q_2 \times 7.76 + q_1 \times (8.11 - 7.76)]$$

$$q_2 = \frac{1}{7.76}\left[\frac{m_0 - m_2}{457{,}317} - q_1\, (8.11 - 7.76)\right]$$

$$= \frac{1}{7.76}\left[\frac{73.10(\overline{1{,}500}^2 - \overline{2{,}000}^2)}{457{,}317} + 36.05 \times 0.35\right]$$

$$(-279.73 + 12.62)$$

$$= -34.42^* \text{ MMCF/D}$$

* If a spacing of 2,640 ft is assumed (640 acres drainage area) the corresponding steady-state injection rate calculates to be approximately 34 MMCF/D.

References

1. Darcy, H. *Les Fontaines Publiques de la Ville de Dijon*, Paris: Victor Dalmont, 1856.
2. King, Hubbert M. "Darcy's Law and the Field Equations of the Flow of Underground Fluids," *Trans. AIME* (1956) 207, pp. 222–239.
3. Klinkenberg, L. J. "The Permeability of Porous Media to Liquids and Gases," *API Drill. and Prod. Prac.* (1941) 200.
4. Fancher, G. H., J. A. Lewis, K. B. Barnes. "Some Physical Characteristics of Oil Sands," Pennsylvania State College, Bull. 12 (1933) p. 65.
5. Tek, M. R. "Development of a Generalized Darcy Equation," *J. of Petro. Tech.* (1957) IX, No. 6, pp. 45–47; *Trans. AIME* (1957) 210, p. 45.
6. Wattenbarger, R. A. and H. J. Ramey, Jr. "Gas Well Testing with Turbulence, Damage and Wellbore Storage," *J. of Petro. Tech.* (1968) pp. 877–878.
7. Muskat, M. *Physical Principles of Oil Production*, New York: McGraw Hill, 1949.
8. Al-Hussainy, R. and H. J. Ramey, Jr. "Application of Real Gas Flow Theory to Well Testing and Deliverability Forecasting," *J. of Petro. Tech.* (1966) pp. 637–642.
9. Aronofsky, J. S. and Jenkins, R. "Unsteady Flow of Gas through Porous Media; One Dimensional Case," Proc., First U.S. National Congress of Applied Mechanics, J. E. Edwards, Inc., Ann Arbor, MI (763) 1952.
10. Jenkins, R. and Aronofsky, J. S. "Unsteady Radial Flow of Gas through Porous Media," *J. Appl. Mech.* (1953) *20*, 210.
11. Bruce, G. D., Peaceman, D. W., Rachford, H. H., and Price, J. "Calculations of Unsteady State Gas Flow through Porous Media," *Trans. AIME* (1953) *198*, 79.
12. Douglas, Jr., J., Peaceman, D. W., and Rachford, H. H. "Calculations of Unsteady-State Gas Flow in a Square Reservoir," *Trans. AIME* (1955) *204*, 190.
13. Forchheimer, P. "Wasserbevegun durch Boden," *Zeitz, Ver Deutsch, Ing.* (1901) *45*, 1731.
14. Hawkins, M. F., Jr. "A Note on The Skin Effect," *Trans. AIME* (1956) *207*, 356–357.
15. Matthews, C. S. and D. G. Russel "Pressure Build-Up and Flow Tests in Wells," SPE Monograph, 1967.
16. Lee, R. L. "Deliverability of Natural Gas," PhD Thesis, University of Michigan, 1986.
17. Van Everdingen, A. F. "The Skin Effect and Its Influence on The Productive Capacity of a Well," *Trans. AIME* (1953) *198*, 171–176.
18. Lee, R. L., Logan R. W., and Tek, M. R. "Effect of Turbulence on Transient Flow of Real Gas Through Porous Media" SPE Formation Evaluation, March 1987.

4

DESIGN AND DEVELOPMENT OF UNDERGROUND STORAGE FIELDS

Having established a perspective on energy as well as an overview on production and storage of natural gas (Chapter 1), the properties of storage systems, rocks, and fluids were discussed (Chapter 2). Chapter 3 was developed in order to document the theory of fluid flow through underground storage systems. The purpose of this chapter is to provide the engineering link between the theory and the practice of storage technology.

The basic elements that constitute the storage systems from the source of gas supply, through the pipeline route, to the storage project near the market, to the city gate and distribution system, to individual meters will be presented to establish the basis for design and real-world development. The concept of design, along with important considerations for optimum service to customers, will be presented next. Some typical engineering data and procedures for design, procurement, construction, testing, and commissioning will be included in support of the design objectives.

CONCEPTS IN STORAGE DESIGN

Underground Storage Systems

Underground storage efficiently combines a constant supply with a variable demand to economic advantage. Figure 4-1 depicts the manner in which the constant capacity of a pipeline is modulated between the low-demand summer months and high-demand winter months when a storage environment is available. Storage facilities are usually classified as *market* or *field* storage. Market storage is near major consuming areas where the variable demand resulting from weather is serviced by a proper combination of pipeline and storage gas. In field storage it is the variable supply to the major market pipelines that is supplemented by the availability of storage gas. Market storage facilities near Chicago, such as Herscher and Manlove, and field storage facilities in Texas and Oklahoma, such as North Lansing and Sayre fields, are typical examples. While most underground storage facilities have been developed in depleted gas or oil reservoirs, some minor but substantial storage is developed on naturally water bearing sands called aquifers. Some minor storage facilities have been developed on dissolved salt caverns. Eminence dome in Alabama and Morton Salt caverns in St. Clair, Michigan are examples of such cavern storage.

Storage reservoirs are occasionally classified by the nature of service they provide. Some are in constant use with relatively small fluctuations of their rates during withdrawal. They are called *base load* type storage facilities.

Others are specially designed to meet very high market demand over relatively short periods of time. These are called *peak shaving* or *peak load* type facilities.

Figure 4-2 shows the major elements of an underground storage facility. These include:

1. A storage reservoir proper, in the subsurface. Shown in cross section, including the storage sand and its various rock layers in the underburden and overburden.
2. A typical I/W (injection/withdrawal) well connected to a surface gathering system.
3. A compression and dehydration plant.

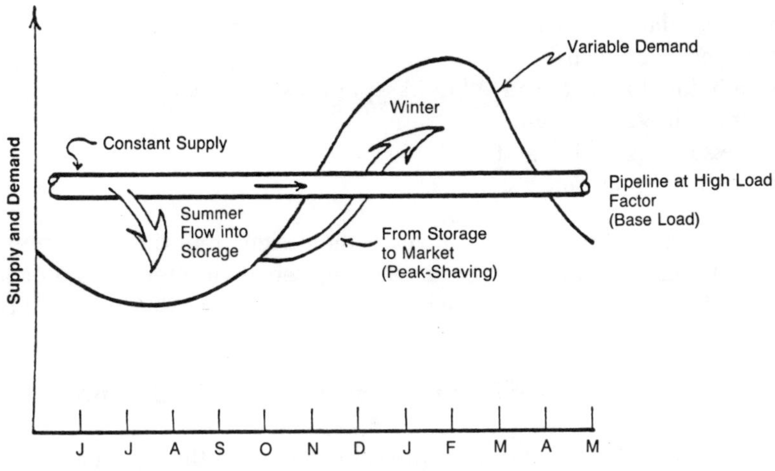

Figure 4-1. Role of underground storage and pipeline meeting market demand.

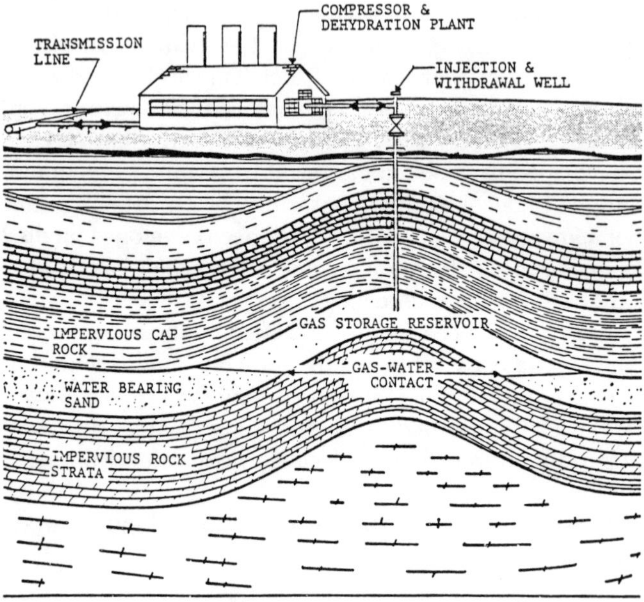

Figure 4-2. Underground storage of natural gas in an aquifer.

4. A transmission line which connects the storage field both to its supply upstream and its market downstream.

In addition to what is shown in Figure 4-2, the storage facilities include:

- Individual well heaters
- Observation wells
- Individual and collective liquid separators
- Brine disposal system
- Pressure regulation and metering
- Methanol injection system

The design of a storage facility includes the design of all the above components. Storage facilities sometimes include their own intermediate or low pressure transmission lines to the city gate, and odorant injection facilities.

Concepts and Main Considerations in Design

Before an underground storage project evolves into the design and development phase, a long-range planning study must determine and document market requirements. Such a study must include the influence of weather on gas consumption, and seasonal studies on space heating loads as affected by such variables as degree-days, degree-day lags, cloud cover, snow, sunshine, and other factors. The degree-day deficiency is defined as the difference between 65°F and the average temperature for that particular day. Once degree-day deficiency is determined for a particular day, the degree-days are summed for a whole year to correlate against the gas sendout to market. In predicting weather in order to forecast the gas sendout to market, one often utilizes 50-year (or some specified number of years) normal temperature records. These are obtained from the statistical records on weather.

The domestic markets usually include home space heating, cooling, cooking, lighting, and other industrial loads. One important aspect of forecasting the market relates to the prediction of *peak day* and *peak hour* sendouts. The objective of marketing surveys is the determination of annual sales forecasts and peak day requirements. In each particular area, consumption factors are determined per each customer meter for residential, commercial, and industrial markets. These consumption factors are usually expressed in terms of MCF/year or MCF/DDD, where DDD denotes degree-day-deficiency.

Knowing the total number of customers projected, using pre-established consumption factors, degree-days and load factors for various markets, the "average day" MMCF and the "peak day" MMCF are determined to set up the two major design parameters. Once total market requirements are determined for each month by class of service (such as residential, commercial, space heating, industrial—both firm and interruptible), a schedule can be prepared for pipeline supply and storage requirements.

Selection of Prospects for Storage

Once the need for storage and a storage schedule is established, the search and screening of various prospects for storage follows. Depending upon size, location, access to major pipeline supplies, and other factors, one considers:

1. Depleted dry gas fields
2. Depleted condensate fields
3. Depleted oil fields
4. Aquifers
5. Possible salt caverns

for evaluation as storage prospects in the indicated order of preference. The depleted dry gas field is the best prospect if available simply because it is closest to the natural environment for dry pipeline quality storage gas.

The depleted condensate field is also desireable, but it will have some engineering problems related to two-phase flow and retrograde behavior in the projected storage cycle.

The depleted oil field is also usually satisfactory, but it too will have some problems related to residual oil.

The aquifer prospect would have many additional characteristics to be evaluated against the specific requirements and is treated separately in the following section.

In the initial screening, if the above four categories do not yield a prospect, the possibility of storage in solution mined salt and/or soft or hard mined caverns are occasionally entertained. Certainly, if one such prospect is available at a strategic location, one could not afford to overlook it.

Screening Prospects for Aquifer Storage

The initial screening for potential structures as prospects for aquifer storage involves surveys of data from federal agencies, oil and gas companies, or geological surveys available on records.

The preliminary work is in general directed to listing and comparing the prospects and then to promptly eliminating those which do not possess the minimum essential requirements.

The absolute minimum requirements for an aquifer storage project are:

1. Sufficient closure to provide the structural volume required.
2. Sufficient porosity for necessary storage capacity.
3. Sufficient permeability to provide the necessary minimum deliverability.
4. Caprock of adequate integrity to ensure against leak and migration.

Surely, if there exists no closure, or if the porosity or permeability are too low, or the protection of an adequate caprock is not available, there would be no need or justification to further evaluate a particular prospect.

There exist specific procedures, tests, and other evaluative techniques for all the four essential requirements listed above. It must also be recognized that, even before the initial preliminary work, one must develop some idea of the size of storage operations from market, economic, conservation, or stockpile considerations. That background study which precedes the initial recommendation for further work involves study of degree-days, consumption factors, heating, cooling, and residential and commercial markets. Availability of gas, logistics of pipelines and alternative energy resources for the future are sometimes among the considerations entertained.

Having determined the desirability as well as magnitude and location of storage, the screening process starts first with geological work directed to item 1 above. Figure 4-3 shows the various tasks involved in selecting prospects for aquifer storage.

Whole core analysis reports usually include depth interval, vertical permeability to water (as measured or less than a cut-out value), caprock permeability, and threshold pressure (where applicable).

The routine core analysis on prospective storage formations will usually report horizontal and vertical permeabilities in millidarcies and percent porosity versus depth. Such core data from several wells at several depths are analyzed, identified by particular strata, and averaged to give representative values for permeability and porosity for each well along with the number of samples.

Relative permeability and capillary pressure tests are also helpful in prospective storage formations to determine connate water, residual equilibrium saturations, and relative permeability curves. Sometimes, connate water saturation is determined by using a centrifuge.

Quite often it is desirable to obtain both *drainage* and *imbibition* capillary pressure curves. The drainage-capillary pressure curves on tight caprock cores are helpful in determining the threshold pressure of the caprock.

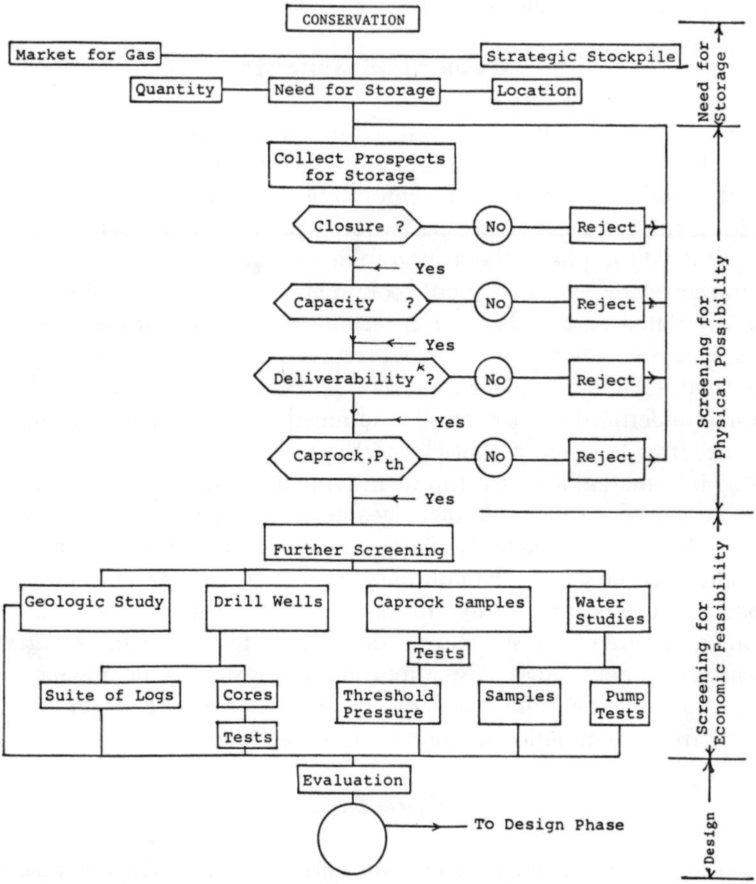

Figure 4-3. Tasks involved in selecting prospects for aquifer storage.

The threshold pressure, however, is usually directly measured from caprock samples by special procedures using core holders and Ruska pumps.

Regional Studies in Water Levels

Before any gas injection is committed to storage, piezometric water levels on all available wells in the area must be recorded. The data on static water levels in wells completed in several zones on both sides of prospective caprocks are important in establishing that the storage zones are not in fluid communication across caprock. The water levels are measured by floats, recorders, sonar, or sonolog type equipment and are usually reported as vertical distance from the stem of the orbit valve. In analyzing the water levels, the surface elevation as well as the elevation of perforations and some other desired datum are noted to translate the liquid-level distances from the surface to psia, absolute pressures at designated datum.

Geological Aspects

The presence of all prospective structures are first searched from regional maps, state records, and drilling records. A number of wells are usually drilled, cored and logged. From information on the regional geology and specific local well records, prospective storage horizons are determined and compared. At this phase, the stratigraphic sequence at each location is studied and correlated. Such composite columnar sections in particular areas are cross-correlated and studied for determining possible storage zones, caprocks, and collector zones. Various prospective strata of different geologic ages, their thickness, homogeneity, their access to overburden, and their relation to underburden are carefully examined. In studying the stratigraphic sequence from logs, self-potential, resistivity, drill-time, gamma ray, and lateralog information are useful in identifying and correlating several strata. An experienced geologist, knowledgeable in reading logs and correlating various data is often essential. Such geologists working with geophysicists who interpret the seismic data ultimately come up with structural maps and isopachs of prospective storage formations. Using core data and dipmeter readings, closure of the structure is determined as well as the *saddles* and potential spill-point areas. Also important at this stage is the presence of insipient fractures and faults which may be identified directly or presumed indirectly from examination of cores and seismic profiles.

Core Data

Cores recovered from drilling are washed, identified, and placed in boxes. Some cores are immediately frozen with their native reservoir fluids and packed in dry ice. Color photographs are taken and each section is described by geologists. Sometimes, cores are placed in plastic bags to prevent drying.

The cores sent to labs are tested for porosity and permeability as a whole or from small "perm-plugs" cut out of them with diamond drills. Plug samples taken from cores give routine permeability and porosity readings by depth intervals.

The water levels observed are subject to local and temporal variations due to possible continuing flow in aquifers and to variations in atmospheric pressure.

Whenever an elastic shock is generated in the crust of the earth as a result of a distant earthquake, the propagation of the elastic wave is also known to affect the water levels, sometimes permanently. When the pressure of the water is sufficient, an *artesian* condition is observed in the static water well. This occurs when the elevation of the water table is above the ground level. In general, however, the water level does not quite reach the well head. In some cases, by filling the tubing with fresh water, one can create an artesian condition artificially by taking advantage of the density difference between the native brine of the aquifer and the fresh water.

In broad regional perspective, the observed water gradients indicative of continuing flow in an aquifer could be of the order of several feet per mile.

Analysis of Water Samples

Waters native to aquifers have widely different characteristics ranging in concentrations from a few hundred parts per million to several hundred thousand parts per million. Typical samples of waters produced from different areas are given in the literature by Wade Watkins.[1]

Variations in specific water analyses and comparisons between the solids content of waters above and below the caprocks are helpful in determining whether direct fluid communication exists across caprocks.

Pumping Tests

Pumping tests are essential and important tasks in evaluation of selected aquifer structures as prospects for storage.

The reasons for conducting pump tests are several:

1. To determine the existence and/or extent of pressure communication across the caprock.
2. To determine the permeability and transmissibility of the storage horizon.
3. To determine the effective value of combined brine-rock compressibility, as will be shown later in this chapter.

Pumping tests take several weeks; 20, 50, 80 days are not uncommon. They are usually conducted at constant terminal rate using a submersible pump, such as a Reda pump, driven by a portable generator on location.

Sequence of Tasks in Underground Storage Design

In converting the suitable prospects for storage or in developing new storage units, the objective of an underground storage company is to provide service to its customers at a minimum possible cost.

Once the market survey relates to the storage needs of the company, the objective indicated above takes on the specific form of a need to provide, say, x MMCF/D for some y heating days in a season. The one additional provision one makes to the requirement stated above is the need to specify the *load factor* of the storage service in terms of the ratio of peak day to average day gas sendout.

In order to meet the design objective so stated, several tasks must be undertaken in a certain sequence. These tasks are graphically depicted in Figure 4-4. They are broadly classified as collection and documentation of input data, engineering design work, and economic analysis.

Input Data. The data used to proceed with design first involve *reservoir description*. These data include all available and pertinent information related to geographic location, geological data, stratigraphic sequence, structural maps, isopachs, logs, and seismic interpretations.

Data on *reservoir performance* are obtained, organized, and documented next. These data relate to the primary depletion of the reservoir whenever conversion of an existing producing facility to storage service is involved. If a new aquifer storage field is to be designed, then the reservoir performance data are generated by simulation.

The next grouping of data essential and important is called *core and fluid data*. Such data must include the permeability, k, and porosity, ϕ, as well as connate water saturation, s_{wc}, from cores and logs. The gas gravity, composition, and viscosity, as well as c_p, k, c_v, and Joule Thomson coefficients are included in the above group. In addition to data from the reservoir rock, reservoir fluids, brine, native gas, and storage gas, one must have data from the overlying caprock. Of particular interest here are the capillary properties of the caprock with respect to determination of the *threshold displacement* pressure. In evaluating the rocks for aquifer storage design, both imbibition and drainage capillary pressure curves become important.

The last set of data which must be organized before design calculations may get underway are *data on existing wells.* These data would include back pressure performance of wells, the completion data, open-flow potential, two-phase flow, and inverse productivity index.

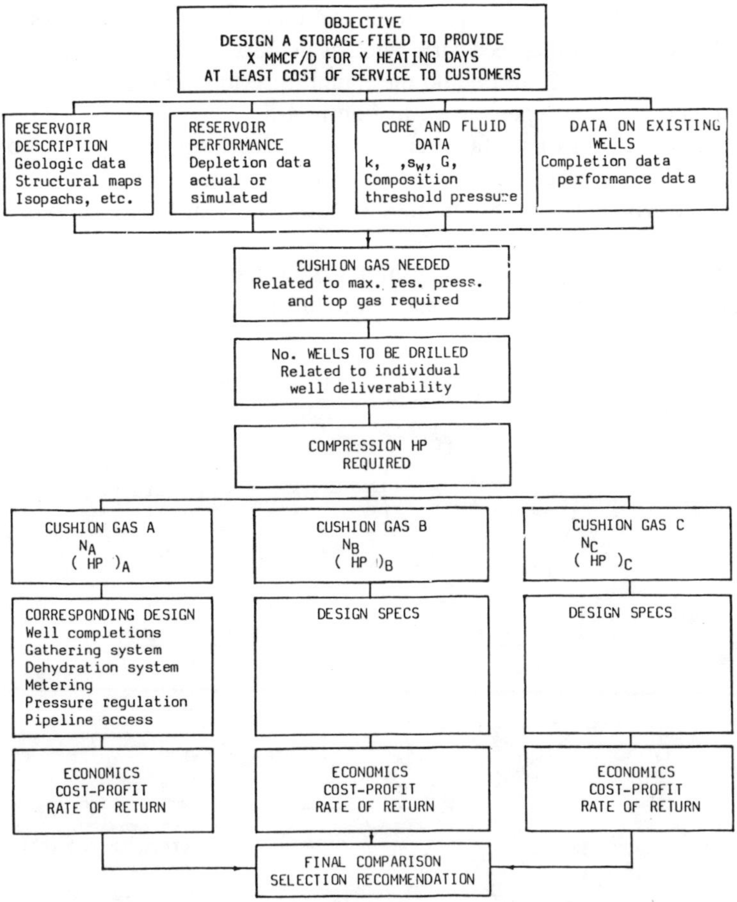

Figure 4-4. Various tasks involved in design of storage fields.

Summary of Major Engineering Design Tasks. Once the seasonal top gas is set forth, the design first focuses on the amount of *base* or *cushion* gas required. This amount is directly related to the maximum pressure of the storage cycle. Since the seasonal *top* gas is usually pre-specified or estimated, the setting of cushion gas fixes both maximum and minimum pressure of the storage cycle at the storage horizon.

Once the storage cycle pressures are so determined at bottom-hole condition, calculations on gas deliverability must be performed before the number of wells required is determined. These calculations on individual well deliverabilities are also necessary to determine the pressures at various points of surface facilities.

Once the well head pressures are determined, it is possible to design the gathering system and its accessories such as heaters, separators, meters, pressure regulators, dehydrators, and methanol injection. The gathering system may be one of three types shown in Figure 4-5.

Depending upon many factors, such as surface topography, number of wells, nature of storage service, pressure, and rates, one may choose a *loop*, *star* or *tree* type network. Sometimes a combination of the above types may be used.

As the design progresses from cushion gas, well location, and well density, the latitude on diameters and particular configurations narrows down. Af-

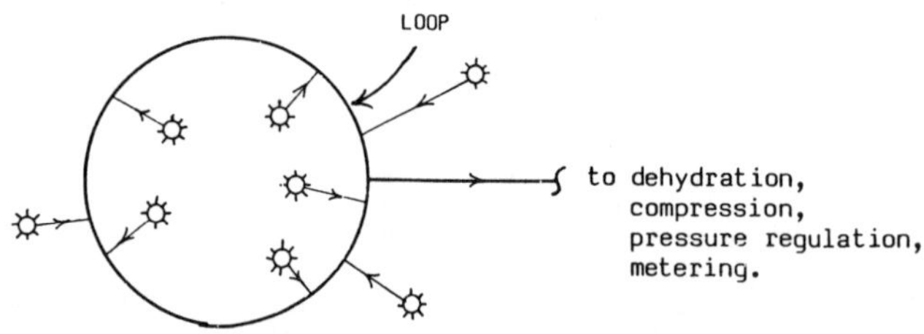

4.5 a. Schematic of a "Loop" Gathering System

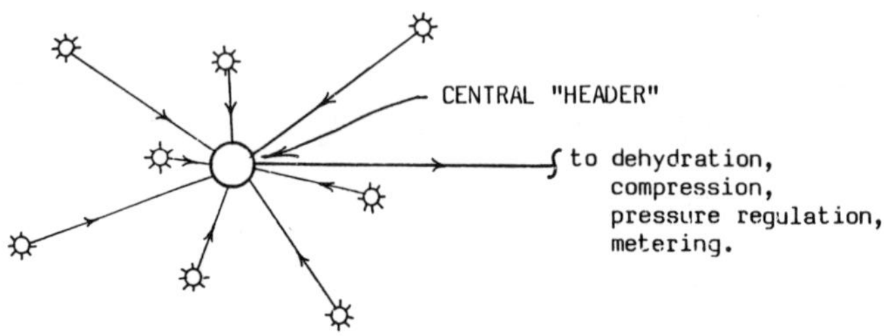

4.5 b Schematic of a "Star" Gathering System"

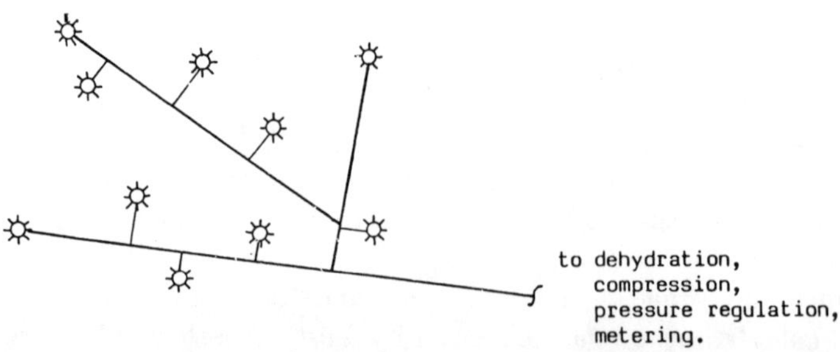

4.5 c. Schematic of a "Tree" Type Gathering System

Figure 4-5. Various types of gathering systems.

ter the task related to the number of wells is completed, the determination of compressor station inlet pressure and the compressor station design must be completed. This includes specific arrangement of types of compression equipment to be used, prime movers, intercoolers, and stand-by capability.

Figure 4-4 depicts the sequence of various design tasks which follow one another. It also shows that, based on three distinctly different cushion gas amounts, the facilities' design may progress downward on three parallel tracks, each consistent with the appropriate cushion gas, number of wells, and compression horsepower. Once the design is finalized on each of the three parallel tracks, A, B and C, an economic analysis is made to determine comparable figures on cost, profit, discounted cash flow, and rate of return.

It is only after the final bottom-line figures are produced by economic analysis that a final comparison yields the recommendation which "freezes" the design.

ENGINEERING PROCEDURES FOR DESIGN OF STORAGE FIELDS

This section of Chapter 4 is devoted to documentation of methods to be used by engineers in their calculations of various components to be designed.

Major elements of the design to be considered will be deliverability into the wellbores, flow and pressure gradients through the wellbores, gradients through the gathering system, the compression plant, and finally, the gas transmission system.

Gas Deliverability

Back Pressure Behavior of Gas Wells. The performance of a natural gas well is determined by the physical properties of the reservoir rock, the extent and geometry of the drainage area, the properties of the flowing fluid, and the pressure distribution in the drainage area. The relation between the quantity of gas delivered and the pressure drop throughout the reservoir is characteristic of the behavior of each well and is often referred to as the *back-pressure* behavior of the well. This relationship is experimentally determined and is usually given in the following form:[1]

$$Q = c(P_f^2 - P_s^2)^n \qquad (4\text{-}1)$$

where c is called the performance coefficient, n is the exponent corresponding to the reciprocal slope of the straight-line relationship between Q and $(P_f^2 - P_s^2)^n$ plotted on logarithmic coordinates, P_f is closed formation pressure, and P_s the flowing sand face pressure.

The determination of a well's performance by the back-pressure behavior permits the production or storage engineers to analyze many reservoir problems essential to the development of gas fields. A few examples of such problems are: calculation of gas deliverability into a pipeline at a predetermined line pressure; design and performance analysis of gas gathering lines; determination of spacing and number of wells to be drilled during field development to meet future market requirements or contract obligations, etc. All depend at least in part on the availability and use of reliable back-pressure curves.[2]

In the past, the behavior of gas wells was evaluated by open flow tests. These tests were wasteful of valuable natural gas and sometimes caused troublesome cave-ins and water coning. The need for better testing methods was first recognized about 20 years ago. For many years, the United States Bureau of Mines, Monograph 7[3] has served as a guide for evaluating the performance of gas wells by back-pressure tests. Since Monograph 7, various methods of testing gas wells have been established and put into practice. These methods, also called "Flow after Flow," "Isochronal Performance Method," etc., have all been based on experiments designed to determine the exponent (n) and the coefficient (c) directly from flow tests.

When a gas well is producing from a high-permeability formation, the drainage radius evolves quickly to the boundaries of the drainage area and only a short period of time is required for semi-steady-state flow conditions to be approximately established throughout the drainage matrix. When a well is stabilized, its back-pressure/flow relation satisfies a definite back-pressure equation. For wells flowing single-phase in the reservoir, the back-pressure curve is a straight line determined by two characteristic quantities: the slope (n) and the performance coefficient (c).

The studies on the isochronal testing methods of gas wells have definitely shown that the steady-state flow conditions are not necessary to determine the slope (n). However, when and so long as transient conditions prevail through the reservoir, the performance coefficient (c) does not become fixed at a stabilized value. This behavior is graphically demonstrated in the shifting of back-pressure curves toward the left while keeping their slope constant.

An experimental determination of the stabilized back-pressure curve requires at least three flow tests, each starting with nearly static pressure distributions and each of proper duration to assure stabilization. Field experience gained throughout the recent years in the development of gas fields such as San Juan, and theoretical studies on the unsteady flow of fluids through porous media indicate that the stabilization time becomes long when the permeability is small. Therefore, direct back-pressure testing of tight wells becomes wasteful of gas and time consuming if the stabilized performance coefficient is to be determined experimentally.

In the development of gas fields with low permeability, it becomes exceedingly desirable to develop a method which would predict the stabilized back-pressure behavior without having to resort to long stabilization tests. Such a method was first suggested, at least theoretically, by A. Houpeurt of French Petroleum Institute and later reduced to engineering practice by M. R. Tek et al.[2]

Prediction of Performance Curves for Gas Wells. The prediction of a stabilized performance curve of a gas well becomes necessary particularly when it is not feasible to perform very long duration back-pressure tests on wells producing from tight formations. In order to determine the back-pressure curve, one must first have the slope, and next the intercept of the back-pressure curve. The slope of the back-pressure curve is usually determined by a short duration test procedure developed by M. H. Cullender, called *Isochronal Testing Procedure.*[4] In measuring both the flowing bottom-hole pressure and the flowrate as a function of time for different well head choke settings, the object of Isochronal Test method is to determine the common slope when data on pressures and flowrates for equal flow duration are determined.

The field test data on a large number of natural gas wells indicate that during the unstabilized performance of a gas well, the back pressure curves for 3 hours, 24 hours, 3 days, 10 days, etc. will constantly shift toward the left without changing their slopes.

Once the slope of the back-pressure curve is obtained, one can calculate the performance coefficient either from field or laboratory determined reservoir parameters.

Computation of Steady-State Performance from Core Data. Laboratory data usually provide some average values for the porosity ϕ and the permeability k. If the composition or the gravity of natural gas is known along with the reservoir temperature, formation thickness, and well spacing, in the absence of actual test information one may use the radial turbulent flow equation to prepare a synthetic log $(P^2_1 - P^2_2)$ vs. log Q curve.

As we have seen earlier, for radial flow the following equation holds in steady state:

$$P^2_1 - P^2_2 = \frac{1{,}424\ \mu\ zTQ\ \ln(r_1/r_2)}{hk} + \frac{3.161 \times 10^{-12}\ GQ^2\ \beta\ zT(1/r_2 - 1/r_1)}{h^2} \qquad (4\text{-}2)$$

With the field units discussed earlier, Equation 4-2 permits the prediction of Q vs. (P^2). It must be noted, however, that the equation only holds for steady-state with constant pressure at the exterior boundary of the drainage area.

Computation of Steady-State Performance from Field Data. For single phase natural gas wells, the stabilized back-pressure curve will be determined if the slope and the performance coefficient are known.

M. H. Culender's *Isochronal* method of determining the slope of back-pressure curves has been in use for many years as the proper way of evaluating the slope (n).[4]

Calculation of the Performance Coefficient. Theoretical studies on the flow of natural gas through porous media indicate that the performance coefficient of a natural gas well is, in general, a function of the following parameters:

k = Effective permeability of the formation
h = Thickness of the formation
μ = Viscosity of the gas
G = Specific gravity of the gas
z = Compressibility factor of the gas
T_f = Temperature of the formation
r_d = Radius of drainage
r_w = Radius of the wellbore
ν = A function of the range of production rates and slope of the back-pressure curve n
b = ρ/p (gr/cc × atm)

Without going into any mathematical or theoretical justification of A. Houpeurt's studies, the relation between the gas production rate and the overall pressure drop between the reservoir boundary and the wellbore radius is:

$$Q = \frac{2.49}{G} (4.659)^n \, \nu \left[\frac{\pi kh}{\mu} \frac{b}{\ln(r_d/r_w)}\right]^n (p^2_f - p^2_s)^n \tag{4-3}$$

If a developed field is considered, the radius of drainage can be obtained directly from the spacing. In the case of a single well, the steady-state radius of drainage becomes equal to the radius of reservoir boundaries. Under the conditions of transient flow, the magnitude of the radius of drainage is not constant. When the well is shut-in, the radius of drainage may be considered to coincide with the wellbore radius. When the well is opened and production resumed, the drainage radius evolves toward the boundaries of the drainage area. An approximate method for the instantaneous evaluation of the drainage radius in the unsteady flow is included in the following.

The term (r_w) designates the effective wellbore radius. Ordinarily, (r_w) can be taken as equal to the radius of the drill bit. In cases where the special-completion technique used on the well may effectively change the value of the wellbore radius beyond the sand face value, a method credited to Houpeurt is developed for calculating the effective value of the wellbore radius. These cases are usually the result of acidization, fracturing, or both.

The preceding equations indicate that, if the test index $\pi kh/\mu$, b, $\ln(r_d/r_w)$ and ν are known, the performance coefficient c_p can be estimated. It is established that the most reliable value of nkh/μ would be the one based on the actual behavior of the well. Although h and μ are usually well known or determined from logs and correlations, the permeability values obtained from core analysis are only point measurements of a general variable, the magnitude of which depends very much on depth, location, and direction. While the permeability obtained in the laboratory by individual core measurements, or by averaging a large number of core measurements, may not be truly representative of the reservoir behavior, it is believed that the effective permeability defined and measured from actual production-pressure behavior of the formation should be used in calculating c_p. This is done by interpreting data obtained from pressure build-up tests.

Supplementary Formulations

The Test-Index. In order to calculate the test-index, a draw-down and a build-up test are required. The well is produced undisturbed through a given choke setting for a flow period long enough to assure that the well behavior is representative of formation characteristics. While true steady-state conditions are never strictly attained, it is, in general, possible to continue production until the gas rate and flowing pressures become approximately constant. Pressure and flow data during the draw-down and pressure data during the build-up are collected.

When the pressure build-up data are plotted vs. the time in log scale, the pressure build-up curve will, in general, exhibit a curved section at small values of time and a curved section flattening out at even larger values of shut-in time. The curved section at the beginning of the pressure build-up curve is due to after production into the wellbore, while the straight-line section represents the theoretically predictable transient flow behavior. The curved section at the end is caused either by reservoir limit or drainage area boundary interference effects.

If S represents the slope of the straight-line section of the pressure build-up curve, Q the production rate previous to shutting the well in, P_s the flowing bottom-hole pressure at the time of shut-in, and T_f absolute the temperature of formation; the following equation can be used to calculate the test-index $\pi kh/\mu$:

$$\frac{\pi kh}{\mu} = 78.1 \frac{QT_f}{P_f S} \tag{4-4}$$

It should be noted that if the well is produced at a low rate prior to build-up, the rate, Q, and the flowing pressure, P, can be considered approximately constant with respect to time.

It will be recalled, consistent with the earlier definition of k and Equation 4-3, the test-index is purposely defined so it will be representative of the formation in the far zone. The units in Equation 4-4 are specifically chosen for direct use in Equation 4-3, which permits the calculation of the performance coefficient.

The Drainage Radius. The drainage radius is the radius beyond which no natural gas is flowing toward a producing well, r_d in Equation 4-3.

If a developed field is under consideration for a condition of semi-steady-state gas production with all the offsets producing the same quantity of gas, the drainage radius r_d becomes about equal to one-half of the spacing between adjacent wells. If a transient condition is prevailing over the drainage area, the drainage radius is not fixed. For no flow condition, it can be assumed to coincide with the wellbore radius and, upon resuming production, it starts propagating toward the reservoir or drainage area boundaries. Under these circumstances, and by mathematically evaluating the drainage radius as the radius beyond which only one percent of the well production is flowing at any instant, an approximate formulation is derived for engineering calculations.[2]

$$r_d = 0.0704 \left(\frac{4kt}{\phi\mu c}\right)^{1/2} \tag{4-5}$$

Since the test-index ($\pi kh/\mu$) is usually determined for calculating the performance coefficient, it will suffice to know the formation thickness (h); the viscosity can be estimated from correlations available in the literature based on bottom-hole pressure and temperature conditions.

The compressibility of the natural gas may be calculated approximately as reciprocal of mean reservoir pressure. When the pressure drop between static reservoir and flowing bottom-hole pressure is small, a more realistic pseudo-compressibility can be calculated from:

$$c = \frac{14.65}{P_0}\left[1 - \frac{P_0}{z_0}\left(\frac{\partial z}{\partial P}\right)_T\right] \tag{4-6}$$

where

P_0 = Static reservoir pressure, psia

z_0 = Compressibility factor of the natural gas under static reservoir pressure and temperature conditions

$\left(\frac{\partial z}{\partial P}\right)_0$ = The slope of $z = z(P,T)$ curve at the point corresponding to the static reservoir conditions.

The Effective Wellbore Radius r_w. A mathematical study of the equation relating the bottom-hole build-up pressures to the production rate preceding the pressure build-up test, the hydraulic diffusivity,* and the shut-in time indicates that the effective wellbore radius can be calculated by the following equation:

$$r_w = \frac{1}{30.5}\left[\frac{kt_n}{\phi\mu c \text{ Antilog}\left(\frac{P_n - P_s}{S} - 0.3515\right)}\right]^{1/2} \tag{4-7}$$

Usually, the slope (S) and the flowing bottom hole pressure (P_s) are available from the calculation of the test-index ($\pi kh/\mu$); the other variables, k and μ are calculated as indicated in the preceding sections. The porosity ϕ is determined from core analysis. P_n and t_n are read from the pressure build-up curves with respect to one another at an arbitrary point on the straight-line section of the build-up curve.

In the case of straight open-hole completions and in the absence of any production stimulation techniques such as fracturing, acidizing, use of explosive charges, etc., the effective wellbore radius may provide useful information on estimating ln r_d/r_w. However, it should be clearly understood that, since this calculation is based on the determination of variables such as k, ϕ, and c, it will be approximate and will depend a great deal on the values of k, ϕ, and c, and on the proper interpretation of pressure build-up data.

It is also interesting to note that a tenfold variation of (r_d/r_w) will only affect the term $\ln(r_d/r_w)$ by approximately 2.3.

Example 4-1: Free No. 4 Hugoton Field

The following data were collected from drawn-down and build-up tests on the well Free No. 4, Hugoton Field:

Q = 5,165 MCF/D approximate stabilized flowrate prior to shut-in
P_f = 424 psia, 5 days shut-in pressure

* The hydraulic diffusivity is defined as the ratio between the permeability and the product of the porosity, viscosity, and compressibility.

T_f = 550 degrees Rankine
P_s = 357.9 psia

Figure 4-6 indicates the pressure build-up data with the slope S = 13.8 psia/cycle.

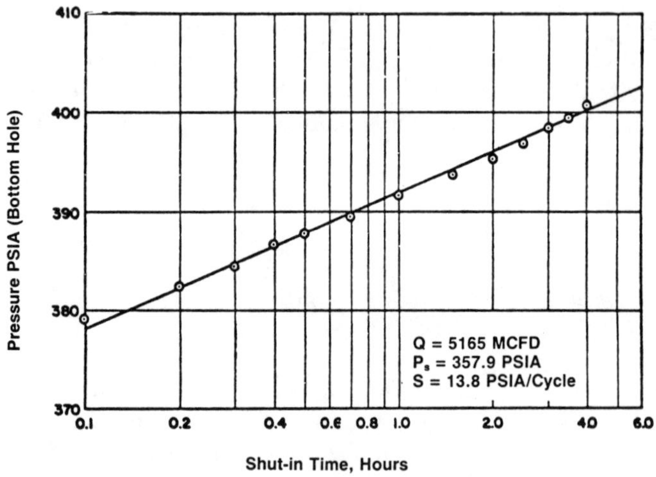

Figure 4-6. Pressure build-up curve on Free No. 4.

1. **Determination of the Slope of the Back-Pressure Curve**
   The slope of the back-pressure curve was determined by back-pressure tests and transmitted by the Natural Gas Department as n = 0.846.
2. **Calculation of the Test-Index**
   Using Equation 4-4:

$$\frac{\pi kh}{\mu} = \frac{78.1 \times 5{,}165 \times 550}{357.9 \times 13.8} = 4.495 \times 10^4 \text{ darcies} \times \text{cm/cp}$$

3. **Calculation of (b) Factor**
   Using Equation 8 from Reference 2:

$$b = \frac{29G}{zR'T_f'}$$

R' = 82 (atm × cc/mole × °K)

Gas gravity G = 0.712 (from transmitted data)

T_c = 388°R

p_c = 664 psia

$$p_r = \frac{424}{664} = .638 \text{ (reduced pressure)}$$

$$T_r = \frac{550}{388} = 1.416 \text{ (reduced temperature)}$$

$$z = 0.925$$

$$T_f = 550° \text{ Rankine, } 90°\text{F, } 32.2° \text{ centigrade, or } 32.2 + 273.15$$
$$= 305.35°\text{K absolute}$$

$$b = \frac{29 \times 0.712}{0.925 \times 82 \times 305.35} = 8.92 \times 10^{-4} \text{ gr/cc} \times \text{atm}$$

4. **Estimation of ln** $\left(\frac{r_d}{r_w}\right)$

For steady-state, the drainage radius may be considered equal to 1/2 of well spacing; r_d = R, and from well spacing information on maps:

R = 1,316 ft

Calculation of r_w

Using the pressure build-up data, Equation 4-7 gives the value of effective wellbore radius as follows:

$$r_w = \frac{1}{30.5}\left[\frac{kt_n}{\phi\mu c \text{ Antilog}\left(\frac{P_n - P_s}{S} - 0.3515\right)}\right]^{1/2} \quad (4\text{-}7)$$

According to the data transmitted from the Natural Gas Department, Free No. 4 has 120 ft of net gas pay:

$$h = 120 \text{ ft}$$
$$\mu = 0.012 \text{ cp}$$
$$\frac{\pi kh}{\mu} = 4.495 \times 10^4 \text{ Darcy} \times \text{cm/cp}$$

thus:

$$k = \frac{4.495 \times 10^4 \times 12 \times 10^{-3}}{3.14 \times 120 \times 12 \times 2.54} = 46.9 \times 10^{-3} \text{ darcies}$$

ϕ = 0.07 (averaged from available core data)

c = Approximate compressibility of the gas (calculated using Equation 4-6)

$$c = \frac{14.65}{424}\left[1 - \frac{424 \times 0.09}{0.92 \times 0.9 \times 664}\right]$$

$$= 3.225 \times 10^{-2} \text{ vol per vol per atm.}$$

Now, going back to pressure build-up Figure 4-3, we read the point:

t_n = 10 hrs, for which P_n = 405.8 psia
p_s = 357.9 psia
s = 13.8 psia per cycle

Substituting all numerical values in Equation 4-7, we calculate

$$r_w = \frac{1}{30.5} \times$$

$$\left[\frac{46.9 \times 10^{-3} \times 10 \times 3{,}600 \times 10^2}{0.07 \times 0.012 \times 3.225 \text{ Antilog}\left(\frac{405.8 - 357.9}{13.8} - 0.3515\right)}\right]^{1/2}$$

$$r_w = 7.14 \text{ ft}$$

hence:

$$\ln \frac{r_d}{r_w} = \ln \frac{1{,}316}{7.15} = 5.21$$

Calculation of Gas Deliverability Using the Concept of *Pseudo-Pressure* *

Any solution related to gas deliverability must include skin effect, wellbore storage, and turbulence in order to represent the real-world conditions prevailing in the drainage area as manifested in terms of field data.

If one treats the lumped parameter $\phi\mu c/k$ as nearly constant, an approximate solution is obtained in dimensionless terms in cylindrical coordinates for Darcy Regime with no-skin as follows:

$$m[p(r,o)] - m[p(r,t)] = 5.0304 \times 10^7 \frac{p_{sc}}{T_{sc}} \frac{Tq_{sc}}{kh} [1/2 \ln t_D + .405] \qquad (4\text{-}8)$$

* Reference 2, Chapter 3

Equation 4-8 is for constant terminal rate and for uniform initial pseudo-pressure distribution $m_i[p(r,o)]$.

where:

$$t_D = \frac{2.637 \times 10^{-4} kt}{\phi\ (\mu c)_i r^2} \qquad (4\text{-}9)$$

A typical plot of the pseudo-pressure for unsteady-state and constant terminal rate is given in Figure 4-7. The effects of skin and turbulence are added to Equation 4-8 in the following manner:

Effect of Skin. The area in the immediate vicinity of the wellbore, having usually lower but sometimes higher permeability than the far zone, is usually recognized as the *near zone* or *skin zone*. The extra pressure drop due to *skin effect* was first studied in 1953 by Van Everdingen[5] who introduced the dimensionless skin factor.

In terms of the pseudo-pressure, m, which was introduced more than a decade later, the skin effect may be expressed as follows:

$$\Delta m(p)\mid_{skin}(r_w,t) = 5.0304 \times 10^7 \frac{p_{sc}}{T_{sc}} \frac{Tq_{sc}}{kh} s \qquad (4\text{-}10)$$

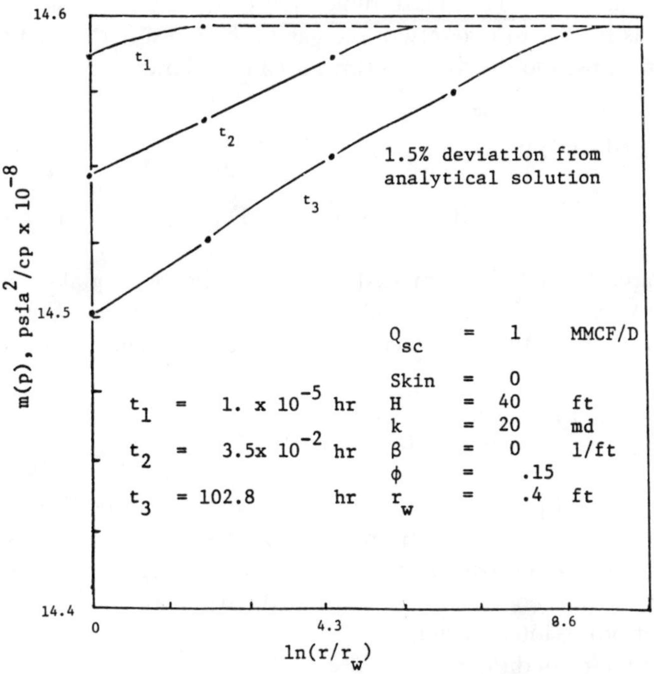

Figure 4-7. Pseudo-pressure profile (without turbulent effect).

Hawkins[6] showed in 1956 that the skin factor s may be related to far and near zone permeabilities as:

$$s = \left(\frac{k}{k_s} - 1\right) \ln \frac{r_s}{r_w} \tag{4-11}$$

From the above, it can be seen that the skin effect can be added directly to Equation 4-8 by inserting the dimensionless s term in the bracket.

Effect of Turbulence on Deliverability. Procedures to include the effect of turbulence on deliverability of gas wells have been published by Smith[7]; Tek, Coats and Katz[8]; Swift and Kiel[9]; and Carter et al.[10]

For an infinite reservoir in radial flow, the effect of turbulence on the pseudo-pressure drop in the wellbore at constant terminal rate and unsteady-state flow may be evaluated by the following equation:

$$m[p(r_w,o)] - m[p(r_w,t)]$$
$$= 5.0304 \times 10^7 \frac{P_{sc}}{T_{sc}} \frac{Tq_{sc}}{kh} [1/2 \ln t_D + .405 + s + Dq_{sc}] \tag{4-12}$$

The term D has the inverse dimension of q_{sc}, (T/L^3). The value of D changes as related to viscosity of the gas as the pressure drops. For a finite reservoir, in pseudo-steady-state flow, it can be shown that

$$m[p(r_e,t)] - m[p(r_w,t)] = 5.0304 \times 10^7 \frac{p_{sc}}{T_{sc}} \frac{T}{kh} q_{sc}$$
$$[\ln r_e/r_w - 3/4 + s + D(\mu_w,r_w)q_{sc}\ (1 - r_w/r_d)] \tag{4-13}$$

When Equation 4-13 is compared to the Forchheimer equation integrated between r_w and r_d for steady-state with constant pressure boundaries one can show that the turbulence coefficient D may be calculated from:

$$D(\mu_w,r_w) = 7.675 \times 10^{-13} \frac{G}{h} \int_{r_w}^{r_d} \frac{\beta k}{\mu r^2} dr \tag{4-14}$$

Equation 4-14 permits calculation of the turbulence coefficient, D, from the radius of drainage r_d and from the properties of natural gas. As a reasonable engineering approximation for the finite range of pressures involved, μ can be treated as constant at its value at the wellbore and, as such, can be taken out of the integral sign.

Under these conditions:

$$D(\mu_w, r_w) = \frac{7.675 \times 10^{-13} G\beta k}{h\mu_w}\left(\frac{1}{r_w} - \frac{1}{r_d}\right) \tag{4-15}$$

Equation 4-15 suggests that, when a gas well is produced at constant terminal rate q_{sc}, as soon as $1/r_d << 1/r_w$, or $r_d >> r_w$, D becomes constant. In reality, however, as the well produces in pseudo-steady-state, the pressures will continue to drop, causing D to become larger and larger. This effect, however, is small (see Figure 4-8).

Figure 4-9 shows the pseudo-pressure profile with turbulence effects calculated by finite difference methods on a main frame computer.*

In the equations for calculating gas deliverability using the concept of pseudo-pressure, the nomenclature of various parameters used is as follows:

A = Coefficient, appropriate units
B = Coefficient, appropriate units
c = Compressibility, vol/vol/psi
D = Turbulence coefficient, appropriate units
G = Gas gravity, dimensionless

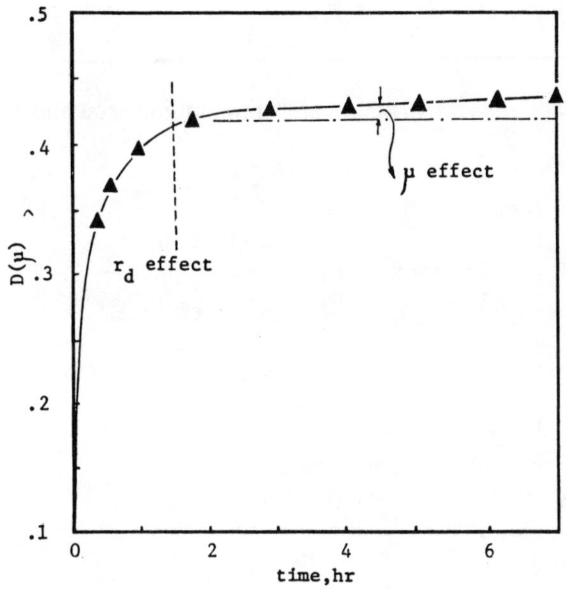

Figure 4-8. Temperature coefficient, $D(\mu_w\ r_w)$ vs. drawdown time.

* For details on this study, refer to recent papers by Tek, Logan, and Lee (SPE 14205) and by Oren, Lee, and Tek (SPE 15525).

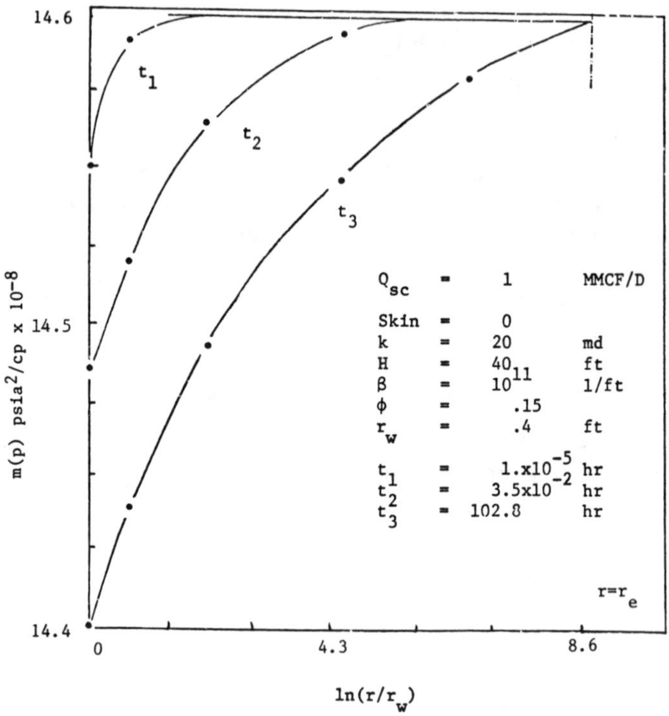

Figure 4-9. Pseudo-pressure profile (with turbulence effect).

h	=	Formation thickness, ft
k	=	Permeability, md or appropriate unit
m[p(r,o)]	=	Pseudo-pressure at r = r, t = o, $psia^2/cp$
m[p(r,t)]	=	Pseudo-pressure at r = r, t = t, $psia^2/cp$
p	=	Pressure, psia
p_{sc}	=	Pressure base, psia (in general = 14.73 psia)
q_{sc}	=	Rate of production, MMCF/D
r	=	Radius, ft
r_s	=	Radius of the skin zone, ft
r_w	=	Radius of the wellbore, ft
s	=	Skin factor, dimensionless
t	=	Time, days or seconds
T	=	Formation temperature, °R abs.
T_{sc}	=	Temperature base, °R abs. (in general = 520 °R abs)
u	=	Velocity, ft/sec
x	=	Distance, ft or approximate unit
z	=	Compressibility factor for gas

WELLBORES, WELL HEADS, GATHERING SYSTEMS, DISTRIBUTING SYSTEMS

After the gas enters the wellbore, its condition, properties, and progress are affected by flow through man-made equipment. In its progress from the perforations, ultimately through the residential/commercial meter and onto the burner or furnace, the route includes flow through well tubing or production string, well head Christmas tree, separators, heaters, surface gathering system, dehydrators, compressors, pressure regulation, metering pipeline, city gate and low pressure distribution networks.

This section will include a review of engineering procedures related to design and performance of major equipment listed above.

Static and Flowing Gradients in Wellbores

The calculations involving wellbore almost invariably start with understanding and interpretation of static pressure gradients. Reservoir pressure data often collected by deadweight testers as psig at well heads must often be converted to bottom-hole pressures expressed as psia. Such static gradient calculations all depart from the fundamental equation of absolute fluid statics derived from the principle of conservation of momentum:

$$dp = \frac{g}{g_c} \rho \, dh \tag{4-16}$$

In Equation 4-16, p = pressure, generally a function of depth (usually measured as vertical distance from the stem of the orbit valve) in psfa.

- ρ = Density of the fluid, pounds mass/ft^3
- g = Acceleration of gravity, 32.17 ft/sec^2
- g_c = Conversion factor, 32.17 (lb m/lb$_f$) × ft/sec^2
- h = Depth in feet measured as vertical distance from the stem of the orbit valve, ft

Static Liquids. When Equation 4-16 (a total, linear, separable, differential equation) is integrated for liquids such as oil, brine, or condensate, one can readily calculate the pressure change due to depth in a static column of liquid, as follows:

$$\Delta P = \rho g/g_c \times \Delta h \tag{4-17}$$

An immediate application of Equation 4-17 is the gradient of static hydraulic pressures prevailing in formations around the wellbore.

If ρ = 62.4 lb m/ft^3 (for pure water)
for h = 1 ft
$\Delta p/\Delta h$ from 4-17 given above
$\Delta p/\Delta h$ $62.4 \times 32.17/32.17 \times 1/144$
= .433 psi/ft

The brines encountered around the wellbores in the crust of the earth are in general heavier than pure water, depending upon their salinity. Accordingly, it is not uncommon to obtain earth gradients of 0.465 to, say, 0.502 psi/ft, because of densities larger than 62.4 lb mass/ft^3.

Static Gases. The equation for calculating the static pressure distribution in gas wells is derived from Equation 4-16 by substituting for the density of gas

$$\rho = \frac{29G}{zRT}\, p \tag{4-18}$$

where G = Specific gravity of the gas, $\left(\frac{\text{Molecular wt}}{29}\right)$

$29G$ = Molecular wt, $\frac{\text{lb mass}}{\text{lb-mole}}$

p = Pressure $p = p(h)$, lb_f/ft^2 abs.

R = The gas constant = 1,544 $\frac{ft \times lb_f}{lb\ mole\ °R}$

T = Average, or constant, temperature, °R abs.

z = The compressibility factor, function of reduced pressure and temperature, dimensionless (as such unitless). With above units the density ρ comes out with units of lb_m/ft^3

In substituting from Equation 4-18 into 4-16 and integrating from h = 0 (well head) to h = x [desired datum depth from the orbit valve stem], one obtains:

$$p_x = (p + 14.7)\, e \frac{(0.01874\ GX)}{T_a z_a} \tag{4-19}$$

where p_x = Pressure at datum, psia

p_{Top} = Pressure at surface, psig

G = Gas gravity

x = Depth of datum, ft

T_a = Average temperature

z_a = Average compressibility factor

Example 4-2

Problem. Draw a curve of the pressure difference between well head and reservoir pressure for a shut-in well 2,000 ft deep, containing natural gas of 0.6 specific gravity. The mean wellbore temperature is 60°F.

Solution. Using Equation 4-19, the z-factor from our charts, and selected well head pressures ranging from 300 to 1,300 psig, we get:

$$0.01874\,\frac{GX}{T_a z_a} = \frac{0.01874 \times 0.6 \times 2{,}000}{520 \times 0.793}$$

$$= 0.0546$$

$p_x = (p_T + 14.7)e^{0.0546}$
$\quad = (p_T + 14.7)1.0561$
For $p_T = 1{,}300$ psig
$\Delta p = 1{,}314.17(1.0561 - 1.0) = 72.9$ psi

Bottom-hole pressure = 72.9 + 1,314.7 = 1,387.6 psia. Mean wellbore pressure = 1,351.6 psia. Repeating calculations for other well head pressures gives the results plotted in Figure 4-10.

Calculation of Flowing Gradients

The prediction of the pressure distribution and pressure gradients along a flowing well is just as important as the prediction of the static gradients in a non-flowing well.

Applications. Analysis of pressure build-up or drawdown behavior, determination of effective reservoir permeabilities, evaluation of the deliverability of natural gas from a well to an existing gathering network, and many other typical reservoir engineering problems require the prediction of flowing gradients.

In any problem where the flowing sand-face pressure is needed, one must either directly measure the bottom-hole pressure by a suitable pressure gage or simply measure the flowing well head pressure and calculate the bottom-hole pressure from pipe flow calculations such as in back pressure calculations. The two methods for calculations of flowing pressure gradients follow.

Calculation of Flowing Bottom-Hole Pressure in a Gas Well. The formula for calculating the flowing bottom-hole pressure from flowing well

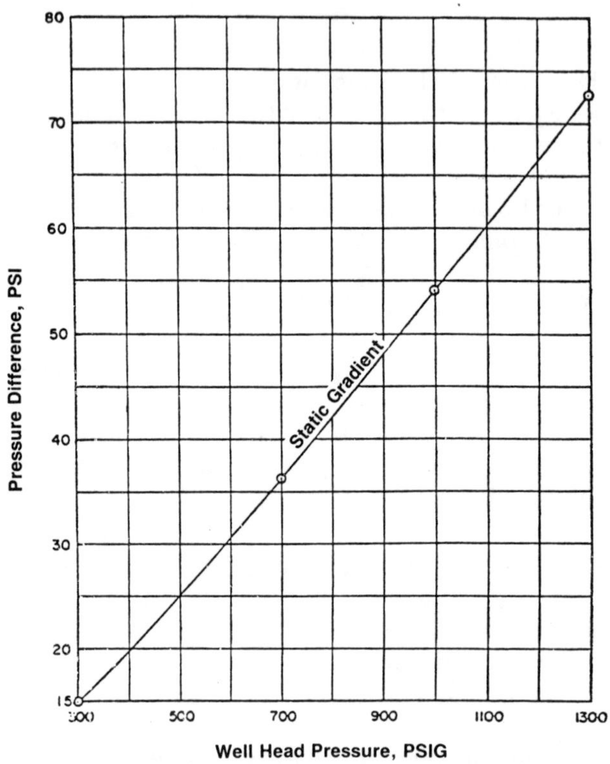

Figure 4-10. Pressure gradient in static wells 2,000 ft. deep for 0.6 gravity gas.

head pressure is derived by combining the static pressure gradient formula with the formula for friction occurring for gas flow in pipes. The static pressure gradient is often a large part of the pressure drop. This is especially true if liquids should be present in the flowing column to increase the average fluid density.

The combined formula in which the static gradient is represented by the e^s term is:

$$P_s^2 - e^s P_w^2 = \pm 0.000667 \, f \, \frac{Q^2 \, T_a^2 \, z_a^2}{d^5} (e^s - 1) \qquad (4\text{-}20)$$

(± for gas production, upward flow)
(– for gas injection, downward flow)

where P_s = Flowing sand face pressure, psia
P_w = Flowing well head pressure, psia

$s = 0.0375\ GL/T_a z_a$
G = Gas gravity
L = Depth of well, feet
T_a = Average flowing temperature
z_a = Compressibility factor of gas at average conditions
e = 2.718
f = Friction factor c.f. Figures 4-11 and 4-12 (Moody)
Q = Gas flowrate, Mcf/day
d = Pipe diameter, ID inch
μ = Gas viscosity, centipoise

The Cullender and Binckley version of this formula permits the elimination of a friction factor as such by relating it to the diameter of the pipe, viscosity of the gas, and gas gravity; they also use P-square values in thousands and Q in MMCF/day.

$$p_s^2 - e^s p_w^2 = (F_d Q)^{1.935}(e^s - 1) = R^2 \tag{4-21}$$

For this formula p_s^2 and p_w^2 are psia^2 in thousands.

$$F_d = \frac{0.137(T_a z_a)^{1.0336}}{d^{2.614}} \left(\frac{\mu}{G}\right)^{0.0336} \tag{4-22}$$

When $T_a = 540°R$, $z_a = 1.0$, $\mu = 0.0109$ cp, and $G = 0.720$, F_d

$$= \frac{65.71}{d^{2.614}}$$

The method given above includes two major assumptions which somewhat restrict the generality of the solution. These are the constant average compressibility factor and constant average wellbore temperature. Actually, the compressibility factor changes as a function of both the pressure and the temperature. The temperature changes as a function of the depth. While the effects of variable compressibility factor and variable temperature on the calculated bottom-hole pressure are not large, the expected accuracy of the methods is enhanced when these two parameters are treated as variables. A method given by Cullender and Smith treats the temperature as a variable. Another method, presented by F. H. Poettman,[66] takes into consideration the variation of the compressibility factor down the flow string.

See Table 4-1 for calculation of back pressure test using Moody friction factors, and Table 4-2 for calculation using Cullender and Binckley factors.

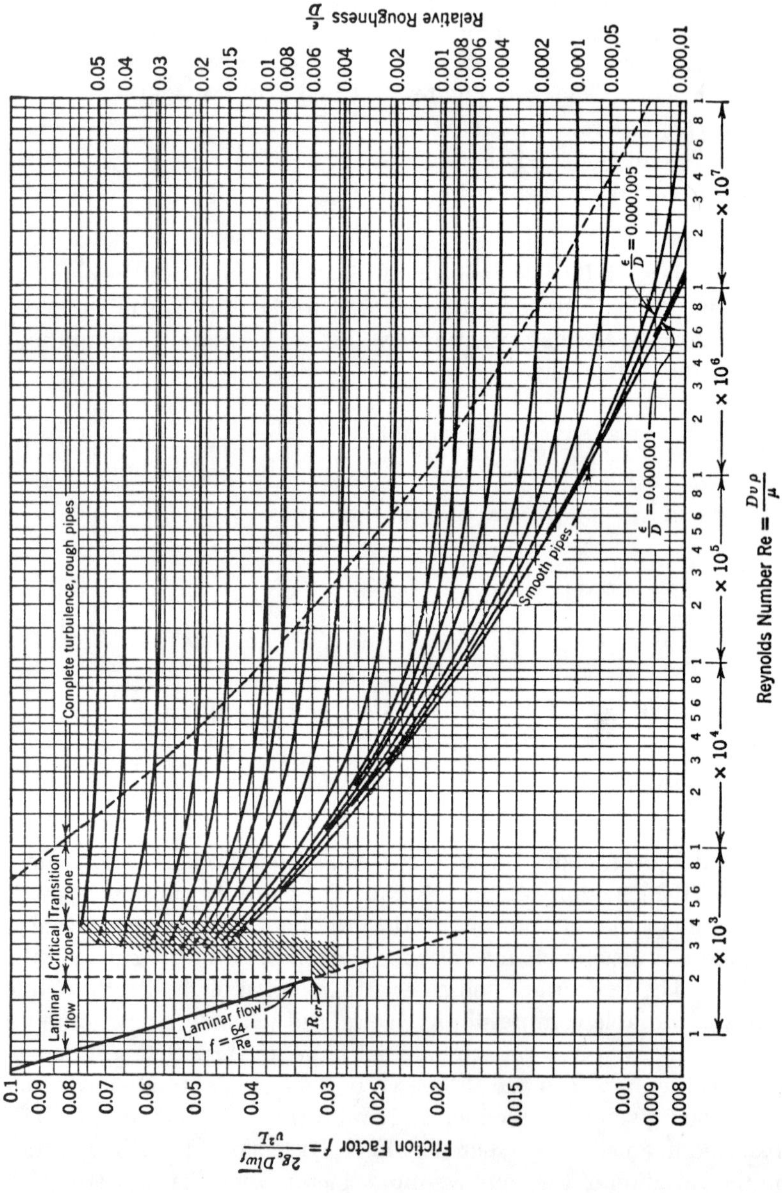

Figure 4-11. Moody friction factor chart (*Fluid Flow and Heat Transfer,* Tek and Wilkes[12]).

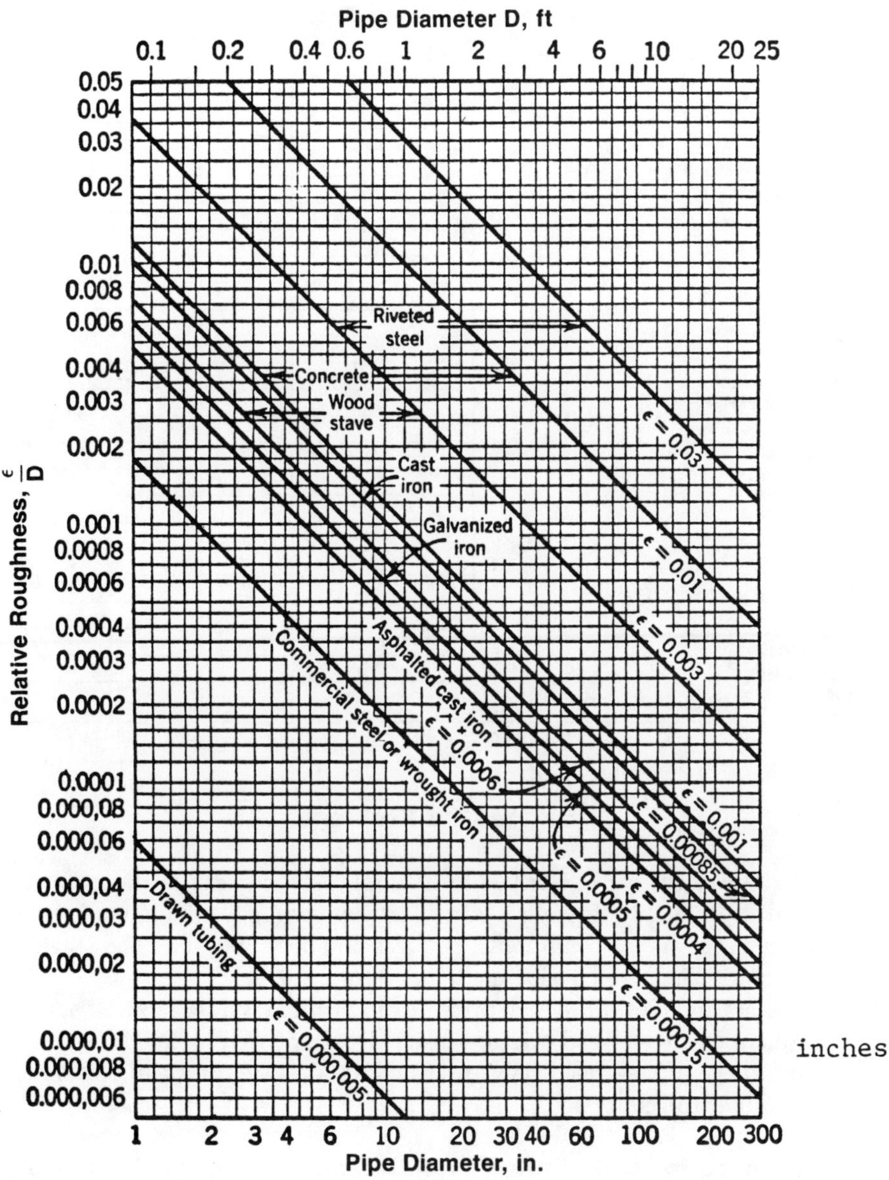

Figure 4-12. Relative roughness chart.[12]

Table 4-1
Calculation of Back Pressure Test Using Moody Friction Factors

	Static Case				
1. Q flowrate MCF/day					
2. p_w well head press., psia					
3. T_a average temp., 460 + °F					
4. p_a assumed psia					
5. $p_a/p_c = p_r$					
6. $T_a/T_c = T_r$					
7. z_a					
8. μ centipoise					
9. G Q/μ					
10. Moody factor					
11. $s = 0.0375\ GL/T_a z_a$					
12. e^s					
13. $e^s - 1$					
14. $(QT_a z_a)^2$					
15. $6.67 \times 10^{-4}(QT_a z_a)^2(e^2 - 1)f/d^5 = b$					
16. p_w^2					
17. $e^s p_w^2$					
18. $p_s^2 = b + e^s p_w^2$					
19. p_f^2					
20. p_f psia					
21. p_s psia					
22. $p_f^2 - p_s^2$, psia^2, thousands					

p_c = psia; T_e = °R; L = ft; G = ; d = in.; d^5 =

Suggested Problem

The natural gas well described in Example 4-2 on static gradients for a gas well is producing through a 2 inch ID flow string 4 million cubic feet (measured at standard conditions 14.73 psia and 60°F) per day. The gravity of gas is 0.6. The average flowing temperature is 60°F.

Calculate and plot the overall pressure drop (from sand face to well head) as a function of well head pressure (psig). Depth of well = 2,000 feet.

Use a range of well head pressures from 300 to 1,300 psig and plot on a graph similar to Figure 4-10.

Procedure for Calculating Well Deliverabilities or Pressure Gradients

A. For Flow Through Casing Tubing Annulus Use:

Table 4-2
Calculation of Back Pressure Test Using Cullender and Binckley Factors

	Static Case					
1. Q flowrate MMCF/day						
2. p_w well head pressure, psia						
3. T_a average well temp., 460 + °F						
4. p_a assumed psia						
5. $p_r = p_a/p_c$						
6. $T_r = T_a/T_c$						
7. z_a						
8. μ centipoises						
9. G						
10. μ/G						
11. $(\mu/G)^{0.0336}$						
12. $(T_a z_a)^{1.0336}$						
13. $F_d = \frac{0.137(T_a z_a)^{1.0336}}{d^{2.614}}\left(\frac{\mu}{G}\right)^{0.0336}$						
14. $s = 0.03754\ GL/T_a z_a$						
15. e^s						
16. $e^s - 1$						
17. $R^2 = (F_d Q)^{1.935}(e^s - 1)$						
18. P_w^2 psia² thousands						
19. $e^s P_w^2$ psia² thousands						
20. $P_s^2 = R^2 + e^s P_w^2$						
21. P_f^2 psia² thousands						
22. P_f psia						
23. P_s psia						
24. $P_f^2 - P_s^2$, psia² thousands						

P = psia; T = °R; L = ft; G = ; d = in.; $d^{2.614}$ =

$$p_s^2 - e^s p_w^2 = 0.000667\ f\ \frac{Q^2 T_a^2 z_a^2\ (e^s - 1)}{(D_1 + D_2)^2\ (D_1 - D_2)^3} \tag{4-23}$$

where D_1 = Inside diameter of casing, in.
D_2 = Outside diameter of tubing, in.
p_s = Sand face pressure, psia
p_w = Well head pressure, psia
$S = 0.0375\ \frac{GL}{T_a z_a}$, dimensionless
G = Gas gravity
L = Length of tubing – casing annulus ft
T_a = Average temperature of flow string °R abs

z_a = Average value of compressibility factor in the flow string
f = Friction factor (Moody), dimensionless

$$f = f(R_e, \epsilon/D)$$

$$R_e = \frac{Q \times 0.0764 \times G\ 1{,}000 \times (D_1 - D_2)}{86{,}400\ \frac{\pi}{4}\ \frac{(D_1^2 - D_2^2)}{144} \times \mu \times 0.000672} \quad (4\text{-}24)$$

Q = Gas rate, MCF/D

Example of Calculation of Sandface Pressure for Flow Through Casing Tubing Annulus

D_1 = 6.456 in.

D_2 = 1.900 in.

p_w = 1,000 psia

p_s = ?

Q = 275,000 MCF/D

L = 6,500 ft

T_a = 460 + 97 = 557°R abs.

p_a = 2,250 assumed

$$p_r = \frac{2{,}250}{671} = 3.35$$

$$T_2 = \frac{550}{360} = 1.53$$

z_a = .785

$d_{eq} = D_1 - D_2 = 6.456 - 1.900 = 4.556$ in.

$\epsilon/d_{eq} = 0.0004$; $G = 0.6$; $\mu = 0.0158$ cp

$$R_e = \frac{275{,}000 \times 1{,}000 \times 0.0764 \times 0.6 \times 4 \times 1.44 \times 4.556}{86{,}400 \times \pi \times (\overline{6.456^2} - \overline{1.9^2})\ 12 \times 0.0158 \times 0.000672}$$

$$= 25{,}126{,}349.93$$

f = 0.016

$$s = \frac{0.0375 \times 0.6 \times 6{,}500}{557 \times .785} = .33$$

$$e^s = 1.40$$

$$e^s - 1 = 0.40$$

$$p_s^2 = 1.40 \times 10^6 + \frac{0.000667 \times 0.016 \times (\overline{275{,}000})^2 \times \overline{557}^2 \times \overline{0.784}^2 \times 0.40}{(8.356)^2(4.556)^3}$$

$$= 1.4 \times 10^6 + 0.323 \times 10^6$$

$$= 1.072 \times 10^7$$

$$\mathbf{p_s = 3{,}275\ psia}$$

For up-flow through tubing or casing, use:

$$p_s^2 - e^s p_w^2 = 0.000667\ \frac{f\ Q^2 T_a^{\,2} z_a^{\,2}(e^s - 1)}{D^5}$$

where D = Inside diameter, in.

$f = f(R_e, \epsilon/D)$

$$R_e = \frac{1{,}000\ Q \times 4 \times 144 \times 0.0764 \times G \times D}{86{,}400 \times \pi D^2 \times 12 \times \mu \times 0.000672}$$

μ = Gas viscosity, cp
f = From Moody chart
Q = Gas rate, MCF/D
z_a = Average value of compressibility factor in the flow string
T_a = Average temperature of flow string, °R abs.
L = Length of tubing – casing annulus, ft
G = Gas gravity

$s = 0.0375\ \dfrac{GL}{T_a z_a}$, dimensionless

p_w = Well head pressure, psia
p_s = Sand face pressure, psia

Design of Surface Facilities

Gas Gathering and Transportation. In the succession of events between the discovery of oil, gas, or condensate field, and the actual delivery of natural gas to city gate, of the crude oil to refineries, or of the petrochemicals to various feedstock markets, the development of the field depends upon the design of piping networks to carry the effluent. Design of a gathering system at the field, of separation, absorption, sweetening, dehydration, natural gasoline processing, low temperature processing—all involve plant piping systems. Large-scale movement of crude oil, natural gas, and products as well as crushed, ground, and cleaned coal, or manufactured high BTU or low BTU gas, liquid fuels—all these involve pipelining through horizontal, vertical, or inclined pipes. Some applications, such as offshore drilling or production, even involve curved pipes. In this section we shall review the basic engineering design tools available to practicing storage engineers for both design and performance evaluation of surface facilities.

Table 4-3 shows the various systems encountered in pipeline technology. Whether single-phase liquid, gas, or multiphase flow, the design of a particular conduit involves horizontal, vertical, inclined, or curved pipes. When the gas must be gathered from various sources and delivered to a central facility, a network of pipe sections is involved. The gathering system networks are either *looped* or *tree* type. In addition, certain installations still use the *star* type gathering system, where each well is directly and independently connected to a central header.

In general, most if not all of the above-indicated specific applications of engineering design equations start with energy balance between two points on the pipe:[12]

Table 4-3
Various Systems Encountered in Pipeline Technology

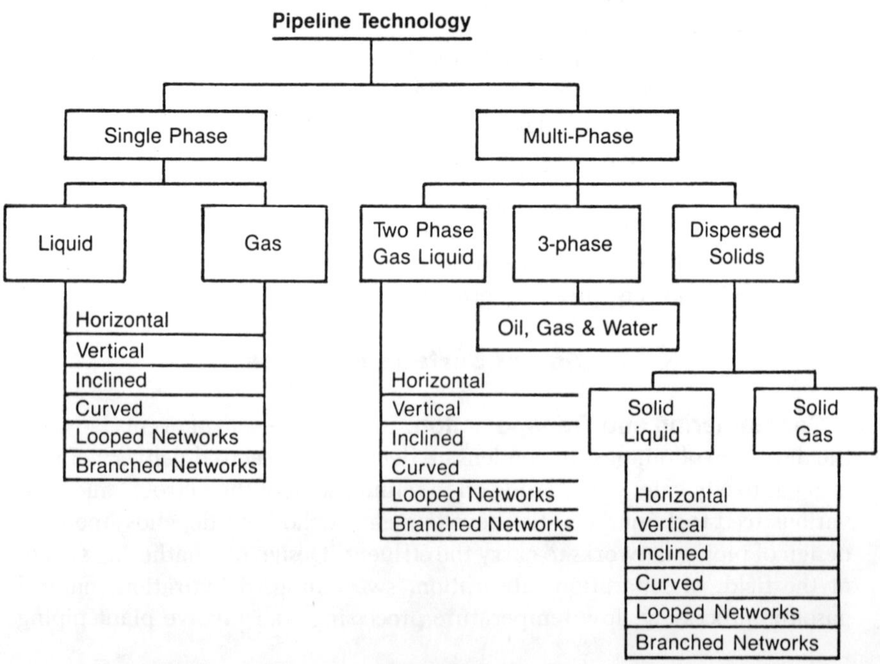

$$\frac{\Delta p}{\rho} + \left(\frac{\Delta V^2}{2g_c}\right) + \Delta\left(\frac{g}{g_c} z\right) = -\overline{w} - \overline{lw} \tag{4-25}$$

Equation 4-25 is for steady-state and for an incompressible liquid. It expresses the energy balance between any two points were quantities refer to point functions of flow work p/ρ, kinetic energy per unit mass flowing $V^2/2g_c$, and potential energy above an arbitrary datum per lb_m flowing. If the flow is compressible, then the first term $(\Delta p)/\rho$ is no longer a point function and becomes path dependent. It is expressed as:

$$\int_{p_1}^{p_2} dp/\rho$$

The term $\overline{w}$ represents the energy crossing the thermodynamic boundary as *shaft-work* expressed in ft $\times$ lb_f/lb_m. It is equal to zero if there are no pumps or compressors between the two considered points relating to the Δ quantities.

If there exists a piece of equipment between points (1) and (2) (with the flow always considered to be going from Point (1) upstream to Point (2) downstream), then if energy is supplied to the fluid, as in the case of a pump or a compressor, then w is a negative quantity. On the other hand, if the fluid is giving energy to the surroundings in the form of shaft work, like in the case of a turbine drive, then w is considered by convention to be a positive quantity.

Whether the flow is single-phase or multiphase, inclined or horizontal, etc., the term $\overline{lw}$ is equated to the following by a Fanning-type friction factor equation:

$$\overline{lw} = f \sum_{i=1}^{N} L_{e_i} \frac{V_i^2}{2g_c D_i} \tag{4-26}$$

In Equation 4-26, f is a friction factor, dimensionless and function of the Reynolds number; ϵ/D is relative roughness; V is the average velocity in the pipe; L_e is the total equivalent length of pipe for the considered section between Points 1 and 2; and D the inside diameter of the pipe.

Pipeline Pressure Drop Calculations—Single-Phase Gas Flow. Several equations are available in the *Main Reference Handbook of Natural Gas Engineering*[66].

1. Weymouth Equation (Modified):

$$Q = 3.22 \frac{T_o}{p_o} \left[\frac{d^5(p_1^2 - p_2^2)}{GTL\, fz_a}\right]^{1/2} \tag{4-27}$$

where Q = flowrate, SCF/hr
T_o = base temperature, °R absolute (520°R)
d = line diameter, in. ID
p_1 = upstream pressure, psia
p_2 = downstream pressure, psia
p_o = pressure base = 14.7 psia
G = gravity of the gas = (M/28.97)
T = flowing temperature, °R absolute
L = length of the pipeline, miles
z_a = average compressibility factor
f = Moody friction factor, a function of the Reynolds number and relative roughness

2. Panhandle Equation:

$$Q = 435.87E\left(\frac{T_o}{p_o}\right)^{1.07881}\left(\frac{p_1^2 - p_2^2}{L\,T}\right)^{0.5394}\left(\frac{1}{T}\right)^{0.4606} d^{2.6182} \tag{4-28}$$

where E is an efficiency factor, usually = 0.92
Q is in SCF/D

There are other equations, such as the Ford, Bacon, and Davis Equation, Clinedinst Equation, and U.S. Bureau of Mines Equation.

Calculation of Compressor Requirements. In storage when the gas must be injected during the summer, and in production of natural gas from a producing reservoir, compression is frequently required. Isothermal work required to compress the gas is:

$$-\overline{w} = \frac{53.241}{G}\,T \ln \frac{p_2}{p_1} \tag{4-29}$$

where $-\overline{w}$ = Isothermal work required to compress an ideal gas from p_1 to p_2 psia, ft × lb_f/lb_m, T °R abs

For a real gas, Equation 4-29 is modified as follows:

$$-\overline{w} = 53.241\ T\left[\int_{0.2}^{p_{r,2}} \frac{z}{p_r}\,dp_r - \int_{0.2}^{p_{r,1}} \frac{z}{p_r}\,dp_r\right] \tag{4-30}$$

where $p_r = p/p_c$ = pseudo-reduced pressure, dimensionless
T = °R absolute, the isothermal flowing temperature

Adiabatic theoretical horsepower required to compress 1 MMCF/d at 60°F and 14.65 psia:

$$HP = 0.08531 \frac{K}{K-1} T_1 \left[\left(\frac{p_2}{p_1} \right)^{\frac{z(K-1)}{K}} - 1 \right] \tag{4-31}$$

where HP = Horsepower

$K = \frac{C_p}{C_v}$ for the gas at suction conditions

z = compressibility factor at suction conditions

p_1 = suction pressure, psia

p_2 = discharge pressure, psia

T_1 = suction temperature, °R abs.

Theoretical adiabatic work required to compress 1 MMCF/D at 60°F and 14.7 psia:

$$-\overline{w} = 0.0432 \, \Delta H \tag{4-32}$$

where Δ H = change in enthalpy, Btu per pound-mass of gas

$-\overline{w}$ = necessary work, ft × lb_f/lb_m

With multistage compression with intercoolers, the equations above may be used on a stage-by-stage basis.

TWO-PHASE FLOW*

There exist over 5,000 technical papers, articles, monographs, published notes, and theses on two-phase flow. The method recommended in this book for two-phase flow calculations is based upon the thesis of Dr. T. L. Gould, highlights of which were published in both SPE[13] and AIChE Journals.

The paper has been recognized as combining the best available correlations for predicting pressure gradients for each of the flow regimes encountered in most gas production or storage wells. The method also has reasonably reliable capability when applied to surface gathering lines and topography-hugging pipelines. It has been evaluated against literature data as well as direct field data from offshore wells. The paper and its summary were published by special permission in SPE *Petroleum Technology*, and AIChE journals.

* The section on two-phase flow is a condensed version of "Two-Phase Flow through Vertical, Inclined or Curved Pipe" by T. L. Gould, M. R. Tek, and D. L. Katz, excerpted by permission. See page 146 of this text for nomenclature.

Introduction and Background on Two-Phase Flow

Problems in two-phase flow are often encountered in design and operation of storage reservoirs.

The basic engineering problem in calculating the pressure distributions in conduits subject to two-phase flow may be spelled out as follows: knowing the geometry of the conduit, physical properties of the two-phase system, and conditions prevailing at one end, predict pressure profile along the pipe.

Through the years, a number of investigators in vertical two-phase flow chose to correlate both slippage and friction losses by a unique and single energy-loss factor[14–19]. The results of their efforts fell short of their desired goal because their correlations did not include the effects of all pertinent variables, and, more importantly, they did not reflect the effect of various flow regimes often co-existing in the pipe.

Another school of thought is represented by other investigators[20–27] who chose to define, measure, and predict slip or holdup as an intermediate parameter leading to the calculation of pressure drop. This approach, along with considerations of energy balance, led to an interpretation of pressure gradient as a sum of three gradients: density, acceleration, and friction.

One of the principal reasons for the failure of most vertical two-phase flow correlations was the coexistence of several flow regimes along the same pipe for a given set of operating conditions. Indeed, if one started at the bottom and proceeded up the tubing in an oil well, one could possibly observe, in succession at various depths (1) single-phase liquid flow, (2) two-phase bubble flow, (3) slug flow, (4) transition (possibly representing annular or froth flow), and finally, (5) mist flow. Any correlation developed specifically for any one of those flow regimes would probably be inadequate to describe flow in the entire pipe.

In 1967, Orkiszewski[28] re-examined all existing correlations for comparison with data he had gathered on 22 heavy-oil wells in Venezuelan fields. He proposed that correlations by Duns and Ros[27] for transition and mist flow, and by Griffith and Wallis[21] for bubble flow could be used with discretion in predicting pressure drops. This approach permitted him to combine the best part of each correlation into one algorithm for pressure drops.

In summary, it may be fairly stated that all work in two-phase flow spanning the period from 1935 to 1972 can be classified into three groups:

1. Work that considers no-slip and no-flow regimes
2. Work that considers slip concept but not flow regime
3. Work that considers both the slip concept (as related to holdup) and the flow regime

The papers discussed above are mostly on vertical two-phase flow systems. There are very few studies on inclined two-phase flow and practically none on two-phase flow through curvilinear pipe, except the papers by Gould and Tek,[29] Beggs and Brill,[68] and some Master's thesis work at the University of Tulsa. On inclined flow, the works of Baker[30] and Flannigan,[31] as well as the thesis by Sevigny,[32] belong to the first group. A number of investigators studied slug flow in inclined pipe.[3–36] Their works deal with only one flow regime and with a limited extent of inclination.

The problem of inclined down-flow has been treated by Sevigny[32] without regard to flow regimes. A study by Greskovich[39] in segregated downflow is concerned with prediction of holdup.

Studies of the effects of physical parameters and operating conditions on the flow regimes in two-phase flow have been extensive.[37–41] Some of these studies discern the flow regimes from trends on pressure-drop curves, others from direct flow pattern observations.

Against the background of a voluminous existing literature only partially cited above, the work presented here involves a computerized method designed and developed first to determine the flow regime likely to prevail at a given point in the pipe; then to use the best correlation available in the literature to predict the density, friction, and acceleration gradients for that particular flow regime before iterating to the next pipe-length increment. The procedure, originally introduced by Orkiszewski,[28] has been generally followed in this investigation. The resulting model, including changes, improvements, and added data on flow regime predictions was cast in finite-difference form, suitable for quick implementation by a computer algorithm. The method so developed predicts pressure drops and holdups for two-phase flow systems in vertical, inclined, or curvilinear pipe.

The generalized model was evaluated on the basis of available laboratory and field data. Maps for predicting flow regime were extended and modified with new direct laboratory data. Also included was a new computer graphics procedure for perspective display of three-dimensional pressure gradient or holdup surfaces, a valuable tool in evaluating correlations and discovering discontinuities.

The Generalized Pressure-Drop Model

In spite of the overwhelming amount of literature on the subject, there exists no perfectly general model capable of representing vertical, horizontal, and inclined flow. However, there are a number of reliable analytical relations or empirical correlations, usually limited to particular flow regimes or certain ranges of geometry, pressure level, and physical properties. The pressure-drop model investigated during this work starts with the thermodynamic energy balance and uses separate and selected correlations for each of the main ingredients of the overall pressure drop for each likely flow regime.

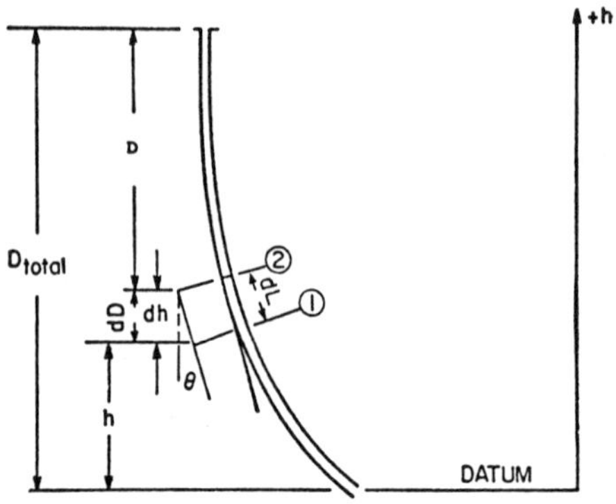

Figure 4-13. Geometry for generalized pressure-drop model.

Figure 4-13 represents the geometry used for inclined and curvilinear pipe to develop the pressure-drop model.

A thermodynamic energy balance, written in differential form between points 1 and 2, gives:

$$\frac{dp}{\rho} + d\left(\frac{v^2}{2g_c}\right) + d\left(\frac{gh}{g_c}\right) = -\overline{lw} \tag{4-33}$$

The lost work term in this equation, $\overline{lw}$, is usually written for pipe flow in terms of a friction factor, as follows:

$$\overline{lw} = \frac{fv^2}{2g_c d}\, d\ell \tag{4-34}$$

For flow in a curvilinear pipe, the length is simply related to the height above datum by $dh = dL \cos\theta$. In oil and gas operations, the well head is usually taken as the point of reference that requires the definition of depth $dD = -dh$. Substituting these relations into Equation 4-33 gives:

$$\frac{dp}{dD} = \frac{\bar{\rho}_m \dfrac{g}{g_c} + \dfrac{f\,\bar{\rho}_m\, v^2}{2g_c\, d \cos\theta}}{1 - \dfrac{wt\, q_G}{g_c\, A^2 \bar{p}}} \tag{4-35}$$

The denominator of this equation assumes that $d\left(\dfrac{v^2}{2g_c}\right)$ is significant in nearly all gas or mist flow only at conditions of low pressure. Wallis[41] has shown that the denominator is actually $1 - M^2$, where M is the Mach number, or ratio of flow velocity to sonic velocity.

For computational purposes, Equation 4-35 is often approximated by taking a finite difference of the pressure gradient. In addition, the friction gradient can be represented as a single term, τ_f, which yields the following:

$$\frac{\Delta p}{\Delta D} = \frac{\bar{\rho}_m (g/g_c) + \tau_f/\cos\theta}{1 - M^2} \tag{4-36}$$

Although Equation 4-36 was developed by considering a homogeneous fluid, we are left with two gradients, $\bar{\rho}_m$ and τ_f, that could represent the flowing properties of a multiphase mixture. Bridging the gap between homogeneous flow and nonhomogeneous or two-phase flow must then be an empirical process.

The friction gradient has not received a great deal of attention recently for two-phase flow. This gradient is dominant only at high flowrates, which are difficult to obtain in laboratory-scale equipment. Most investigators simply base the friction gradient on the single phase that is in continuous contact with the pipe wall. Of course, Equation 4-36 shows that under inclined-flow conditions, this gradient will be magnified. In a truly horizontal pipe, the two-phase friction gradient has been shown to be as much as two to three times the magnitude of the homogeneous flow gradient.[47,48] One wonders if this would also be true for vertical two-phase flow. However, to date the vast majority of work has been concerned with the empirical correlation of the term $\bar{\rho}_m$ in Equation 4-36.

The term $\bar{\rho}_m g/g_c$ may be referred to as the density gradient, where $\bar{\rho}_m$ is average mixture density. The density of the mixture may be computed as:

$$\rho_m = \rho_l \epsilon_\ell + \rho_g(1 - \epsilon_\ell) \tag{4-37}$$

where ϵ_ℓ is the actual flowing volume fraction liquid referred to as holdup. It can be shown that the holdup is directly related to the relative or *slip* velocity between the phases. Based on the superficial velocity of each phase, the slip velocity is given by:

$$v_s = \frac{v_{sg}}{1 - \epsilon_\ell} - \frac{v_{sl}}{\epsilon_\ell} \tag{4-38}$$

In liquid-phase continuous flow, or bubble flow, the slip velocity may be taken as the terminal velocity of bubble rise. For most two-phase flow problems in the bubble regime, it is recognized that small bubbles coalesce into larger ones and large bubbles, which do not survive the usual turbulence, split into smaller ones, resulting in a size-invariant bubble-rise velocity. In this investigation, the following equation (given by Harmathy[43]) has been used in calculating the bubble-rise velocity, v_∞:

$$v_\infty = v_s = 1.53 \left[\sigma g\, g_c \left(\frac{\rho_l - \rho_g}{\rho_l^2} \right) \right]^{1/4} \tag{4-39}$$

During the late stages of this investigation, another method was also successfully implemented to estimate holdup in the bubble regime. That method, credited to Wallis,[41] also gave answers comparable to those obtained by Equation 4-38. It is called the *Drift-Flux* method.

When a large number of bubbles is present, the swarm does not always rise with the terminal velocity of the single bubble. The factor for correcting the value of v_∞ for a single bubble for the condition of hindered ascent is in turn a function of the holdup, ϵ_ℓ. There are several correlations available in the literature to evaluate this effect, such as those of Marucci[44] and Bhatia.[45] This was not quantitatively evaluated in the work reported here.

Referring to Equation 4-36 and Figure 4-13, we can develop an iterative finite-difference algorithm to calculate length of depth increments corresponding to selected values of Δp, provided that the holdup and friction gradients can be evaluated at the pressure, temperature and flow conditions in the increment. The procedure is applied in successive steps up or down the pipe until the sum of the increments equals the total depth. This iterative calculation permits the evaluation of the flow regime in the next pipe segment. If the flow regime is known, the best correlations available for ρ_m and τ_f can then be used. This procedure at the same time allows for the presence of several different flow regimes at various depths in the same pipe.

Determination of Flow Regimes. Using the physical properties of vapor and liquid phases determined *in-situ*, the individual flowrates, pipe geometry, and pressure level, one can predict the flow regime likely to occur at that particular point. One of the principal shortcomings of most vertical two-phase flow models in the literature has been their failure to account for the coexistence of several flow regimes along the pipe for a given set of operating conditions.

Using some quantitative data for cases known to apply to mist flow condition, Gould and Tek[38] proposed an extension of the flow regime map earlier published by Duns and Ros.[27] On the basis of their results, observations on literature data, and direct laboratory experiments, a vertical flow regime map was developed (Figure 4-14). The region designated as *heading* might also be classified as both phases continuous, as discussed below. Figure 4-14, which has been developed by analyzing direct laboratory and field data, is used to arrive at the best possible locations of liquid-phase-continuous, alternating phases, and gas-phase-continuous boundaries.

Effect of Inclination on Flow Regimes. Throughout the investigation reported here, the basic geometry has been that of the vertical pipe. How-

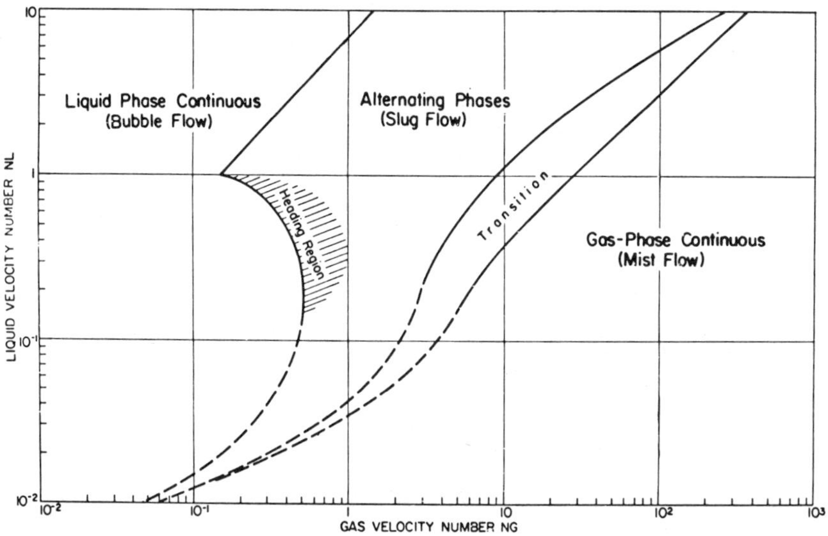

Figure 4-14. Extended flow regime map for vertical two-phase flow.

ever, as the model developed for pressure drop was extended and generalized to oblique and even curvilinear pipe, there arose the question of the effect of inclination on the flow regime map. This effect was investigated by direct observations on a two-phase flow apparatus designed to operate at various angles of inclination.

The results of the observations are shown in Figures 4-15 through 4-17. Figure 4-15 is for totally vertical flow and Figures 4-16 and 4-17 are for 45° and 90° inclinations from vertical, respectively. The flow pattern data have been classified on the basis of the observed continuous phase.[37]

As the flow regime quadrant is traversed in both directions, various observations on liquid-phase-continuous (L), alternating phases (A), transition from liquid-phase continuous to alternating (AL), and from alternating to gas-phase continuous (AG), have been spotted at appropriate locations. The letter B designates the transition between both continuous and alternating phases. Note that the data for Figure 4-15 bear out the same approximate locations of previously established bubble/slug, slug/transition, and transition/mist boundaries for vertical flow.

The region where both phases are continuous is newly defined and is a region for vertical two-phase flow where no known pressure-drop correlation exists. In this region, gas is separating from the liquid and allowing liquid fallback. Even though the net mixture flow is upward, individual pipe segments contain countercurrent liquid downflow for short periods of time. The phenomenon of liquid fallback was called *heading* by Ros.[27] Figure

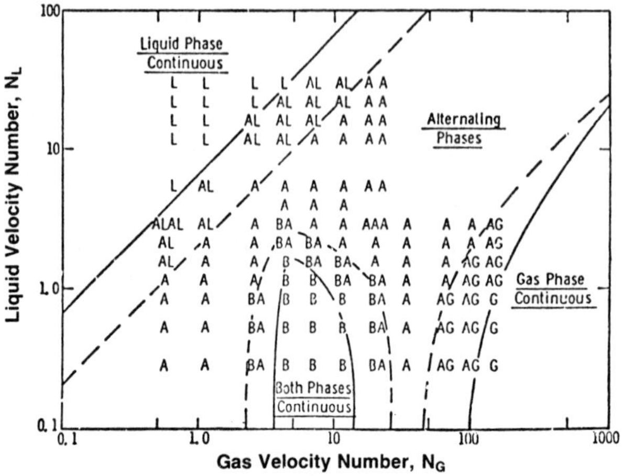

Figure 4-15. Flow regime map for inclination of 0° (vertical).

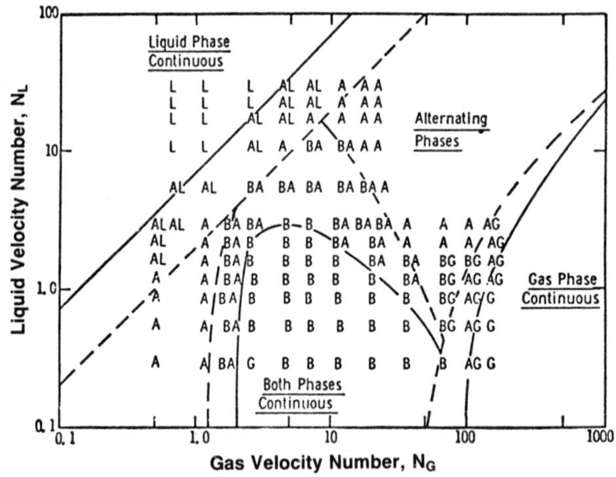

Figure 4-16. Flow regime map for inclination of 45°.

4-16 shows that this region occupies a significant portion of the flow regime quadrant.

For horizontal flow, the region where both phases are continuous occupies a significant portion of the flow regime quadrant, as shown in Figure 4-17. The region corresponds to the conventional classification of stratified flow for horizontal pipe. This conclusion is substantiated when the data of Hoogendoorn[42] for air and water are replotted on a flow diagram (Figure 4-18).

By combining Froth and Slug in the alternating-phase classification and combining Bubble and Plug in the liquid-phase continuous classification, there is almost perfect agreement between Figures 4-17 and 4-18.

Comparing Figures 4-15, 4-16, and 4-17 leads to a conclusion that will be important in the development of pressure-drop correlations for inclined flow. The liquid-phase continuous flow regime boundaries do not appear to vary significantly with inclination. This implies that it should be possible to

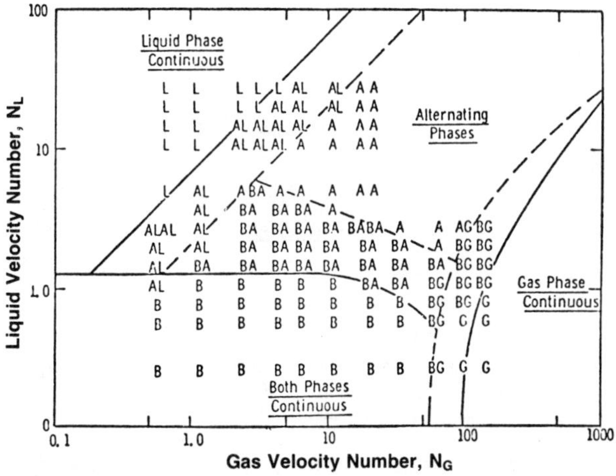

Figure 4-17. Flow regime map for inclination of 90° (horizontal).

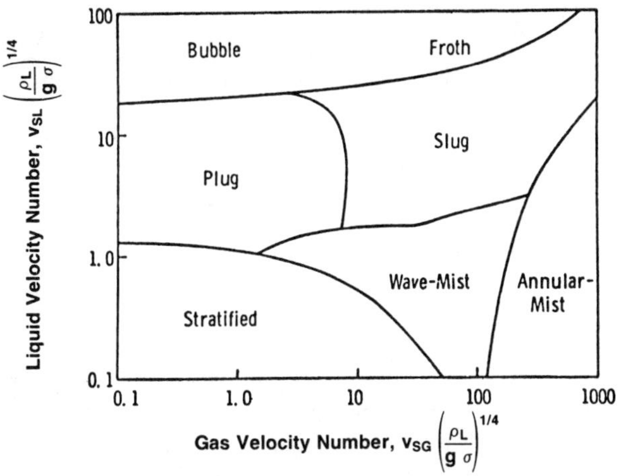

Figure 4-18. The horizontal two-phase flow regime map (Hoogendoorn data[42]).

develop an alternating-phase pressure-drop correlation that is generalized for all inclinations and that is a continuous function at the flow regime boundaries.

Calculation of Acceleration Gradient. The acceleration gradient given previously in Equation 4-19 as a fraction of total pressure gradient has been derived for the case $q_l << q_g$ or $q_g \cong q_T$ (ℓ for liquid, T for total):

$$\left(\frac{dp}{dL}\right)_A = \frac{W_T q_g}{g_c A^2 p}\left(\frac{dp}{dL}\right)_T \tag{4-40}$$

In this study, for all two-phase flow regimes, Equation 4-40 has been used to evaluate the acceleration gradient. (A thermodynamic analysis of two-phase flow leading to the derivation of the two-phase pressure gradient equation is included in Appendix E.) This implies a tacit assumption that for bubble, slug, or transition regimes, the acceleration gradient, which is small anyway, may be computed from a procedure more rigorously applicable for mist flow. Dukler[48] gives a procedure that does not require the assumption of homogeneous mist flow, but he does assume a constant ratio between the liquid and gas velocities.

Since Equation 4-36 was implemented by finite difference techniques on the computer, the size of Δp increments as related to the current value of p occasionally resulted in computational instability wherever the numerical value of the coefficient $\left(\frac{W_T q_g}{g_c A^2 p}\right)$ was close to unity. Further examination of these cases disclosed that under those conditions the sonic velocity was being approached by the two-phase mixture in the pipe. In fact, if this coefficient is exactly unity, the pressure gradient from Equation 4-36 is infinite, corresponding to a shock front or choked flow.

Logically then, we can determine directly the flowing pressure at which sonic flow occurs by:

$$p = \frac{W_T q_g}{g_c A^2} \tag{4-41}$$

as long as the volumetric gas flowrate is much greater than the liquid flowrate. Figure 4-19 shows a plot of this pressure in field units for various flow conditions. Operating at pressures lower than those predicted from the curves will result in choked flow in the wellbore. This type of analysis can be crucially important to safety equipment design.

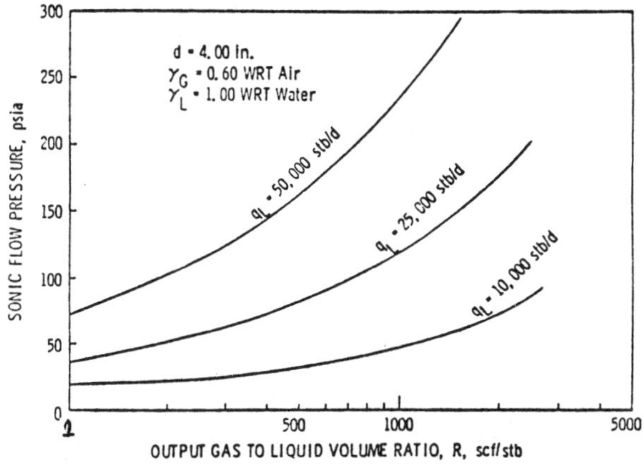

Figure 4-19. Sonic flow boundaries for various flowrates in a 4.0 in. diameter pipe.

Evaluation of the Pressure-Drop Model. Field and laboratory data available from various literature sources have been examined, computed, and compared with results of procedures developed in this work.

Using the literature data available from five sources, the authors have compared statistically the procedures developed during this investigation with reported results of the Orkiszewski and Duns and Ros methods. This comparison is given in Table 4-4. On the basis of 148 data points so sampled, this study shows good agreement with Orkiszewski, as expected. The slug-flow correlation used was essentially the same, but the bubble-flow holdup was modified from his method. Since the data used in this comparison are primarily located in the slug-flow regime, one would not expect the results to be too different from those reported by Orkiszewski. Calculated and measured pressure drops are plotted in Figure 4-20. Note that there appears to be more deviation at the lower pressure drops than at higher values. This may indicate possible measurement errors. Because the sample size is too small, any quantitative generalization of the results of Table 4-4 may be hazardous. On the other hand, the magnitude and the range of average and standard deviations observed confirm the overall validity of the methods used.

Table 4-5 shows the range of variables involved in this study. The most restricted variable was diameter, whereas the gas/liquid ratios available covered a fairly wide range. Since there were not a great deal of data available on large diameter two-phase flow, it is interesting and relevant to report the few large-diameter points where the procedure has been tested against data. (See Table 4-6.) The comparison is indeed encouraging from the standpoint of hydrodynamic scale-up, with average error being within acceptable limits for all three cases.

Table 4-4
Statistical Comparison of Predicted vs. Measured Pressure Drops

Data Source	This Method	Orkiszewski Method	Duns and Ros Method
Baxendell-Thomas[1] (25 points, $d = 2.992$ and 2.441 in.)			
Average error, percent	−1.5	−2.1	+2.3
Standard deviation, percent	7.5	11.1	20.0
Fancher-Brown[2] (20 points, $d = 1.995$ in.)			
Average error, percent	+4.4	+0.3	+1.7
Standard deviation, percent	12.9	11.8	32.1
Hagedorn-Brown[9] (32 points, $d = 1.38$ in.)			
Average error, percent	−3.9	+0.1	−16.9
Standard deviation, percent	12.7	8.2	36.6
Orkiszewski[15] (22 wells, $d = 2.992$ in.)			
Average error, percent	+1.3	−1.2	+22.7
Standard deviation, percent	10.7	10.4	18.7
Poettmann-Carpenter[5] (49 wells, $d = 1.995$ to 2.992 in.)			
Average error, percent	−1.1	−1.0	+5.8
Standard deviation, percent	10.4	12.0	12.4
Over-all results (148 data points)			
Average error, percent	−0.096	−0.8	+2.4
Standard deviation percent	11.3	10.8	27.0

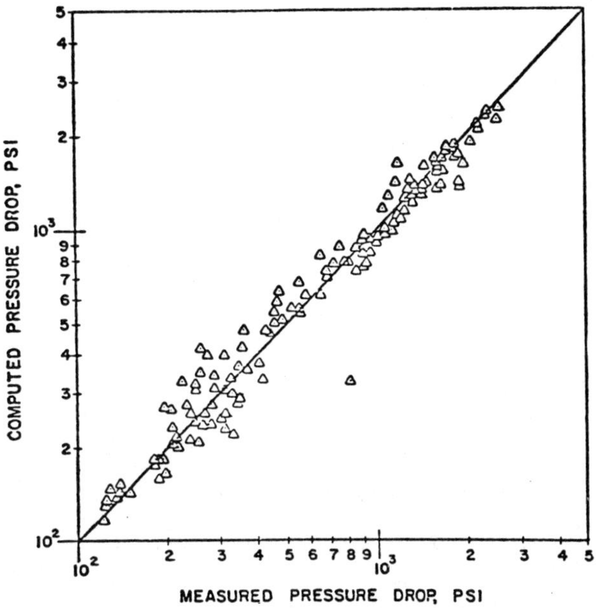

Figure 4-20. Comparison of the pressure-drop model with published field data.

Table 4-5
Range of Variables Tested against Field Data for the Pressure-Drop Model

Variable	Range
Diameter (d), in.	1.380 to 2.992
Viscosity (μ_1), cp	0.7 to 2,000
Gas-Liquid ratio (R), SCF	200 to 100,000
Pressure level, psia	30 to 4,000
Length of flow (L), ft	1,000 to 10,000
Flow regime	slug froth
Inclination, degrees from vertical	0

Table 4-6
Statistical Comparison of Predicted vs. Measured Pressure Drops for Large Diameter Pipes

d_h (in.)	Number of Data Points	Average Error (percent)
10.05	3	−9.8
6.398	4	+11.0
6.0	2	+1.5

Nomenclature

The nomenclature for the section on two-phase flow is given below:

A = Cross-sectional area of pipe, ft^2
D = Depth from surface, ft
d = Diameter, ft
f = Moody friction factor, dimensionless
g = Acceleration of gravity, ft/sec^2
g_c = 32.17 $(lb_m/lb_f) \times ft/sec^2$
ϵ_ℓ = Volume fraction liquid (holdup)
h = Height above datum, ft
L = Length of pipe, ft
$\overline{lw}$ = Lost work, $ft \times lb_f/lb_m$
M = Ratio of flow velocity to sonic velocity, Mach number
N_{gv} = $V_{sg}\sqrt{\rho_l/gg_c\sigma}$, gas velocity influence number, dimensionless
N_{lv} = $V_{sl}\sqrt{\rho_l/gg_c\sigma}$, liquid velocity influence number, dimensionless
p = Pressure, lb_f/ft^2 (abs.)
$\bar{p}$ = Average pressure, lb_f/ft^2
q = Volumetric rate of flow, ft^3/sec
v = Velocity of mixture, ft/sec
v_s = Slip velocity, ft/sec
v_∞ = Terminal velocity of single bubble, ft/sec
W_T = Total mass flowrate, lb_m/sec
Δ = Increment or decrement
γ = Fluid gravity
ρ = Density, lb_m/ft^3
ρ_m = Mixture density, lb_m/ft^3
σ = Surface tension, lb_f/ft
τ_f = Friction gradient, lb_f/ft^3
θ = Angle with vertical, radians

Subscripts:

A = Acceleration
b = Bubble
g = Gas
l = Liquid
S = Slip
sg = Superficial gas
sl = Superficial liquid
T = Total

TEMPERATURE DISTRIBUTIONS IN STORAGE WELLS

As development of storage fields continued, two new ideas emerged during 1978 and 1979[57,69]. One relates to the control of flowrates with bottom-hole chokes preempting use of pressure regulation at the surface. The other concerns flow through casing or casing-tubing annulus to take advantage of geothermal heat of the earth around the wellbore warming the gas so it does not cool sufficiently to form hydrates at the well head before free-flowing into the pipeline. The second idea precludes the necessity of installing equipment such as dehydrators and methanol injection systems. With appropriate temperature and pressure control, it was suggested that underground storage service could be provided with minimum heating, dehydration, and pressure regulation at the surface.

Design concepts related to temperature-pressure gradients through wells, chokes, and associated equipment are also of continuing interest for applications in geothermal wells, compressed air storage, production from Arctic regions, and underground storage of heat.

The literature on temperature distributions includes papers on unsteady-state heat transfer,[58] temperature distribution in water injection wells,[59,60] temperatures in gas producing wells,[61] and an analytical method of calculating theoretical temperature distribution.[62] More recently, temperature effects became of particular interest for estimating pipe movements and thermal stresses.[63] Development of storage in salt domes and operation of deep solution cavities with stored fluids interacting with environment temperatures also provide excellent examples of the need for predictive methods in temperature gradients. More recently, a paper on simultaneous computer solutions of fluid-flow heat transfer equations gave several examples of matching temperature and pressure profiles in wellbores.[64]

The purpose of this section is to document practical engineering procedures for analytical and finite difference calculation of temperature and pressure gradients.

Thermodynamics of the Wellbore

Figure 4-21 is a schematic representation of the wellbore thermodynamics. The gas enters the tubing or casing-tubing annulus at the bottom and flows upward between sections 1 and 2. As the gas expands in rising in the wellbore it acquires kinetic and potential energies.

A thermodynamic energy balance implementing the first law, applied to steady-state flow between points 1, upstream and 2, downstream gives:

$$U_1 + \frac{1}{J} p_1 V_1 + \frac{m}{J}\frac{v_1^2}{2g_c} + \frac{1}{J}\frac{m\,g}{g_c} h_1 + q - \frac{w}{J} = U_2 + \frac{1}{J} p_2 V_2 + \frac{m}{J}\frac{v_2^2}{2g_c} + \frac{m}{J}\frac{g}{g_c} h_2 \quad (4\text{-}42)$$

Equation 4-42 simply states that total energy entering the system through section 1, plus the heat added across the thermodynamic boundary, minus energy removed as work across the same thermodynamic boundary, must be equal to the energy leaving at section 2. Note that irreversible energy losses are implicitly included in the difference between the total internal energy terms U_2 and U_1. As the flow goes from point 1 to point 2, using the usual delta convention:

$$\Delta U = U_2 - U_1 \quad (4\text{-}43)$$

Equation 4-42 may be written in a different form as:

$$\Delta U + \frac{1}{J}\Delta(PV) + \Delta\left(\frac{m}{J}\frac{v^2}{2g_c}\right) + \Delta\left(\frac{m}{J}\frac{g}{g_c} h\right) = q - w \quad (4\text{-}44)$$

In general for problems involving gas wells, the shaft work term $w = 0$ and Equation 4-44 may be written in terms of enthalpies per pound mass of flowing gas, as follows:

$$\Delta H + \frac{1}{J}\Delta\left(\frac{v^2}{2g_c}\right) + \frac{1}{J}\Delta\left(\frac{g}{g_c} h\right) = \bar{q} \quad (4\text{-}45)$$

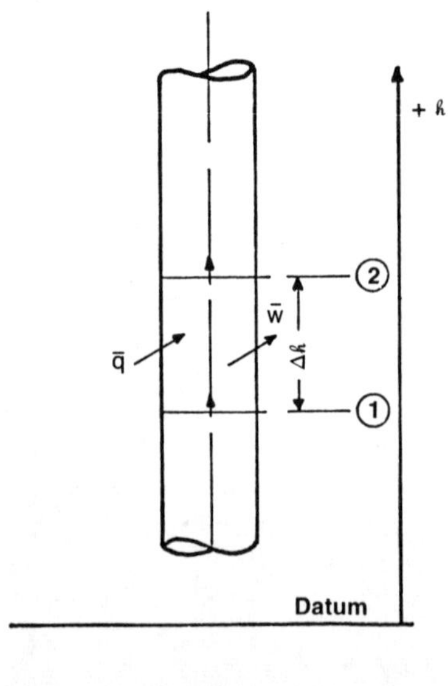

Figure 4-21. Model for energy balance.

Energy Balance for Gas Wells in Steady Flow

Equation 4-45 may be written symbolically as:

$$\Delta H + \Delta(KE) + \Delta(PE) = \bar{q} \tag{4-46}$$

Equation 4-46 simply states that in steady-state flow the changes in enthalpy, kinetic energy, and potential energy per unit mass flow between any two points going from (1) upstream to (2) downstream must be equal to the net-heat transferred to the system across the thermodynamic boundary.

Equation 4-46 is calculated stepwise and incrementally up the wellbore, starting at the upstream terminal at the bottom, term-by-term, to give the temperature distribution. This requires knowing both pressure and temperature at the bottom and the pressure distribution from bottom to top.

The pressure distribution itself is obtained from the mechanical energy balance and is nearly independent of the temperature distribution.

The Mechanical Energy Balance

The mechanical energy balance is also derived from the *first law*, written explicitly in terms of internal energy U and flow work pV.

By expressing the change in internal energy as a sum total of heat effects $\int_1^2 TdS$, compression/expansion effects $\int_1^2 p(-dV)$, surface effects $\int_1^2 \gamma d\sigma$, and chemical effects $\int_1^2 \mu_A dm_A + \int_1^2 \mu_B dm_B$ + etc. we get:

$$\Delta U = \int_1^2 TdS + \int_1^2 p(-dV) + \int_1^2 \gamma d\sigma + \int_1^2 \mu_A dm_A + \int_1^2 \mu_B dm_B + \ldots \tag{4-47}$$

and currently:

$$\Delta(pV) = \int_1^2 pdV + \int_1^2 Vdp \tag{4-48}$$

Substituting the above into Equation 4-44:

$$\int_1^2 TdS + \int_1^2 Vdp + \int_1^2 \gamma d\sigma + \int_1^2 \mu_A dm_A + \int_1^2 \mu_B dm_B + \ldots$$

$$+ \int_1^2 pdV + \int_1^2 Vdp + \Delta(KE) + \Delta(PE) = q - w \tag{4-49}$$

Now:

$$\int_1^2 TdS = q + \overline{lw} \tag{4-50}$$

where q is the net-heat transferred across thermodynamic boundaries and $\overline{lw}$ represents the sum total of all irreversible energy loss processes occurring between sections 1 and 2. In the absence of chemical and surface effects, Equation 4-50 written intensively, per lb_m flow gives:

$$\int_1^2 \frac{dp}{\rho} + \Delta\left(\frac{v^2}{2g_c}\right) + \left(\frac{g}{g_c} h\right) = -\overline{w} - \overline{lw} \tag{4-51}$$

which is the mechanical energy balance. Equation 4-51 simply states that, in a steady flow system between 2 points, (1) upstream and (2) downstream, the sum total of changes in flow-work, kinetic and potential energies per-pound-mass flow plus energy loss, plus shaft work per-unit-mass flow added to the system, must be equal to zero.

Equation 4-51 gives the pressure distribution for the gas flowing through the well.

Temperature and Pressure Gradients Observed in Storage Wells

Figure 4-22 shows several typical temperature and pressure gradients in relation to the temperature and pressure prevailing in the environment near the wells. Curve A on the right hand side represents the hydraulic gradient prevailing in the formation around the wellbore. The curves C, D, and E represent the flowing gas gradients in the wellbore. Of these, curve D depicts flow through casing or casing-tubing annulus with little pressure drop. Curve E represents high flowrate through the wellbore originating from the same bottom-hole draw-down. Finally, the line GD represents the stepped-down pressure across the sand face due to a bottom-hole choke followed by upflow through the casing with little frictional pressure drop. Curve B shows the static gas gradient.

The same figure also depicts three temperature gradients of interest. These are:

1. The straightline geothermal gradient F
2. I, the gradient for flow with a bottom-hole choke
3. H, gradient for flow through the tubing without a bottom-hole choke

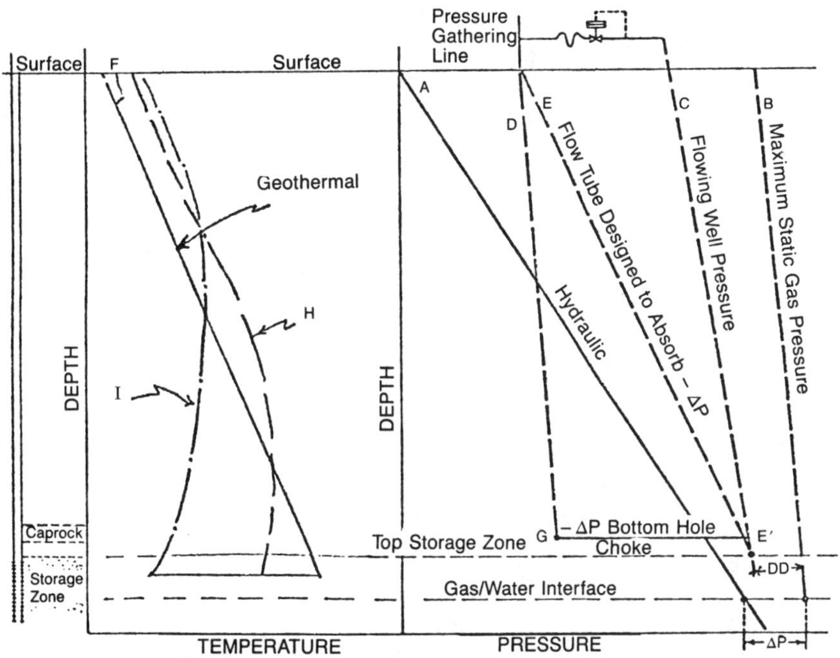

Figure 4-22. Temperature and pressure gradients in storage wells.

Procedures for Calculating Temperature and Pressure Gradients

The stepwise procedure for calculating the temperature distribution in the well is presented in detail in SPE 7495.[57] It starts at the bottom, inside the wellbore where the pressure is known or measured, and the temperature results from isenthalpic cooling of the gas as it enters the wellbore, having taken a pressure drop through perforations or, where applicable, through a bottom-hole choke.

Knowing the temperature, pressure, and therefore the enthalpy at the bottom, the procedure involves calculation of ΔH from Equation 4-46, first using the usual heat transfer correlations for q and chosen height for the potential energy term. Having calculated ΔH, the pressure drop is calculated next, resulting in a new pressure and a new enthalpy at the downstream end of the incremental section just calculated.

Knowing either the composition or gravity of the gas at the calculated pressure and updated enthalpy, the computer program next updates the temperature before the entire algorithm is repeated. The equations used for pressure drop calculations and those for various heat transfer coefficients were given in the cited SPE paper.[57] Note that the choice of small ΔH increments and the fact that the pressure drop calculation only depends upon the absolute value of the temperature, and is therefore rather insensitive to small temperature changes, makes the calculation possible in the prescribed sequence.

Calculation of Pressures. Knowing $T = T_1$ and $p = p_1$ (at the bottom) and the specific gravity of the gas G, select a Δh, vertical distance up the pipe and calculate p_2 using the well known equation of gas flow through vertical pipe (assume for the increment Δh, the flow is isothermal at $T = T_1$). The following equation may be used for this calculation:

$$p_2 = \frac{1}{\sqrt{e^s}} \sqrt{p_1^2 - \frac{fQ^2 T_a^2 Z_a^2 (e^s - 1) 10^6 \times 0.000667}{d_i^5}} \tag{4-52}$$

$$s = 0.0375 \frac{G\Delta h}{T_a Z_a} \tag{4-53}$$

f = The Moody friction factor, a function of the Reynolds number R_e and the relative roughness. The Reynolds number may be calculated from the gas rate and diameter of pipe by:

$$R_e = \frac{Q \times 10^6 \times 0.0764 \times G \times d_i}{12 \times \dfrac{86{,}400 \times \pi\, d_i^2}{4 \times 144} \times \mu \times 0.000672} = 2.014 \times 10^4 \frac{4QG}{d_i \mu} \tag{4-54}$$

In the case of flow through casing-tubing annulus, equivalent diameter is used for calculating R_e.

Calculation of Temperatures. Once p_2 has been calculated, the next step is to calculate the change of enthalpy H between points 1 and 2 from potential, kinetic energy changes and heat transfer into the flowing gas from the outside environment.

Equation 4-45 or 4-46 may be used for calculating the change in enthalpy. While the kinetic energy change between the inlet and outlet of an incremental section is usually negligible, under conditions of low pressure level and high velocities it may be appreciable. In order to check or verify the magnitude of it the following equation may be used:

$$\frac{1}{J} \Delta \left(\frac{v^2}{2g_c} \right) = \frac{1}{J} (v_2^2 - v_1^2) \frac{1}{2g_c} \tag{4-55}$$

$$v_{1,2} = Q \times 10^6 \times \frac{p_{ST}}{p_{1,2}} Z \frac{T}{T_{ST}} \frac{1}{86{,}400 \dfrac{\pi\, d_i^2}{4 \times 144}} \tag{4-56}$$

The heat transfer into the gas flowing through the well is calculated from the following equation:

$$q = \frac{(T_o - T_i)}{\left(\dfrac{1}{UA}\right)_i} \tag{4-57}$$

The temperature difference $(-\Delta T)$ is the average temperature difference prevailing throughout the h section considered. T_o is, in general, a linear function of depth for the particular area.

The term $(1/UA)_i$ represents the overall thermal resistance to the flow of heat across the thermodynamic boundary. It includes, for flow through tubing, the convective film resistance inside the tubing, the conduction resistance across tubing, the outside convective film outside tubing, convection across the particular fluid in casing-tubing annulus, the convective resistance on the inside of casing, the conduction resistances across the casing wall, and successive conduction resistances across the cement and the formation beyond:

$$\left(\frac{1}{UA}\right)_i = \frac{1}{h_i A_i} + \frac{\Delta r_t}{K_t A_{mt}} + \frac{1}{h_o A_{ot}} + \frac{\ln \dfrac{d_c}{d_t}}{2\pi \Delta h K_f} + \frac{1}{h_{ic} A_{ic}} + \frac{\Delta r_c}{K_c A_{mc}} + \frac{\ln \left(\dfrac{d_{c_2}}{d_{c_o}}\right)}{2\pi \Delta h K_c} + \frac{\ln \left(\dfrac{d_{f_2}}{d_{c_2}}\right)}{2\pi \Delta h K_f} \tag{4-58}$$

Several terms on the right hand side of the equation are, in general, small and may be neglected. In flow through tubing, from several calculations performed, it appears that conduction through the casing-tubing annulus controls the rate of heat transfer. The fluid in the casing-tubing annulus is natural gas or sometimes brine or condensate.

The principal variables in heat transfer from gas to pipe wall are gas flow velocity, pipe diameter, and fluid properties. Equation 4-59 (commonly known as the *Dittus-Boelter Equation*) applies for gases being cooled when the Reynolds number > 10,000:

$$\frac{h_i d_i}{K_g} = 0.023(R_e)^{0.8}(Pr)^{0.3*} \tag{4-59}$$

* = 0.4 for gases receiving heat

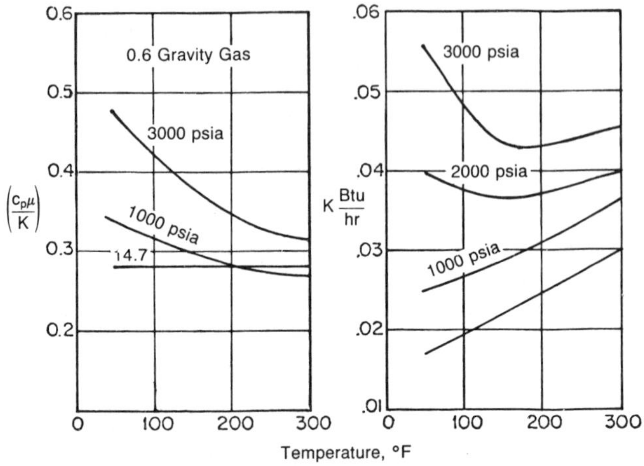

Figure 4-23. Charts for Prandtl no. and thermal conductivity for natural gas of 0.6 gravity.[57]

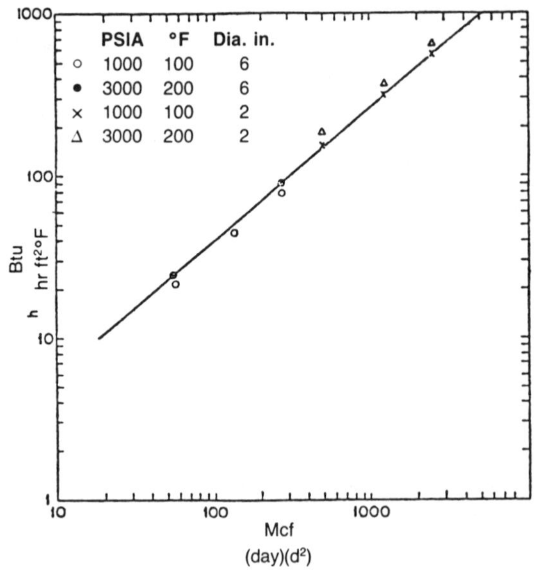

Figure 4-24. Chart for estimating film coefficient of natural gas.[57]

where R_e = Reynolds number, Equation 4-54

Pr = Prandtl number = $\mu C_p/k$, dimensionless; values of $\mu C_p/k$ and K have been prepared for 0.6 gravity natural gas,[52] Figure 4-23. A quick approximate solution is given for h from gas flowrate and pipe diameter squared as shown by Figure 4-24.[52]

The thermal conductivity data on gases at high pressure have been given by Lenoir *et al.*[65] Once the rate of heat transfer q into the flowing gas is determined, then $\bar{q}$ to be used in Equation 4-45 or 4-46 can be calculated as:

$$\bar{q} = \frac{q}{3{,}600\ m} \tag{4-60}$$

where m represents the mass flowrate of gas in lb_m/sec.

$$m = \frac{Q \times 10^6 \times 0.0764 \times G}{86{,}400} \tag{4-61}$$

Having respectively calculated q and ΔH, one can readily determine the enthalpy of the gas at the outlet of the section considered:

$$\overline{H_2} = \overline{H_1} + \Delta\overline{H}$$

Having just calculated H_2 and previously p_2, one can determine the temperature T_2 from a Mollier chart available in the *Handbook of Natural Gas Engineering*[66]. Once p_2 and T_2 are determined, the calculation is repeated algorithmically in the direction of flow. The prediction of temperatures for real gas behavior in wellbores was improved subsequently using computer subroutines which predict the enthalpies from the Soave-Redlic Kwong equation of state.[67]

Temperature Changes of Natural Gas in Flow through Perforations—Chokes—Valves

It can be shown that any throttling process would cause the gas to expand along an isenthalpic path. Original Mollier Charts given by Brown[66] for 0.6 and 0.7 gravity gas can be used to predict temperature changes by interpolation. In SPE 7495,[57] a Joule Thomson cooling chart was given by D. L. Katz incorporating recent experimental data.

Practical Use of Joule-Thomson Cooling Coefficients

The Joule-Thomson cooling coefficient permits calculation of the number of degrees cooling of the gas from thermodynamic coordinates and the isenthalpic pressure drop in appropriate units. The Joule-Thomson coefficient is thermodynamically defined as:

$$\mu_j = \frac{T\left(\frac{\partial V}{\partial T}\right)_p - V}{c_p} \tag{4-62}$$

Consider 1 lb_m of gas; let M be the molecular weight:

$$pV = z \frac{1}{M} RT$$

$$V = \frac{zRT}{pM}$$

$$\left(\frac{\partial V}{\partial T}\right) p = \frac{zR}{pM} + \frac{RT}{pM} \left(\frac{\partial z}{\partial T}\right) p$$

$$\frac{T}{T_c} = T_r$$

$$dT = T_c \, dT_r$$

$$\left(\frac{z}{T}\right) p = \frac{1}{T_c} \left(\frac{\partial z}{\partial T_r}\right) p$$

Substitute everything in (1):

$$\mu_j = \frac{\frac{RT^2}{pM}\left(\frac{\partial z}{\partial T}\right)}{C_p}$$

$$= \frac{RT^2}{C_p \, p \, M \, T_c} \left(\frac{\partial z}{\partial T_r}\right) p_r$$

$$= \frac{1{,}544}{778} \frac{RT^2}{C_p \, p \, M \, T_c} \left(\frac{\partial z}{\partial T_r}\right) p_r$$

$$\mu_j = 1.98 \frac{T^2}{C_p \, p \, M \, T_c} \left(\frac{\partial z}{\partial T_r}\right) p_r,$$

°R/psi or °R/psf (depending on unit of p) (4-63)

Method for Practical Design Calculations of Wellbore Temperatures

When natural gas is relatively dry and flow takes place through casing, or casing-tubing annulus, one does not encounter large pressure drops in the wellbore. Under these circumstances it is reasonable to neglect the effect of pressure on the enthalpy and express both H and q as functions of temperature of the gas in the wellbore and the geothermal temperature prevailing outside around the wellbore. The resulting differential equation is linear, first-order, non-homogeneous; and gives an analytical expression for wellbore temperatures similar to those published earlier by Kunz and Tixier in 1955,[61] by Ramey in 1962,[58] and more recently by Romero-Juarez in 1979.[63] The derivation of the differential equation, as well as its solution, is given in Appendix F.

There have been several papers published during the last two years on calculating well temperature gradients stepwise and with reasonable accuracy and precision using numerical procedures implemented on computers. It must be emphasized that all of these procedures combine steady-state energy and momentum balances in the wellbore with assumed steady or unsteady effects occurring around and away from the wellbore. That is why, while reasonably accurate and reliable, these methods are still not quite perfect predictors. In the past several years, investigators have recognized that if the gas is ideal, or the process isobaric, one can calculate wellbore temperatures from direct analytical solutions of differential equations describing the thermodynamic energy balance. The pressure distribution equation in the wellbore depends only upon the absolute value of average temperature prevailing in the wellbore. As such, it is rather insensitive to temperature variation in the wellbore. Furthermore, the pressure drop becomes small when the gas is allowed to flow through casing or casing-tubing annulus.

The procedure developed for calculating wellbore temperatures has two distinct steps:

1. Calculation of the temperature from an analytical equation developed for an isobaric path (equally well and valid for ideal gas).
2. Correcting the temperature for the effect of actual pressure drop using a current estimate of the Joule-Thomson cooling coefficient.

The comparison on a Mollier Chart of various thermodynamic paths involved in the calculation of wellbore temperatures is shown in Figure 4-25.

Starting at point A (temperature T_1, pressure p_1), if temperatures dropped isobarically one would follow the path AB. If the gas flowed isothermally and the final pressure was p_3, the path AT would be isothermal. By contrast, a constant entropy path (isentropic) would bring the temperatures down, along AS, if such a path occurred. The isentropic path represents what would physically happen if the gas was being cooled by removing shaft work from it via a frictionless expansion device. Likewise, an isenthalpic path at point A would lead to AI. That particular path represents what would physically happen if the gas was expanding quickly through a throttling device without much time available for heat transfer to take place across the thermodynamic boundary ($\Delta H = 0$).

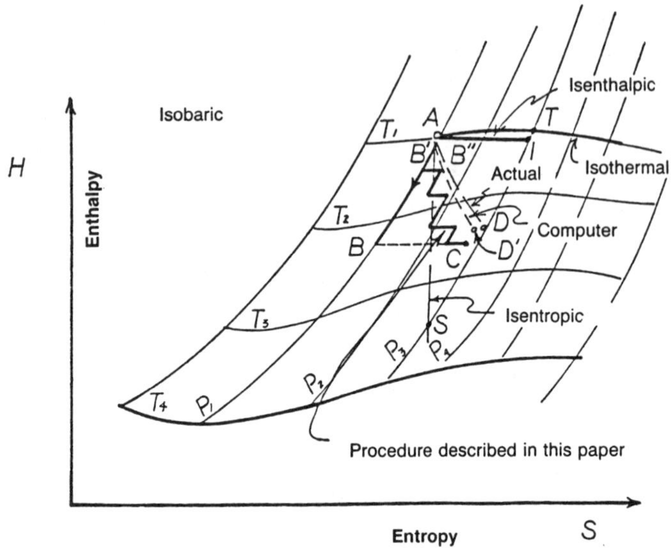

Figure 4-25. Comparison on Mollier chart of various thermodynamic paths involved in calculation of wellbore temperatures.

As the gas flows through the wellbore as a real gas, with some inside friction occurring while some heat energy is being exchanged with the environment, the actual thermodynamic path may be represented by AD. This is called *polytropic path.*

During the course of this research, stepwise computer solution to energy mass and momentum balances in the wellbore is shown to give a path AD′ reasonably close to AD.

The stepwise computational procedure is represented by the stepwise path from A to C. It involves first calculating the temperature drop in the wellbore from bottom A to point B′ some Δh feet above, from analytical solutions developed for upflow at constant pressure.

The actual pressure drop would be calculated for the considered Δh increment next. Knowing the pressure drop, the path B′ B″ would be calculated next, correcting the temperature at B′ (precalculated isobarically) for the pressure drop via the Joule-Thomson expansion process.

Our calculations on several examples and field measurements have so far shown that assumption of an isobaric step followed by an isenthalpic one predicts a final temperature reasonably close to actual wellbore temperature.

The solution just described is put in the form of an equation giving the temperature of the gas as a function of height above datum h.

Equation for Estimating the Actual Wellbore Temperature

The equation developed for estimating the wellbore temperature is derived in detail in Appendix F. The results in appropriate units are given as follows:

$$T(h) = T'(h) + \frac{\overline{m}c_p}{2\pi aKg_t} - \frac{\overline{m}}{J}\frac{g}{g_c}\frac{1}{2\pi aK}$$

$$- \left[T_o' - T_o + \frac{\overline{m}c_p}{2\pi aKg_t} - \frac{\overline{m}}{J}\frac{g}{g_c}\frac{1}{2\pi aK}\right]e^{-\frac{2\pi ak}{\overline{m}C_p}h} - \mu_j(-\Delta p)_{o\rightarrow h} \quad (4\text{-}64)$$

where $m = \dfrac{Q \times 10^6 \times 0.0764\ G}{24}$

For flow through casing:

$$\frac{1}{2\pi aK\ L} = \frac{1}{h_iA_l} + \frac{\Delta r}{K_cA_m} + \frac{\ln\dfrac{d_{cem}}{d_c}}{2\pi K_{cem}\ L} + \frac{\ln\dfrac{d_{earth}}{d_{cem}}}{2\pi K_{earth}\ L} \quad (4\text{-}65)$$

$$\frac{h_id_i}{K_g} = 0.023\ (R_e)^{0.8}\ (P_r)^{0.4*} \quad (4\text{-}66)$$

$$\mu_j = \frac{1{,}544}{778}\ \frac{T^2}{P\ C_p\ M\ T_c}\left(\frac{\partial z}{\partial T_r}\right)_{P_r}$$

For the nomenclature and units involved in the design calculations related to temperature distributions, an example problem is given in step-by-step detail at the end of this chapter.

Unsteady-State Effects

In storage, the gas that is injected into the porous formation exchanges heat readily and effectively with native gas in the pore space and the reservoir rock.

During the flow through porous media the gas remains at constant formation temperature. This is basically due to having a very large surface-to-volume ratio between the gas and the porous rock, a very large source or sink for heat. As the pressure drops near the wellbore and the gas flows into the wellbore through perforations, the temperature of the gas drops at constant

* if the gas is cooling use 0.3

enthalpy. The continuous production of gas originating at equilibrium formation temperature convectively and cumulatively removes heat from the producing formation at the bottom. This causes a gradually enlarging area around the wellbore at the bottom subject to temperatures lower than the initial geothermal. As the gas flows up through the wellbore, at some intermediate depth its temperature becomes higher than geothermal. From that point upward the effect of gas flow on the prevailing temperatures around the wellbore is such that it gradually heats the near-zone areas from original geothermal to higher temperatures. As a result, in long-time flow, the geothermal temperature distribution near the wellbore, which is nearly linear at the bottom and somewhat curved very near the top, gets distorted to become warmer at the top and cooler at the bottom. The opposite occurs in response to injection.

Since underground storage involves a producing withdrawal season not longer than 100 to, say, 120 days, the cooling effect near the bottom in response to production has been estimated to affect a radius of 10 to 15 feet at most. Using the unsteady-state conduction by line source equations and thermal properties of the formations, the radius subject to temperature drop was estimated to grow to 12 ft in 100 days. Accordingly, it was concluded that during a typical storage withdrawal season unsteady temperature effects would not appreciably change the findings based on steady-state equations.

Furthermore, the computer program that calculates the heat transfer through successive thermal resistance zones was used to simulate the effect of an enlarging region around the wellbore where the heat flow is by radial conduction. The results showed that, on production, the assumption of original geothermal temperature prevailing at a diameter equal to or more than 20 inches around the wellbore did not appreciably affect the temperature of the gas arriving at the surface. In flow through the tubing, the controlling resistance for heat transfer is natural convection of stagnant gas in casing-tubing annulus. In flow through casing or casing-tubing annulus, the largest resistance to heat is radial conduction out in the formation from the edge of the cement to a point in the formation where the temperature is equal to geothermal. That point gets farther and farther as production proceeds, but contribution of this unsteady-state effect is through a term proportional to natural log of (d_{earth}/d_{cement}). It is rather insensitive to d_{earth}, as the radius around the wellbore affected rarely exceeds 10 ft.

Design Concepts Related to Control of Temperature and Pressures

During 1978 three gas fields in northern Michigan were considered for conversion to storage service. The total working storage contemplated was 55 BCF.

Maximum reservoir pressures will be approximately 4,600 psi at a depth of 6,500 ft. A total of some 30 wells would be used to deliver the storage gas during a 100-day season. Maximum peak well flowrates considered were as high as 50 MMCF/D. Static reservoir temperature was 115°F. Wellbore drawdowns would be relatively small due to the high permeability of the reservoirs and good completion techniques.

In an attempt to reduce surface regulation requirements, thereby reducing gas consumption in surface heaters, many of the wells would be completed with equipment designed to permit pressure drops in the wellbore rather than the surface. Design requirements called for flexible field flowrates to meet peak demand, while always maintaining a 1,000 psia well head/gathering system pressure. This would be done by equipping some wells with down-hole chokes and others with various sizes of tubing-casing strings.

The down-hole choke wells would be used at the start of the season when shut-in well head pressures are near 4,000 psia. The temperature drop occurring due to gas expansion across the down-hole choke was expected to be offset by geothermal heating as the gas rose in the wellbore. The gas would reach the surface at 1,000 psia and a temperature above the hydrate point, thereby requiring no further regulation or heating. Peak flow demands at this early stage of the withdrawal season would be met by flowing unrestricted wells and making limited temporary use of surface regulation and heating.

As the field pressure declined, daily rates would be met by using the variously-sized tubing-casing wells in combination. All pressure drop would occur due to friction in the wellbore. It was important, therefore, that these wells not be used until field pressures were low enough (and pressure drops in the flow string were low enough) to result in surface temperatures above the hydrate threshold.

Ultimately, the large-casing unrestricted wells would be called upon to provide high flowrates at low field pressure conditions. It was anticipated that by this time the resultant flowing well head temperatures would be well above the hydrate point even at high flowrates.

The Joule-Thomson cooling caused by pressure regulation can allow downstream hydrate formation at pipeline pressure conditions unless the gas stream is heated prior to regulation. Figure 4-26 illustrates a typical withdrawal flow schematic for Northern Michigan Niagaran Reef Storage, showing pressure and temperature changes as the gas stream passes from well bottom to pipeline.[69] Gas dehydration is normally not needed for Michigan Niagaran Reef Storage operations, so no dehydrator is included in the schematic. The thermodynamic path taken by the temperature as the gas stream progresses from well bottom to pipeline is observed starting at 107°F at well bottom. The gas is allowed to cool as it flows to the surface and then must be reheated to approximately the reservoir temperature prior to regula-

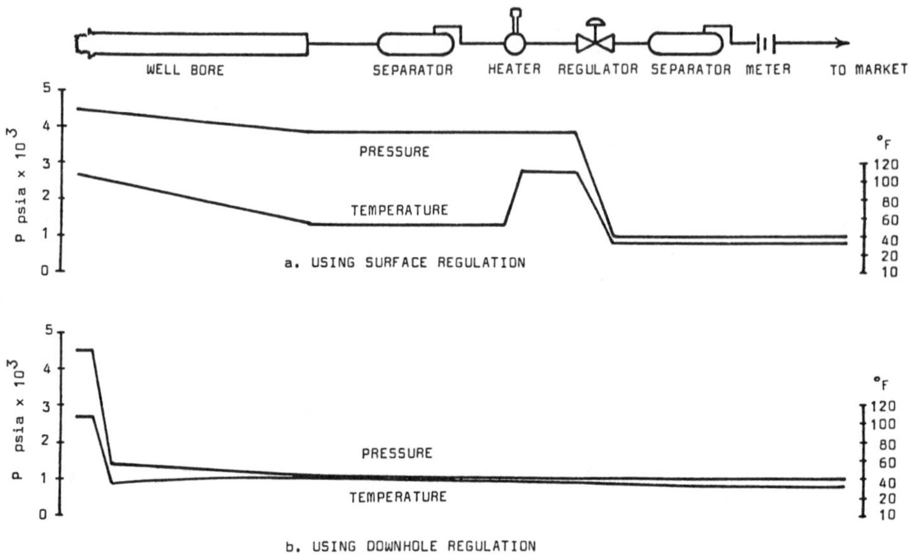

Figure 4-26. Typical northern Michigan gas withdrawal schematic, 50 MMCFD gas rate for a 7 in. OD well.

tion. Considering the high replacement cost of gas, this reheating of the gas stream was a very costly operational step.

Figure 4-26 illustrates a departure from traditional surface regulation by making use of a downhole choke and earth temperature to result in pipeline gas conditions at the wellhead, eliminating the need for further heating and regulation. The mechanics of the process are quite simple, involving the installation of a packer and wireline landing nipple at depth in the wellbore so that a properly sized wireline locking mandrel and choke bean can be located. Another method yet to be tested involves simply sizing the flow string to cause the proper pressure drop at the required flowrate. In all cases, the wellbore is designed to allow optimum heat transfer into the wellbore from the surrounding earth. These concepts were being entertained in the design and operation of future Niagaran Reef Storage for the ANR Storage Company in Northern Michigan.[69] Four reefs were being developed to a total of 51 BCF working gas storage service for six interstate pipeline customers. Two of the reef projects, totalling 38.3 BCF working storage (with an additional 5.9 BCF base gas) were under construction.* The other two reef projects, with 12.8 BCF working storage (with an additional 1.9 BCF base gas), were awaiting F.E.R.C. certification.* Maximum expected withdrawal rate was over 600 MMCFD for the project. The anticipated annual fuel savings

* At the time of publication of references 57, 68, and 69.

from using downhole regulation during withdrawal could exceed 75,000 MCF. This fuel savings, coupled with reduced heater investment and low base gas requirement was thought to help allow the ANR Storage Company to provide relatively inexpensive storage service to the industry.

Application of this process to gas streams containing high water vapor content is also feasible. Gas production operations as well as gas storage operations can benefit from the fuel savings. In areas of the world where cool geothermal gradients result in relatively low reservoir temperatures, a two-step pressure reduction (downhole and surface) can still result in considerable fuel savings. A producing gas well described in previous work by Tek, Katz, et al.,[57] located in Southeastern Michigan and having a reservoir temperature of only 84°F, was fitted with a downhole choke and resulted in hydrate-free production without the use of surface heat. Geothermally warm areas such as California and the U.S. Gulf Coast are prospective locations for using one-step (downhole) regulation with large ultimate fuel savings.

Comparison between Calculated Wellbore Temperatures and Field Data

During the course of this research, a reasonably sufficient amount of data direct from field operations was made available to the author and his coauthors on two previous papers.[57,68]

These data were not always complete in terms of precise static geothermal and flowing wellbore profiles. Sometimes they only referred to bottom-hole and surface conditions, where they only provided checks on end points of flow strings.

They were, however, carefully checked and screened as to their reliability and, as such, provided the background against which calculated values were compared with actual measurements.

Table 4-7 gives a summary of comparison between calculated and measured well head temperatures. A comparison is also shown in Figure 4-27.

Figure 4-28 shows the geometry of the wellbore from surface string to perforations and total depth for one of the wells studied in detail. Figure 4-29 shows the pressure distribution in the tubing while the well was flowing. It may be noted that the pressure appears to be rising as the gas flows upward at the two bottom points immediately above the bottom-hole choke. This is probably due to temperature correction or turbulence effect related to a disturbance created by the pressure bomb lowered to a short distance above the bottom-hole choke. The difference between the measured and calculated surface pressure appears to be less than a few percentage points. Figure 4-30 shows a comparison between measured and calculated temperature distribution on the same well at two different surveys. It may be noted that the agreement between measured and calculated temper-

Table 4-7
Summary of Temperature Tests and Comparison Between Calculated and Measured Wellhead Temperatures

	Field A	Field B	Bruce*	W. Columbus	Aliso Canyon*	Aliso Canyon*
Well	2 A-I/w 16	1 B-I/w 2	Jones 1-30	W. Columbus #3	IW-80	SS-9
Date of Test	8-24-77	8-24-77	5-23-78	1-9-79	1-17-75	1-17-75
Depth of Well (ft)	6842	6818	3606	3050	7600	8450
Size of Casing (in.)	5 1/2	3 1/2	4 1/2	7(6.276 I.D.)	8 5/8	7.0
Size of Tubing (in.)	----------	--------	----------	----------------	2 7/8	2 7/8
Flow Geometry	Casing	Casing	Tubing	Casing	Annular	Annular
Reservoir Temp.						
Geothermal (°F)	220	218	82.50	71	190	205
Flowing (°F)	219	138	62.50	69.4	106	167
Flowing Reservoir						
Pressure (psia)	1635.7	1571.7	930.70	1561.4	804	1495
Flowrate (MMCF/day)	3.536	7.452	3.817	50.0	32.0	13.12
Well Head Temperature						
Measured (°F)	81	98	54	57	104	93
Calculated (°F)	87	95	50	57.47**	102	98
Geothermal (°F)	72	72	50	(48–51)	70	70

* Data were taken from SPE 7495.
** Calculated by approximate analytical method.

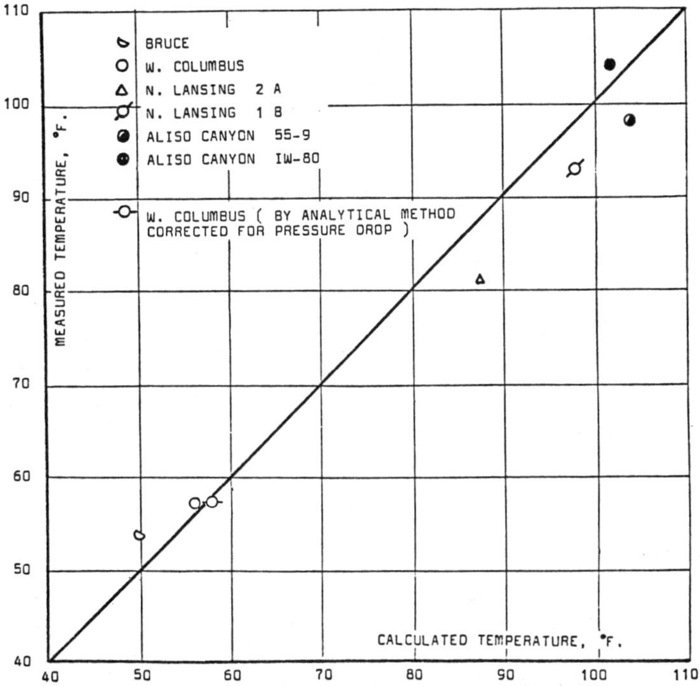

Figure 4-27. Comparison between predicted and calculated temperatures.

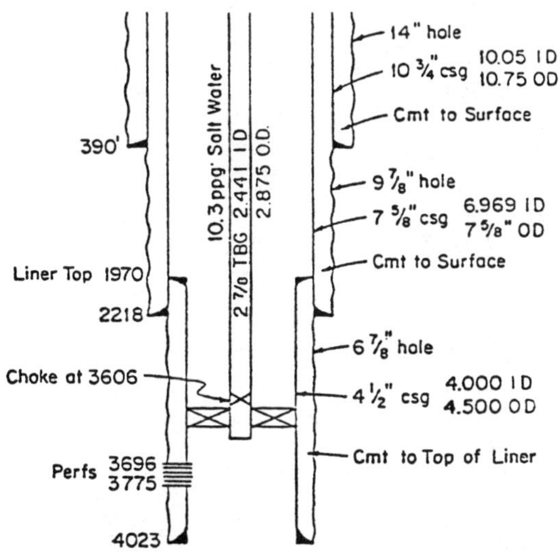

Figure 4-28. Detail of wellbore geometry on Jones I-30 well.

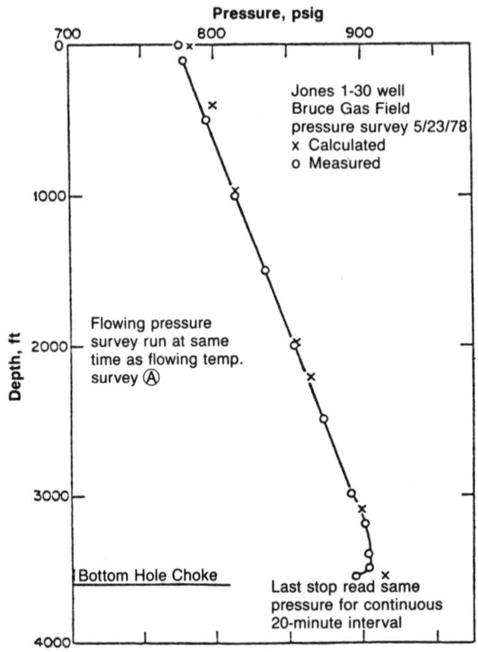

Figure 4-29. Pressure distribution, Jones I-30 well.

Figure 4-30. Comparison between measured and calculated temperature distributions on Jones I-30 well, Bruce Field, Michigan.

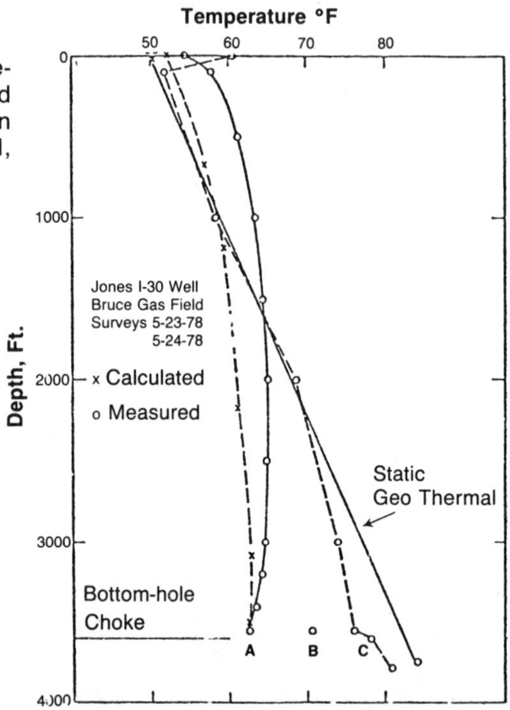

Legend

A. Flowing survey flowrate—3300 MCFD flowtime prior to survey—3 days

B. Shut-in 1 hr 38 min after test A

C. Static survey run 27 hrs after test A. Air temp. at time of test—92°F (bottom-hole choke had been removed before this survey)

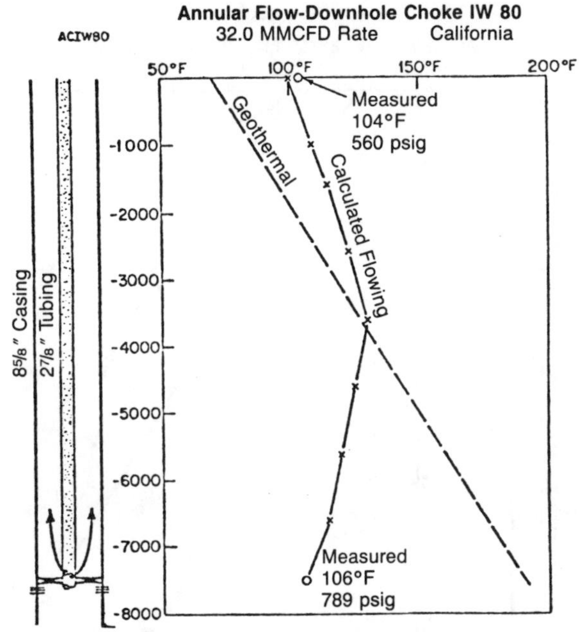

Figure 4-31. Comparison between measured terminal temperatures and calculated temperature distribution for SS-9 Well, Aliso Canyon Field, California.

atures is excellent. Figure 4-31 shows a comparison between measured downhole and surface temperatures and calculated pressure distribution for a casing-tubing annulus flow for the Well IW-80 in Aliso Canyon, California.

Figure 4-32 shows a similar comparison between calculated and measured temperatures for the Well SS-9, Aliso Canyon Field, California. The casing-tubing annulus flow through the bottom-hole choke and details of the wellbore geometry are sketched at the left hand side. The agreement between calculated profiles and measured terminal temperatures appears equally good for both Aliso Canyon Wells.

Figure 4-33 shows a comparison between calculated and measured temperatures on a storage well. Figure 4-27 shows a comparison between calculated and measured temperatures on several storage wells in Michigan and California.

Special Test and Temperature Survey on a Michigan Well

In order to directly evaluate the effect of time on gas temperatures in a Michigan storage well producing at a high rate, a special test was performed

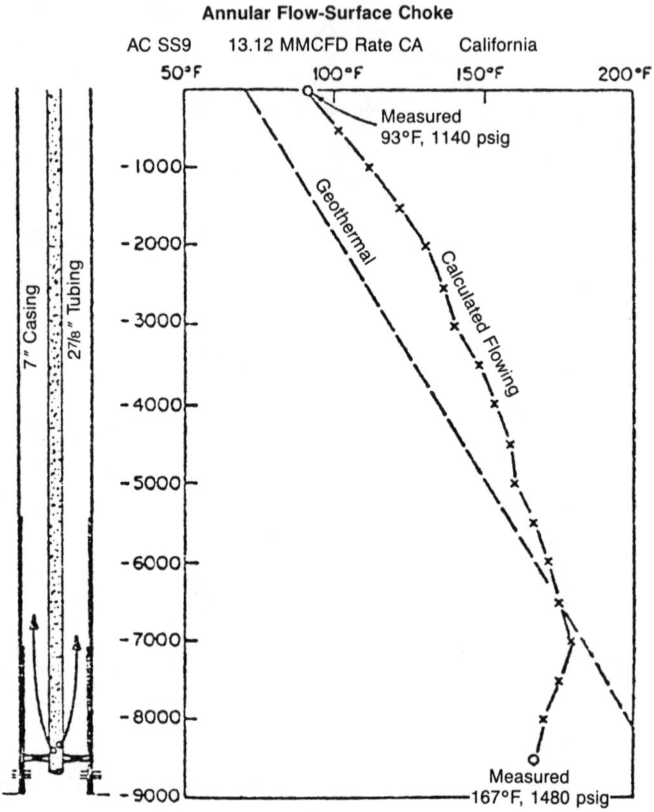

Figure 4-32. Comparison between measured terminal temperatures and calculated temperature distribution for SS-9 Well, Aliso Canyon Field, California.

and a temperature survey was made. The results indicated that the surface temperature of the gas did not vary between 19 hours and 187 hours.

The data obtained during this test and a comparison of wellbore temperatures calculated by the method of this paper and the methods of References 8 and 12 are presented in Table 4-8.

An examination of Table 4-8 indicates that:

1. The temperatures of the gas reaching the surface for 50 MMCF/D flowrate was the same for 19 hours and 187 hours.
2. The computer program based on the equation from SPE 7495 and Equation 4-62 predicted the surface temperature better than the three methods tried.
3. The approximate analytical method of this paper was also used and gave a surface temperature about 0.5° lower than measured.

Figure 4-33 shows a comparison between measured, geothermal, and predicted temperature distributions by Equation 4-62 and two previous publications, hand-calculated.

Figure 4-34 shows a comparison between temperatures measured at 19 and 187 hours and those calculated by the methods of this paper. The surface temperatures predicted by References 58 and 62 are also shown for flow periods of 19 hours and 187 hours.

Summary and Conclusions. On the basis of engineering calculations, field data, and special field tests conducted, one may conclude the following:

1. The temperature distribution of gas in a wellbore can be predicted with reasonable accuracy in underground storage operations involving withdrawal or injection using computer-implemented stepwise energy balances in the wellbore.
2. When the pressure drop in the wellbore is small, the energy balances may be expressed in the form of a differential equation which can be

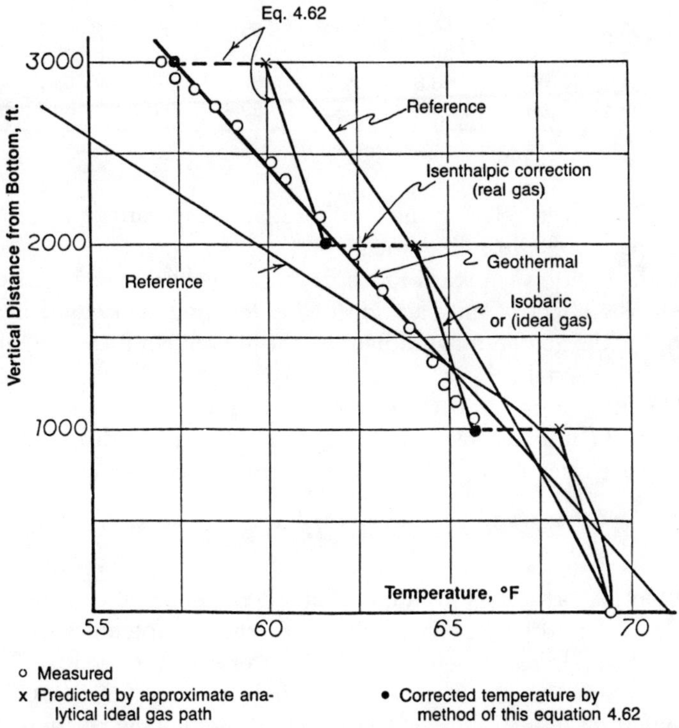

Figure 4-33. Comparison of calculated temperatures with actual field data.

Table 4-8
Comparison of Measured Temperatures from a Michigan Well "W" vs. Temperatures Calculated by Various Methods

Depth (ft)	SPE 7495 and This Paper T, °F		Reference T, °F		Reference T, °F		Measured T, °F	
	19 hrs	187 hrs	19 hrs	187 hrs	19 hrs	187 hrs	19 hrs	187 hrs
0	56.9	56.5	60.6	60.2	52.1	52.5	57.0	57.0
50	57.2	56.9	60.8	60.4	52.5	52.9	57.4	57.4
100	57.4	57.2	61.0	60.5	52.9	53.3	57.9	57.7
200	57.9	57.9	61.4	60.9	53.7	54.1	58.5	58.2
300	58.3	58.5	61.7	61.2	54.5	54.9	59.2	58.5
500	59.1	59.8	62.4	61.8	55.8	56.5	60.1	58.7
600	59.6	60.5	62.7	62.1	56.9	57.3	60.5	59.4
800	60.4	61.7	63.4	62.7	58.5	58.9	61.5	60.7
1000	61.2	63.0	64.1	63.3	60.1	60.5	62.4	61.7
1200	61.9	64.2	64.7	63.9	61.7	62.1	63.2	62.9
1400	62.7	65.4	65.3	64.4	63.3	63.7	63.9	62.9
1600	63.4	66.5	65.9	65.0	63.9	65.3	64.6	63.7
1700	63.7	67.2	66.2	65.2	65.7	66.1	64.9	64.1
1800	64.0	67.8	66.5	65.5	66.5	66.9	65.2	64.4
1899	64.3	68.2	66.8	65.8	67.3	67.7	65.7	64.6
2950	69.4	68.3	69.4	68.3	69.4	68.3	69.4	68.3

readily solved relating the wellbore gas temperatures to geothermal temperatures prevailing outside.

3. As the gas flows upward for short times involved in storage flow periods, the effect of unsteady distortion (cooling at bottom, heating at top) of the geothermal gradient on the gas temperature arriving at the surface is small.
4. The final temperature predicted by the approximate analytical method can be corrected for the effect of pressure drop with reasonable accuracy using data on Joule-Thomson cooling coefficients.

Example 4-4

Problem. A gas storage well has a total depth of 3,050 ft from surface to datum level at mid-perforations. The geothermal temperature near the surface is 58°F and linearly increases 0.435°F every 100 ft of depth. The well is flowing through 7 in. casing (ID = 6.276 in.) at a rate of 50 MMCF/D. The flowing bottom-hole pressure is 1,561.4 psia. The flowing bottom-hole temperature is 69.4°F. Previous design calculations indicate that the overall heat transfer coefficient k between the gas flowing up inside and the formation where geothermal temperature prevails outside is 5 Btu/hr × ft² × °F. Calculate and plot the estimated steady-state temperature distribution for this well.

Solution.

Q = 50 MMCF/D

$$T'(h) = T'_o - \frac{h}{g_t}$$

h = 3,000 ft T′(h) = 58°F

$$g_t = \frac{100}{.435}\ \text{ft/°F}$$

$$= 229.89\ \text{ft/°F}$$

$$T'_o = 58 + \frac{3{,}000}{229.89} = 71.05°\text{F}$$

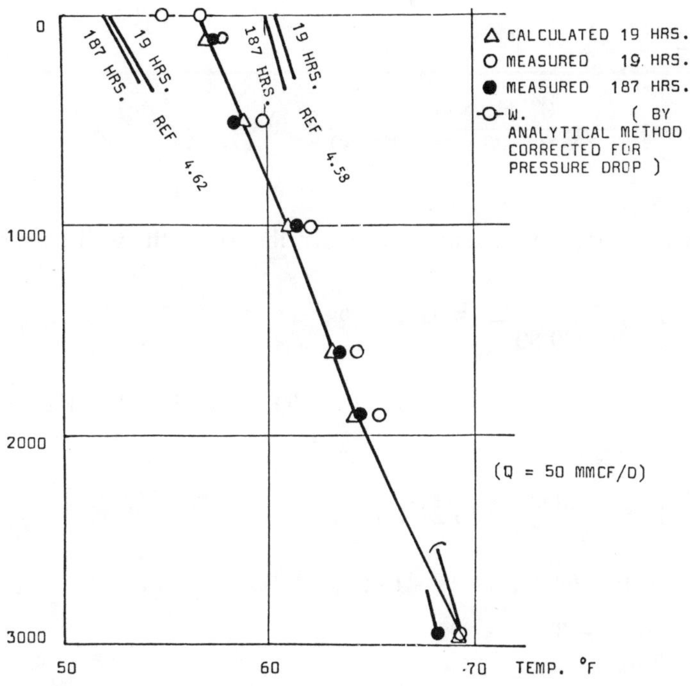

Figure 4-34. Comparison between calculated and measured temperatures, Michigan Well "W."

$$\overline{m} = \frac{50 \times 10^6 \times 0.0764 \times 0.612}{24}$$

$$= 97{,}410 \text{ lb}_m/\text{hr}$$

$$a = \frac{6.276}{12} = 0.26 \text{ ft}; K = 5 \text{ Btu/hr} \times \text{ft}^2 \times {}^\circ\text{F}$$

$$\frac{m}{2\pi a K}\frac{1}{J}\frac{g}{g_c} = \frac{97{,}410 \times 32.17}{2\pi \times 0.26 \times 5 \times 778 \times 32.17} = 15.24^\circ\text{F}$$

Next, calculate c_p for pipeline gas:

$c_p = 0.525$ at $p = 1$ atm (*Handbook of Natural Gas,*[66] p. 131 Figure 4-49)

Pressure Correction $p_{bot} = 1{,}561.4$ psia

$${}_p p_r = \frac{1{,}561.4}{672} = 2.32$$

$${}_p T_r = \frac{529.4}{395} = 1.34$$

$$c_p = .525 + \frac{7.5}{17.3} = 0.95$$

$$\frac{\overline{m} c_p}{2\pi\ a\ K\ g_t} = \frac{97{,}410 \times 0.95}{2 \times \pi \times 0.26 \times 5 \times 229.89} = 49.28$$

$T_o = 69.4$

So now the equation for temperature distribution in the well becomes:

$$T(h) = 71.05 - \frac{h}{229.89} - 15.24 + 49.28 - [71.05 - 69.4 + 49.28 - 15.24]e^{-\frac{2\pi \times 0.26 \times 5h}{9{,}740 \times .95}}$$

Take h = 1,000 ft

$$T(h) = 71.05 - \frac{1{,}000}{229.89} - 15.24 + 49.28 - [71.05 - 69.4 + 49.28 - 15.24]e^{-0.09}$$

$$= 100.74 - 35.69 \times .92$$
$$= 67.91^\circ\text{F}$$

Next, calculate the pressure drop for 1,000 ft from bottom:

Q = 50,000 MCF/D

$$p_{bot} = 1{,}561.4 \text{ psia}$$

$$p_s^2 - e^s p_w^2 = +0.000667 \text{ f } Q^2 T_a^{\,2} Z_a^{\,2} \frac{(e^s - 1)}{d}$$

$$s = 0.0375 \frac{GL}{T_a z_a}$$

$$G = 0.612$$

$$p_r = \frac{1{,}361.4}{672} = 2.32$$

$$T_r = \frac{529.4}{395} = 1.34$$

$$z = 0.7$$

$$s = \frac{0.0375 \times 0.612 \times 1{,}000}{529.4 \times 0.7} = 0.6$$

$$e^s = 1.06$$

$$R_e = \frac{50 \times 10^6 \times 0.0764 \times 0.612 \times 6.276 \times 4 \times 144}{864{,}000 \times 12 \times 0.000672 \times 0.0145 \times \pi \times \overline{6.276}^2}$$

$$= 6{,}760{,}396$$

$$\epsilon/D = 0.0003$$

$$f = 0.015$$

$$p_{1{,}000\,ft} = \left(\frac{\overline{1{,}561.4}^2}{1.06} - \frac{0.000667 \times 0.015 \times \overline{50{,}000}^2 \times \overline{529.4}^2 \times \overline{0.7}^2 \times 0.06}{1.06 \times \overline{(6.275)}^5} \right)^{1/2}$$

$$(2{,}299{,}971.66 - 19{,}968.79)^{1/2} = 1{,}509.97 \text{ psia}$$

Correction of calculated temperature for the actual pressure drop:

$$\mu_j = \frac{1.98 \, T^2}{C_p \, M \, p \, T_c} \left(\frac{\partial z}{\partial T_r} \right)_{p_r} \text{ °F/psi}$$

$$p_r = \frac{1{,}561.4}{672} = 2.32$$

$$\begin{array}{llll} & 1.45 & & 0.776 \\ T_r = & 1.34 & z = & 0.7 \\ & 1.30 & & 0.650 \end{array}$$

$$\frac{\Delta z}{\Delta T} = \frac{.126}{.150} = 0.84$$

$$\mu_j = \frac{1.98 \times (460 + 67.91)^2 \times 0.84}{0.95 \times 17.3 \times 1{,}561.4 \times 395} = 0.046°F/psi$$

$$\Delta T = (1{,}561.4 - 1{,}509.97) \times 0.046 = 2.35°F$$

$$T_{corrected} = 67.91 - 2.35$$
$$= 65.56\ °F$$

Now use same procedure for the next 1,000 ft:

$$T_2 = 66.70 - 1{,}000/229.89 - 15.24 + 49.28 - [66.70 - 65.56 + 49.28 - 15.24]e^{-0.09}$$
$$= 100.73 - 32.37 - 4.45$$
$$= 64.02°F$$

Next calculate p_2 at 2,000 ft from the bottom: $T_{r_2} = 1.34$ $P_{r_2} = 2.25$ $z_2 = .7$

$$p_2 = \left[\frac{\overline{1{,}510}^2}{1.06} - \frac{0.000667 \times 0.015 \times \overline{50{,}000}^2 \times (\overline{525.56})^2 \times \overline{.7}^2 \times 0.06}{1.06 \times (\overline{6.276})^5}\right]^{1/2}$$

$$= (2{,}151{,}037.74 - 19{,}893.42)^{1/2}$$

$$= 1{,}459.92 \text{ psia}$$

$$p_{r_2} = \frac{1{,}510}{672} = 2.25$$

$$T_{r_2} = \frac{524.4}{395} = 1.34$$

$$\begin{matrix} 1.4 & .245 \\ z_2 & = 0.7 \\ 1.3 & .655 \end{matrix} \quad ; \frac{\Delta z}{\Delta T} \cong \frac{.09}{0.1} = 0.9$$

$$\mu_j = \frac{1.98 \times (524.02)^2 \times 0.9}{0.95 \times 17.3 \times 1{,}510 \times 395} = 0.05 = 2.5°F$$

$$T_{2_{correct}} = 64.02 - 2.5 = 61.52$$

Next calculate T_3

$$T_3 = 62 - \frac{1{,}000}{229.89} - 15.24 + 49.28 - [62 - 61.52 - 15.24 + 49.28].92$$

$$= 62 - 4.35 - 15.24 + 49.28 - [62 - 61.52 - 15.24 + 49.28].92$$
$$= 91.69 - 34.52 \times 0.92$$
$$= 59.93$$

Table 4-9
Summary of Temperature and Pressure Distribution Calculations on Example Problem (Same as Michigan well featured in previous section)

h, ft	T′ °F	T(h)	μ_j °F/psi	T(h) corrected	T measured
0	71.05	69.4	0.046	—	69.4
1000	66.71	67.91	0.05	65.56	66.0
2000	62.37	67.91	0.05	61.52	62.25
3000	58.03	59.93	—	57.47	57.0

$$T_r = \frac{519.93}{395} = 1.32$$

$$p_r = \frac{1{,}459.84}{672} = 2.17$$

$$z = 0.685$$

$$s = \frac{0.0375 \times 0.612 \times 1{,}000}{519.93 \times 0.685} = 0.06$$

$$e^s = 1.06$$

$$p_3 = \left(\frac{\overline{1{,}459.92}^2}{1.06} - \frac{0.000667 \times 0.015 \times \overline{50{,}000}^2 \times \overline{519.93}^2 \times \overline{.685}^2 \times 0.06}{1.06 \times \overline{6.276}^5}\right)^{1/2}$$

$$= 1{,}411.48 \text{ psia}$$

$$p_{r_3} = \frac{1{,}411.48}{672} = 2.1$$

$$T_{r_3} = \frac{519.93}{395} = 1.32$$

1.35 .72
z_3 = 0.7
1.30 .68

$$\frac{\Delta z}{\Delta p} = \frac{.045}{.05} = 0.9$$

$$\mu_j = \frac{1.98 \times \overline{519.93}^2 \times 0.9}{0.95 \times 17.3 \times 1{,}459.84 \times 395} = 0.05$$

$$\Delta T = (1{,}459.84 - 1{,}411.48)0.05$$
$$= 2.46°F$$

$$T_{3_{corrected}} = 59.93 - 2.46$$
$$= 57.47\ °F$$

These calculations are summarized in Table 4-9.

References

1. Tek, M. R. and D. L. Katz. "Underground Storage of Gases," course notes, Boyne Mountain Lodge, June 18–25, 1981.
2. Tek, M. R., F. H. Poettmann, and M. L. Grove. "Method for Determining the Back Pressure Behavior of Low Permeability Natural Gas Wells," *Pet. Trans. AIME*, (1952) 210, 302.
3. Rawlins, E. L. and M. A. Schellhardt. "Back-Pressure Data on Natural Gas Wells and Their Application to Production."
4. Cullender, M. H. "The Isochronal Performance Method of Determining the Flow Characteristics of Gas Wells," *Trans. AIME*, (1955) 204:137.
5. Van Everdingen, A. F. "The Skin Effect and Its Influence in the Productive Capacity of a Well," *Trans. AIME* (1953) 198, 171–176.
6. Hawkins, M. F., Jr., "A Note on The Skin Effect," *Trans. AIME* (1956) 207, 356–357.
7. Smith, R. V. "Unsteady Gas Flow into Gas Wells," *J. Petro. Tech.* (1961) 13, 1151–1159.
8. Tek, M. R., K. H. Coats, and D. L. Katz. "The Effect of Turbulence on Flow of Natural Gas through Porous Reservoirs," *J. Petro. Tech.* (July 1962) 799–806.
9. Swift, G. W. and O. G. Kiel. "The Prediction of Gas Well Performance Including the Effect of Non-Darcy Flow," *J. Petro. Tech.* (1962) 14, 791–798.
10. Carter, R. D., S. C. Miller, Jr., and H. G. Riley. "Determination of Stabilized Gas Well Performance from Short Flow Tests," *J. Petro. Tech.* (1963) 15, 651–658.
11. Lee, R. L., R. W. Logan, and M. R. Tek. "Deliverability in Gas Storage," CEEC Summer Conference, University of Michigan, Ann Arbor, 1984.
12. Tek, M. R. and J. O. Wilkes. "Fluid Flow and Heat Transfer," University of Michigan, Ann Arbor, 1974.
13. Gould, T. L., M. R. Tek, and D. L. Katz. "Two-Phase Flow through Vertical, Inclined or Curved Pipe," *Journal of Petro. Tech.* (August 1974) SPE.

14. Baxendell, P. B. and R. Thomas. "The Calculation of Pressure Gradients in High Rate Flowing Wells," *J. Petro. Tech.*, 1023–1028; *Trans. AIME*, 222 (Oct. 1961).
15. Fancher, G. H., Jr. and K. E. Brown. "Prediction of Pressure Gradients for Multiphase Flow in Tubing," *Soc. Petro. Eng. J.*, 59–69; *Trans. AIME*, 228 (March 1963).
16. Gaither, O. D., H. W. Winkler, and C. V. Kirkpatrick. "Single- and Two-Phase Fluid Flow in Small Vertical Conduits Including Annular Configurations," *J. Petro. Tech.*, 309–320; *Trans. AIME* 228 (March 1963).
17. Govier, G. W. and W. L. Short. "The Upward Vertical Flow of Air-Water Mixtures," *Cdn. J. Chem. Eng.* (Oct. 1958) 36, 195.
18. Poettmann, F. H. and P. G. Carpenter. "The Multiphase Flow of Gas, Oil and Water through Vertical Flow Strings with Application to the Gas-Lift Installations," *Drill. and Prod. Prac.*, API 257 (1952).
19. Tek, M. R. "Multiphase Flow of Water, Oil and Natural Gas through Vertical Flow Strings," *J. Petro. Tech.*, 1029–1036; *Trans. AIME*, 222 (Oct. 1961).
20. Ros, N. C. J. "Simultaneous Flow of Gas and Liquid as Encountered in Well Tubing," *J. Petro. Tech.*, 1037–1049; *Trans. AIME*, 222 (Oct. 1961).
21. Griffith, P. and G. B. Wallis. "Two-Phase Slug Flow," *J. Heat Transfer; Trans. ASME*, 307–320 (Aug. 1961).
22. Hagedorn, A. R. and K. E. Brown. "The Effect of Liquid Viscosity in Vertical Two-Phase Flow," *J. Petro. Tech.*, 203–210; *Trans. AIME*, 231 (Feb. 1964).
23. Hughmark, G. A. "Holdup in Gas Liquid Flow," *Chem. Eng.* (April 1962) 58, 62.
24. Hughmark, G. A. and B. S. Pressburg. "Holdup and Pressure Drop with Gas-Liquid Flow in a Vertical Pipe," *AIChE J.* (Dec. 1961) 7, 677.
25. Moore, T. V. and H. D. Wilde. "Experimental Measurements of Slippage in Flow through Vertical Tubes," *Trans. AIME*, 92, 296–319 (Oct. 1930).
26. Moore, T. V. and R. J. Schilthuis. "Calculation of Pressure Drops in Flowing Wells," *Trans. AIME* 92, 170–190 (1933).
27. Duns, H., Jr. and N. C. J. Ros. "Vertical Flow of Gas and Liquid Mixtures from Boreholes," Proc., Sixth World Petro. Cong., Frankfurt II, Paper 22-P.D, 6, 1963.
28. Orkiszewski, J. "Predicting Two-Phase Pressure Drop in Vertical Pipe," *J. Petro. Tech.*, 829–838; *Trans. AIME*, 240 (June 1967).
29. Gould, T. L. and M. R. Tek. "Digital Simulation of Unsteady State Vertical Two-Phase Flow," Proc., Conf. on Natural Gas Research and Tech., AGA-IGT, Chicago, 1971.

30. Baker, O. "Effect of Hills on Two-Phase Pressure Drop," *Oil and Gas J.* (Nov. 1957) 58, 150.
31. Flannigan, O. "Effect of Uphill Flow on Pressure Drop in Design of Two-Phase Gathering System," *Oil and Gas J.* (March 1958) 56, 132.
32. Sevigny, R. "Investigation of Isothermal Co-Current Two-Fluid, Two-Phase Flow in an Inclined Tub," Ph.D. Thesis, U. of Rochester, N.Y., 1962.
33. Bonderson, L. W. "An Analysis of the Pressure Drops and the Void for a Two-Phase Slug Flow in Inclined Pipes," M.S. Thesis, M.I.T., Cambridge, MA, 1969.
34. Bonnecaze, R. H., W. Erskine, and E. J. Greskovich. "Hold-up and Pressure Drop for Two-Phase Slug Flow in Inclined Pipelines," AIChE J. (Sept 1971) 17,1109.
35. Singh, G. and P. Griffith. "Determination of the Pressure Drop Optimum Pipe Size for a Two-Phase Slug Flow in an Inclined Pipe," *J. of Eng. Ind.; Trans. ASME*, 717 (Nov. 1970).
36. Zukoski, E. E. "Influence of Viscosity, Surface Tension and Inclination Angle on Motion of Long Bubbles in Closed Tubes," *J. Fluid Mech.* (1966) 25, 821.
37. Gould, T. L. "Vertical Two-Phase Flow in Oil and Gas Wells," Ph.D. Thesis, U. of Michigan, Ann Arbor, 1972.
38. Gould, T. L. and M. R. Tek. "Steady and Unsteady State Two-Phase Flow through Vertical Flow Strings," SPE paper 2804, SPE-AIME Second Symposium on Numerical Simulation of Reservoir Performance, Dallas, Feb. 1970.
39. Greskovich, E. J. "Holdup Predictions for Stratified Downflow of Gas-Liquid Mixtures," *Ind. and Eng. Chem. Process Design Development* (1972) 11, 81.
40. Hewitt, G. F. and N. S. Hall Taylor. "Annular Two-Phase Flow," Oxford: Pergamon Press, 1970.
41. Wallis, G. B. "One-Dimensional Two-Phase Flow," McGraw-Hill, 1969.
42. Hoogendoorn, C. J. "Gas-Liquid Flow in Horizontal Pipes," *Chem. Eng. Science* (1959) 9, 205.
43. Harmathy, T. B. "Velocity of Large Drops and Bubbles in Media of Infinite or Restricted Extent," *AIChE J.* (1960) 6, 281.
44. Marrucci, G. "Rising Velocity of a Swarm of Spherical Bubbles," *Ind. and Eng. Chem. Fundamentals* (May 1965) 225, No. 2, 4.
45. Bhatia, V. K. "Gas Hold-Up of a Bubble Swarm in Two-Phase Vertical Flow," *AIChE J.* (May 1969) 15, No. 3, 466.
46. Kelmer, E. and G. A. Poole. "A Preliminary Investigation of Flowing Wells," *Drill. and Prod. Prac.*, API 140, 1936.

47. Beggs, H. D. and J. P. Brill. "A Study of Two-Phase Flow in Inclined Pipes," *J. Petro. Tech.*, 607–617; *Trans. AIME*, 255, May 1973.
48. Dukler, A. E. "Research Results, Gas-Liquid Flow in Pipelines," AGA and API Project NX 28, May 1973.
49. Espanol Herrera, J. H. "Comparison of Three Methods for Calculating a Pressure Traverse in Vertical Multiphase Flow," M.S. Thesis, U. of Tulsa, Oklahoma, 1968.
50. U. of Tulsa, Private communication—Data Group 1, unpublished.
51. Camacho, C. A. "Comparison of Correlation for Predicting Pressure Losses on High Gas-Liquid Ratio Vertical Wells," M.S. Thesis, U. of Tulsa, Oklahoma, 1970.
52. Lawson, J. D. and J. P. Brill. "A Statistical Evaluation of Methods Used to Predict Pressure Losses from Multiphase Flow in Vertical Oilwell Tubing," *J. Petro. Tech.*, 903–914; *Trans. AIME*, 257, Aug. 1974.
53. Chierici, G. L., G. M. Ciucci, and G. Sclocchi. "Two-Phase Vertical Flow in Oil Wells—Prediction of Pressure Drop," *J. Petro. Tech.*, 927–937; *Trans. AIME*, 257, Aug. 1974.
54. Moody, L. F. *Trans Am. Soc. Mech. Engrs.*, 66, 671–684, 1944.
55. Tek, M. R. and J. O. Wilkes. "Fluid Flow and Heat Transfer," University of Michigan, Ann Arbor, 1974.
56. Poettmann, F. H. and P. G. Carpenter. "The Multiphase Flow of Gas, Oil and Water through Vertical Flow Strings," *Drill. and Prod. Prac.*, API, 257, 1952.
57. Tek, M. R., D. L. Katz, J. R. Elanbaas, M. Whims, and J. L. Roberts. "Temperature and Pressure Gradients in Gas Wells," SPE 7495, Houston, Texas, Oct. 1978.
58. Ramey, J. J. "Well Bore Heat Transmission," *Trans. AIME*, Vol. 225, p. 427, 1962.
59. Smith, R. C. and R. J. Steffensen. "Computer Study of Factors Affecting Temperature Profiles in Water Inventory Well," *Trans. AIME*, Vol. 249, p. 1447, 1970.
60. Smith, R. C. and R. J. Steffensen. "Interpretation of Temperature Profiles in Water Inventory Wells," *Trans. AIME*, Vol. 259, p. 777, 1975.
61. Kunz, K. S. and M. P. Tixier. "Temperature Surveys in Gas Producing Wells," *Trans. AIME*, Vol. 204, p. 111, 1955.
62. Lesem, L. B. et al. "A Method of Calculating the Distribution of Temperature in Flowing Gas Wells," *Trans. AIME*, Vol. 210, p. 169, 1957.
63. Romero-Juarez, A. "A Simplified Method of Calculating Temperature Changes in Deep Wells," *Trans. AIME*, Vol. 31, p. 763, 1979.
64. Chierici, G. L., G. Sclocchi, and L. Terzi, AGIP S.p.A., Milan. "Pressure and Temperature Profiles are Calculated for Gas Flow," *Oil and Gas J.* (Jan. 7, 1980).

65. Lenoir, J. M., W. A. Junk, and E. W. Conings. "Measurements and Correlation of Thermal Conductivities of Gases at High Pressure," *CEP* (1953) Vol. 49, #10, p. 539.
66. Katz, D. L. et al. *Handbook of Natural Gas Engineering*, McGraw-Hill, 1959.
67. Farrar, M. A. "Prediction of Temperature Profiles in the Well Bore of Gas Storage Wells," M.S. Thesis, U. of Michigan, Ann Arbor, 1979.
68. Tek, M. R., and J. R. Elenbaas, M. Whims, and J. L. Roberts. "Design of Storage Fields with Temperature and Pressure Control of the Surface," SPE 8413, Las Vegas, Nevada, Sept. 1979.
69. Tek, M. R., "Design of Storage Fields—Combining Old Concepts with New Ideas," Proceedings, AGA Natural Gas Transmission Conference, 1980.

5

OPERATION OF STORAGE FIELDS

Operation of underground storage reservoirs becomes the joint responsibility of both the field operations group and several central storage company groups once the storage unit is commissioned and starts to supply the market. The field personnel, under the direction of a superintendent, include field engineers, geologists, and technicians. These people are in charge of day-to-day operation, as well as seasonal and periodic maintenance on wells and surface facilities. Their work includes gathering data and the upkeep of wells, Christmas trees, gathering systems, water disposal, metering, pressure regulation, dehydration, compression, methanol injection, and other facilities. The field group maintains contact with gas control and dispatch as well as reservoir engineers, usually centrally located. The field personnel are also party to longer-range planning and development such as acquisition of protective acreage, drilling of new wells, and recompletions and installation of new equipment. Conduct of routine tests, such as seasonal shut-in pressure surveys and special tests required by the reservoir engineering group, is another responsibility of the field office.

In comparison with the function and responsibilities of the field office in charge of hardware and facilities, the centrally located reservoir engineering group is primarily concerned with software, methods, performance analysis, and planning. Interfacing with the field on one hand and, on the other hand, with marketing, dispatch, and gas control, the reservoir engineering group reports to the management on the state of storage and makes recommendations on both supply and withdrawal planning. This work includes monitoring, testing, and continual evaluation of storage performance.

EVALUATION OF STORAGE RESERVOIR PERFORMANCE

There are three basic requirements recognized in the performance of underground storage reservoirs. These, sometimes called *Performance Attributes*, are:

1. Verification of inventory
2. Containment against migration
3. Assurance of deliverability

These three storage attributes, Inventory-Migration-Deliverability, are related two-by-two as shown in Figure 5-1.

Inventory in Gas Storage

The verification of inventory is probably the most important performance attribute of the storage reservoir.

In the area of inventory, the particular spatial distribution and location of storage gas is of primary interest. Whether the storage reservoir is a depleted gas or oil field, a small reef, or a large aquifer; whether the surrounding environment is tight shale or 100% water-saturated sand; the phase distribution in the porous matrix of the storage system must affect the mobility and recoverability of storage gas. In viewing the storage inventory, it is cus-

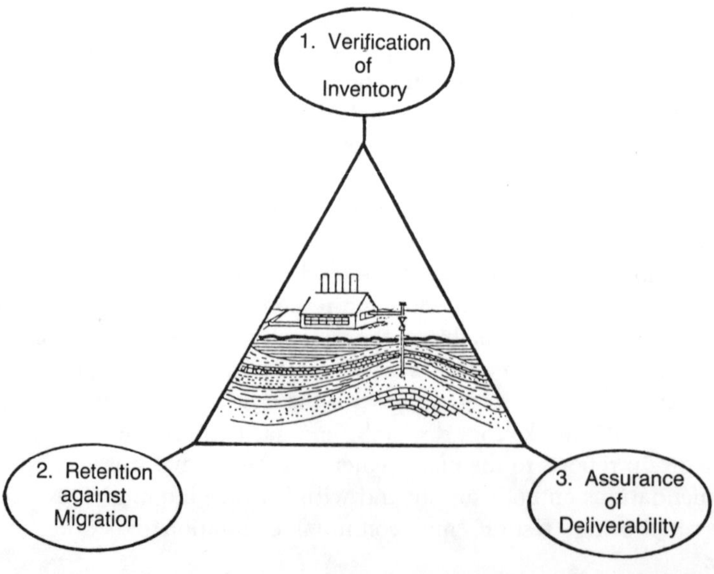

Figure 5-1. The three basic requirements in underground storage: inventory, migration, and deliverability.

tomary to differentiate the gas in and out each season from the rest, which provides the basis for the cyclic process. Accordingly, total gas in storage is divided into two basic components, as shown in Table 5-1. These are:

1. Top gas (or working storage)
2. Base gas (or cushion gas)

Table 5-1
Performance Attributes Related to Capacity, Operating Parameters and Efficiency

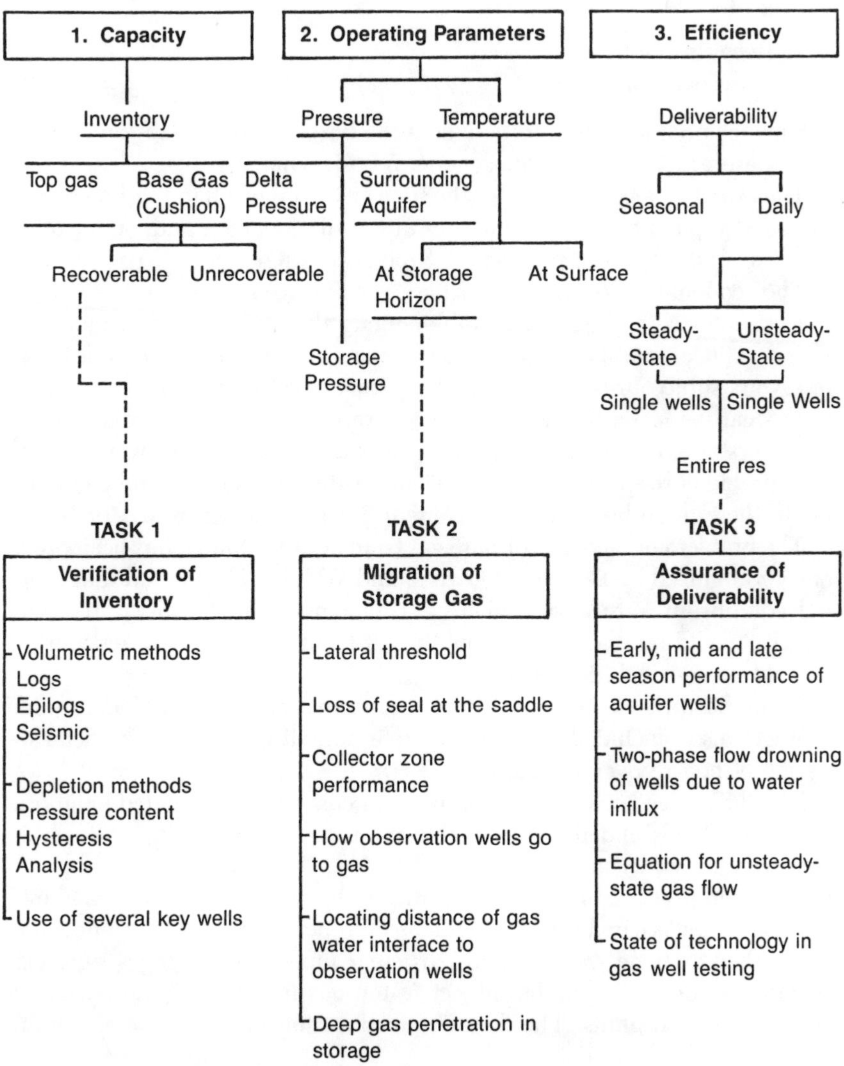

Table 5-1 relates the three performance attributes to these three familiar engineering characteristics: capacity, operating parameters, and efficiency.

Recoverability of Inventory

The top gas that can be produced varies from season to season with the demand imposed upon storage gas by weather. The maximum amount of top gas that can be produced is usually designated *top gas capability*. The base gas, which by virtue of its presence in the reservoir provides the pressure level necessary for the deliverability of top gas, is sometimes viewed as having two components:

1. Recoverable base
2. Unrecoverable base

Some companies prefer to designate the base gas as a direct function of some base pressure, be it atmospheric, 50 psi, whatever.

The extent of recoverability of the base gas is, on one hand, a function of surface equipment, compression, pressure regulation, etc.; and, on the other hand, a function of the properties of the porous reservoir matrix.

The economic recoverability of base gas has been the subject of several research efforts in the past. Notable among these are lab experiments by Cierici, Ciucci, and Long[1] on gas saturations behind the water front in gas reservoirs subject to water drive. They concluded that, while reasonable ranges can be given for residual gas saturations in consolidated sandstone and limestone, no correlation could be found to relate the recoverability to measurable pore-space properties. Their results were similar and comparable to those published earlier by Geffen, Parrish, Haynes, and Morris.[2]

The problem of recovery of base gas from aquifer storage has been previously recognized by Gardner, Downie, and Wyllie.[3] They noted that residual equilibrium saturation, where gas remains immobile in the reservoir matrix, would be that which would result from capillary hysteresis on the first drainage-imbibition cycle.

More recently, an AGA project on "Unrecoverable Gas in Underground Storage" also concluded that ". . . there is no uniform method for determination of non-recoverable gas. . . ."

The difference between recoverability of cushion gas indicated in laboratory core floods and that reasonably projected from field operations has been recognized by many of the investigators discussed above.

In actual practice of storage operations, the need to have a reliable estimate for top gas capability is essential for proper planning. Every now and then when extreme weather and particular pipeline storage gas supply so dictates, a reservoir may be subject to top gas (or working gas) demands near its physical limits. The data obtained during such extreme operations are valuable for extrapolation into ultimate top gas capability. In fact, some reservoirs with which the author has been involved permit constant revision of their top gas ratio capability by analysis of field data obtained during special *season's end* tests. The practice of such season's end tests in actual late season peaking conditions permit meaningful extrapolation on all of the wells turned on to full flow just before the turnaround for injection season. The methodology for this evaluation is directed primarily toward the analysis of such field data. After all, if we know by theory of flow and by actual field tests how wells will perform, the sum total of the performance of all wells should tell us a great deal about the capability in the field.

Aside from direct analysis of storage field data, some critical laboratory experiments may have to be performed to shed light on two-phase flow phenomena taking place in the interpore space.

As aquifer storage reservoirs mature and approach their size objectives, their expected performance under near steady-state conditions becomes important. During the early stages of growth, the development of a cohesive gas bubble, drainage of water between gas saturated streamers and displacement of some water from pore space by injected gas results in an increase of the amount of gas that can be withdrawn during a given season. The gas produced during the winter, commonly referred to as *top gas*, is a definite fraction of maximum inventory before withdrawal.

The past performance of aquifers being developed indicates that during any particular season, their ability to produce a fraction of their maximum inventory is constantly enhanced as the size of the reservoir grows. The physical reason for this is continually increasing permeability to gas, more continuity of flow paths available to gas in the drainage matrix, and gradually larger reservoir thickness available for flow through porous media.

As storage reservoirs are, in general, heterogeneous, the areas between the wells are subject to varying ranges of permeability to gas and, as such, respond variably to the call for gas by the pressure drop imposed upon the field gathering system. As the gas storage bubble grows during the development of the field and the drainage areas become drier and more continuous, it becomes possible for the wells to assist one another in response to demand for production. During each year, as the flow is reversed from injection to production, some gas which fingers out into water-saturated rock gets entrapped and become immobile. This phenomenon occurs less frequently as the gas bubble gets larger, drier, more cohesive, and more homogeneous. On the other hand, some part of the gas penetrates regions of very low permeability during the long injection season and reaches areas from which it becomes impossible for it to return during the relatively short producing season. In addition, some gas tends to be adsorbed at solid surface while some either gets dissolved in the ever present brine or moves and diffuses through cracks, fissures, faults, etc. Some gas penetrates and becomes immobile in dead-end pore volumes. Another part might move into shale environment from which it cannot return. It is also well known that some of the injected inventory migrates away from the main gas bubble long distances down-structure. Such down-structure gas occasionally gets physically separated from the main gas bubble and eventually moves into shallower areas where, by virtue of its inherent delta pressure, it grows in size. The sum total of these various complex processes causes a rather large fraction of total inventory to be present in the reservoir but not readily recoverable. The gas present in the reservoir which by its presence makes the top gas recoverable, is usually referred to as *base gas* or *cushion gas*. In highly permeable, dry, naturally occurring gas reservoirs, most of the base or cushion gas is recoverable. In comparison, in aquifer storage reservoirs, the cushion gas may be only 10% to 30% recoverable. In tight and heterogeneous storage reservoirs, it may be recoverable only in the 10% to 20% range. In usual aquifer storage sands with highly permeable matrices, it may be recoverable to the extent of 30% to 50%. In carbonate reservoirs and tight limestones it is somewhat less recoverable. During each year of reservoir growth, the net injection must not only replenish the previous year's production (removed top gas), but must also provide for the planned growth of both top gas and base gas. Problems related to phase distribution in the pore space also depend critically upon our understanding of connate water, structural geometry, and heterogeneities.

As storage reservoirs become mature and reach a near steady-state, the gas produced every year must be replenished during the injection season. In practice, this is accomplished by balancing the pound × days in and out of the reservoir.

Under these conditions, it is known that zero-growth of the inventory results in some depressurization of the delta pressure donut developed in the vicinity of the storage bubble. This depressurization in turn causes both top gas and cushion gas to expand. When they expand, they invade additional pore volume where some top gas transfers into unrecoverable status. At first glance, this would seem to cause some decline in top gas ratio capability. However, it is also possible that some previously unproduceable cushion gas may be expanding to connect along a path leading to the wellbore. That would transfer some cushion gas back to top status. Understanding of such phenomena by way of simply designed laboratory experiments and interpretation of field data to assess likely changes in top gas ratio capability are among areas of ongoing research.

Sometimes the storage inventory is inevitably shrunk due to our inability to replenish the previous season's withdrawal. When that happens, the net effect on top gas ratio capability becomes crucial. The erosion of ultimate top gas ratio capability due to curtailed growth is an important area of concern to the operating storage engineer.

VERIFICATION OF INVENTORY

The verification of storage inventory has been an area of continuing interest probably for as long as the history of underground storage.

The problems with verification of inventory in storage are analogous to those in estimating reserves in production. As such, they must be approached independently and simultaneously by *volumetric* and *depletion* type methods. Only if and when the two are reconciled, the stored quantities can confidently be expected to be there to perform their intended functions. There have been new methods, tools, and developments on volumetric survey of the storage inventory. These include logs, epilogs, and the recent downhole seismic and color-coded seismic interpretation techniques. In the area of depletion methods, the pressure content hysteresis has been of interest during recent years.* Through proper volumetric integration of pressure data and use of computer data processing techniques, it may even be possible to preempt costly biannual shut-in pressure surveys. More recently, use of pressure content data from several key wells (instead of one) indicated the promise of permitting the calculation of more representative volumetric average pressures. This work, applicable to dry gas reservoirs, has the potential of being applicable to reservoirs with significant amounts of water drive, including the aquifer storage fields.

Specific procedures for verification of inventory are included in Appendixes G and H.

Migration in Storage

The likelihood of migration of storage gas is of paramount interest in all underground storage reservoirs. In the past, considerable attention was given to vertical migration past and possibly across caprocks. The work on threshold pressure in storage sets the industry standards for such migration.[5] What was not done during that study, however, was to pay equal attention to conditions that could lead to horizontal migration, such as "threshold of stratigraphic traps" and possible loss of hydraulic seals at the saddle. The manner in which observation wells go to gas, the nature of deep gas penetration, and the hydraulic performance of collector-zone displacement are among problems of interest to the storage engineer. Of particular interest is the procedure developed for calculating the location of gas/water interface from storage pressure and observation well water level data.[6] The results in actual cases tried so far have been so encouraging that it appears that the classic problem of triangulating leaks in storage reservoirs may be pursued with a reasonable chance of success. The methodology of Reference 6 in-

* Special reference is made here to the AGA Pipeline Research Project on Inventory-Migration-Deliverability at the University of Michigan during the years 1980–1984.[10]

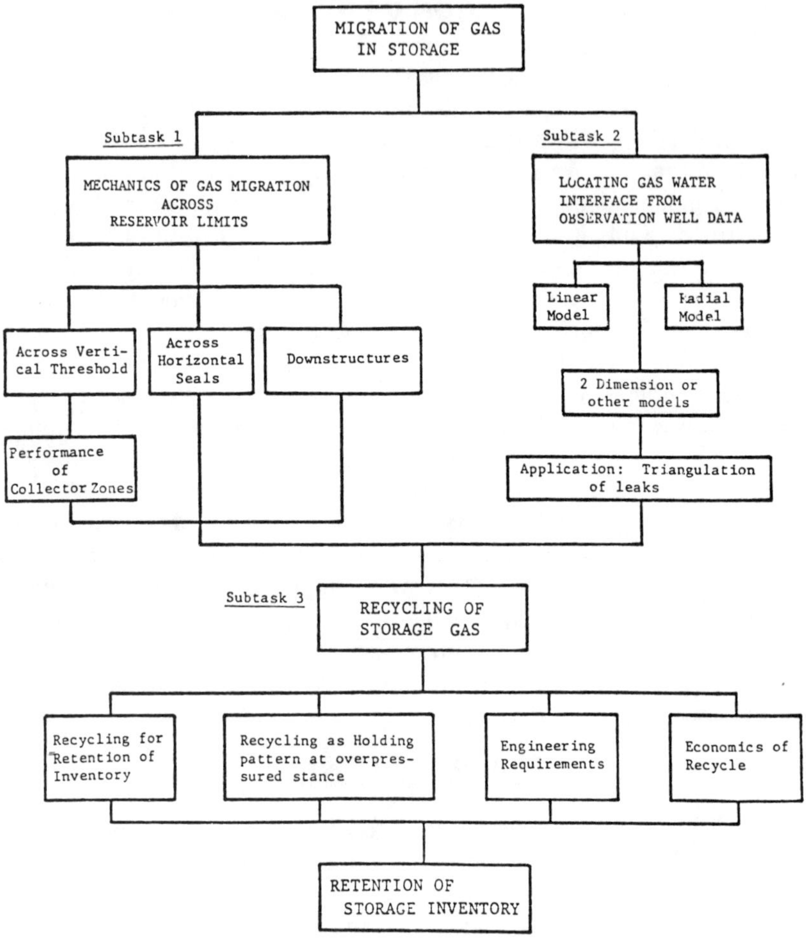

Figure 5-2. Topics in migration/retention.

volves mathematical analysis of field data, models for computer implementation, and evaluation of the procedure against field data.

An overview of tasks and topics in migration/retention is shown in Figure 5-2. Some of these relate to the study of the mechanics of gas migration across reservoir limits. Others are specifically directed to locate the edge of gas storage bubble from water level data on distant observation wells. The top half of Figure 5-2 relates to diagnostic monitoring of gas migration. The bottom task relates to several aspects of recycling as a remedial, prognostic measure.

The work of Udegbunam[6] suggests that observation well data from several locations may be helpful in locating leaks by triangulation.

Equally as important and significant as monitoring are methods related to engineering economics of retention within protective acreage by means of recycling. Recycling of storage gas is practiced in areas where, due to channeling or directional permeability, the gas has a tendency to migrate in a particular direction. Under these circumstances, the purpose of recycling is to retain the inventory against the tendency to migrate.

Sometimes, in the operation of underground storage reservoirs, adverse weather necessitates holding the inventory at overpressured condition during a mild winter. With rather large pressure gradients near the edge of the field, certain areas may be subject to recycle in order to retain the gas from migrating out. Engineering requirements and economic constraints would then provide options and alternatives on how much and where to recycle.

The area of deliverability of gas from storage is a logical extension of the two previously discussed attributes in storage.

The inventory is related to capacity and to thermodynamic equilibrium content. Problems of migration as well as prospects for containment are related to pressures and temperatures as they prevail in the reservoir. Those are operating parameters otherwise referred to as *thermodynamic coordinates*. By analogy, one may perceive the concept of deliverability as a measure of storage reservoir efficiency, since it relates directly to the ultimate intent of underground storage, i.e., cyclic deliverability of gas.

The daily and seasonal deliverabilities of both dry and wet storage wells operating in steady- or unsteady-state mode are both of interest for optimum operation of storage fields.

Deliverability as Related to Inventory and Migration

In overall perspective, the triad of storage performance attributes are each interrelated two-by-two. The "inventory" is, basically, a thermodynamic quantity. It is related to the amount of gas in storage. Its verification is approached by two independent concepts: *volumetric* and *depletion*. *Migration*, by contrast, is a fluid dynamic concept. Its containment and its monitoring may be described as rate processes. Similarly, *deliverability* is a measure of rate at which gas can be withdrawn from storage or returned to storage. It relates to steady- and unsteady-state flow of gas through porous media. Inventory is like money in the bank, while deliverability is comparable to cash flow in deposits and withdrawals. Migration is similar to unearned profit or passive income tax which causes attrition of inventory.

Coming back to the storage reservoir, the deliverability from the wells is related to inventory because it is a function of the pressure level in the reservoir. Obversely, the migration is related to deliverability and vice-versa. As we know, excessive injection rate or high pressure gradient in the gas bubble behavior will cause a migrating finger to develop and carry the gas across the threshold beyond the intended storage horizon. Furthermore, migration is related to inventory because it takes some unmetered inventory away from storage.

In storage, the ability to deliver the contracted gas to market is maximal at the turnaround from injection to withdrawal. Early in the producing season, therefore, deliverability from storage reservoirs usually exceeds the demand imposed upon it by gas dispatch. In contrast, in late winter when inventories are low, peak loads and high demand from dispatch occurs. There are existing procedures and strategies for meeting such an unfortunate combination as described above. Use of supplementary peak shaving facilities and resources, high well density, close spacing, infield drilled areas saved for late season, line packing, and gas swaps are among the alternatives.

From a strictly operational viewpoint, better understanding of problems related to deliverability lead to better design and selection of equipment, as well as procedures optimized to meet the challenge of "peak day."

Aspects of Deliverability in Need of Research

Study of deliverability involves review of fundamental concepts in flow of natural gas through porous media. The major areas of fundamental concepts affecting deliverability of gas, as they relate to each other, are shown in Figure 5-3. Central to fundamental concepts are the phenomenological relations involved in Darcy and non-Darcy flow regimes. While no new

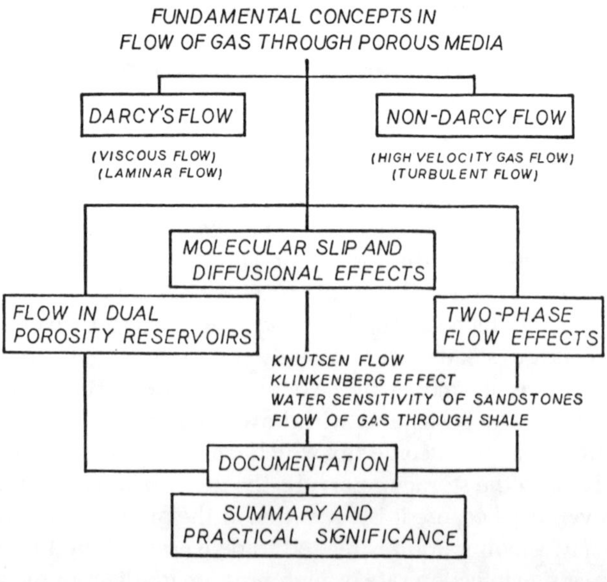

Figure 5-3. Major areas of fundamental concepts affecting deliverability.

knowledge is likely to be developed in the area of fundamental concepts, clear understanding of the various requirements and a documentation of their practical effect in gas well deliverability is important. In addition, such concepts as Knutsen Flow, Klinkenberg Effect, water sensitivity, and presently available information on flow of gas through shale need to be reviewed, investigated, and documented. In studying problems of deliverability, one area that does not appear to have received sufficient attention is gas flow in dual-porosity reservoirs. Whether the dual porosity is attributed to incipient fractures or vugular matrices in limestone or other carbonate rocks, propagation of unsteady-state effects in heterogeneous environment is of continuing interest to the storage engineer who must continue to meet the unsteady and ever-changing demand of dispatch. In underground storage reservoirs subject to water drive, deliverability problems related to two-phase flow are frequently encountered at the end of the withdrawal season. With excessive water encroachment, the wells occasionally load-up and water-out. With the present knowledge and understanding of two-phase flow, the practical implications for storage deliverability must be understood, described, and documented.

Figure 5-4[10] shows the methods and procedures related to deliverability in overview with both injection and withdrawal operations in perspective. The rate of deliverability from a gas well draining a porous and permeable matrix is an unsteady and non-linear process for which a general analytical solution is not available.

Our present understanding of gas well performance involves the concept of *real gas potential*, or pseudo-pressure, which provides an effective transformation to reduce the non-linearities due to pressure dependence of the compressibility, compressibility factor, and the viscosity of natural gas. The methods and procedures for unsteady-state effects are not as well understood or routinely used by storage engineers as the steady-state approximations or back-pressure test oriented empiricism. Once it is recognized that underground storage problems are basically unsteady-state with time varying boundary conditions, particularly in the withdrawal mode, it becomes important to translate the theory of unsteady-state real gas flow to practical field procedures that would be easy to follow and understand.

Of particular interest in aquifer storage is the productivity decline that is encountered as the season progresses. As the water saturation in the drainage matrix changes from nearly dry at turnaround (to withdrawal) to maximum water influx at the end of the producing season, the performance of certain wells becomes variably affected.

Operation of these wells sometimes requires special attention for maximum deliverability. The deliverability of several wells viewed collectively as that of the field must be investigated in both injection and withdrawal cycles. The deliverability from the storage field, time integrated into the so-

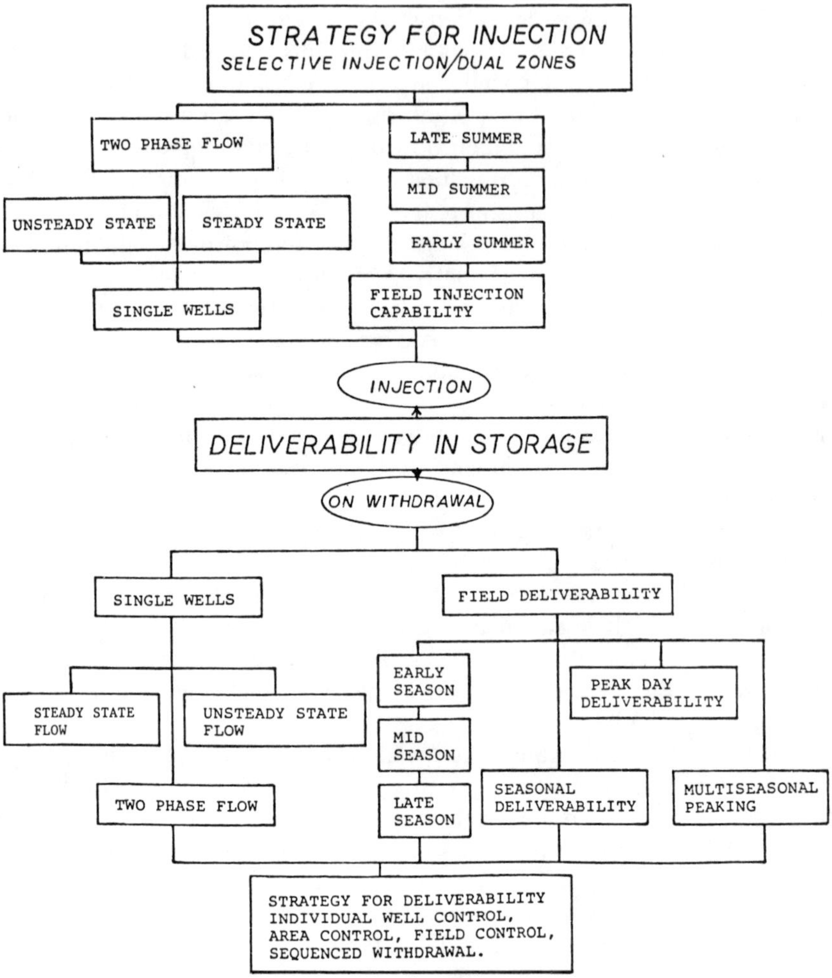

Figure 5-4. Methods and procedures related to deliverability.

called "top gas" performance is important not only for peak day commitments, but also for seasonal budgeting of cyclic gas and multiseasonal peaking requirements.

As depicted in Figure 5-4,[10] the bottom lines on withdrawal and on its mirror image, injection mode, are strategies that would result from understanding field deliverability characteristics. Such strategies, already suggested from experience in the operation of certain fields, are: individual well control, area control, selective injection, optimized pattern injection, sequenced withdrawal, special spacing patterns for edge wells, and in-field drilled areas.

Effect of Surface Facilities on Deliverability from Storage

In underground storage operation, one area that merits interest and attention is the link-up between the subsurface capability and surface facilities. As the flow of gas takes place between the storage environment and the wellbore, before it eventually passes through the residential or commercial meter, the performances of several systems designed and operated by the storage engineers interact with one another. In consecutive sequence these include:

1. Wells
2. Gathering system (with or without brine separation, water disposal, individual heaters)
3. Field separators, slug catchers
4. Dehydration
5. Pressure regulation
6. Compression
7. Metering
8. Pipeline to market
9. Distribution—intermediate and low pressure regulation
10. Customer meters

The succession of the several elements listed above are not necessarily always in series. Parallel loops, sectionalized arrangements, and by-pass provisions often further complicate the picture. Our engineering understanding of the physical performance of each unit-operation involved is well established in theory and practice. Yet, often the surface capability lags behind that of reservoir deliverability.

A smoothly operating, optimized design with adequate fail-safe protection against component malfunction requires analysis of the system as a whole for the purpose of deliverability. The number of wells to be turned on and off in order to adjust the individual well deliverabilities to ever-changing dispatch request from gas control requires decision making based on well productivity as well as wellbore deliverability. Combining individual well productivities with wellbore to well head pressure drops involves a large number of nonlinear algebraic equations with many unknowns. Applications of numerical techniques through the use of computer programs need to be entertained to obtain solutions and analyses for optimized deliverability.

Another area where there is a coupling wellbore capability with variable gas demand is compressible flow and unsteady-state effects in surface equipment. Problems and prospects in line packing to anticipate demand and real-time computer control of gas dispatch are areas of interest for further research.

Recently, new contributions were made to our understanding of wellbore temperature distributions in both injection and production. On withdrawal, as the gas cools from bottom hole to well head, the desirability of flow through casing-tubing annulus with surface controlled bottom-hole chokes were suggested in Chapter 4 as a possible means to preempt some surface equipment such as heaters, regulators, and dehydrators. The matter received substantial interest but not sufficient attention, and the idea has not been sufficiently explored for definite evaluation as of this writing.

Verification of Inventory Using Pressure-Content Hysteresis

Verification of inventory is similar to estimating reserves.[10] It can be accomplished by volumetric calculations or by analysis of production-pressure depletion data. The two methods often provide cross-checks on one another.

The volumetric calculation involves estimates of the net pore volume occupied by storage gas and either calculation or direct measurement of equalized reservoir pressure. Pressure surveys are conducted in general at times corresponding to turnaround to injection and/or withdrawal seasons.

If one designates the equalized reservoir pressure as p, then the number of moles present in the storage unit can be calculated from:

$$pV = znRT_R \tag{5-1}$$

where p = Pressure, psfa
z = Corresponding compressibility factor, dimensionless
V = Net pore volume occupied by storage gas
n = Number of moles present in storage horizon (1 mole = 379 SCF)
R = Gas constant, 1,545 ft × lb_f, lb-moles × °R abs
T_R = Reservoir temperature, °R abs.

When data on individual pressures are available from several wells, the p, thermodynamically meaningful equilibrium pressure reflecting the inventory may be calculated as follows:

$$p = \frac{\sum_{i=1}^{N} p_i h_i A_i}{\sum_{i=1}^{N} h_i A_i} \tag{5-2}$$

or

$$p = \frac{\int_A phdA}{\int_A hdA} \tag{5-3}$$

p is also called the *volumetric average pressure* representing the inventory. In Equations 5-2 and 5-3, h denotes the net formation thickness and A represents the drainage area around each well. The integer, i, represents each well for which data are available. N is the total number of such wells.

Use of Single "Key" Indicator Well

Frequently, based on past practical experience, one well is designated as a *key well* or *indicator well* because its surface (or bottom-hole) reading correlates with the prevailing average reservoir pressure. Such a well, preferably located high and dry on structure and kept at static condition, tracks the inventory pressure data through its p/z readings:

$$\left(\frac{p}{z}\right) = \frac{RT}{V}\,\frac{10^9}{379}\,Q \tag{5-4}$$

In Equation 5-4, Q is the total inventory present in BCF, and V the actual pore volume in cubic feet.

Tracing p/z vs. Q in the pressure-content quadrant permits verification and auditing of the inventory directly by tracing the same path during different cycles. If the storage unit performs volumetrically; i.e., V = constant, the path becomes a straight line which passes through the origin when extrapolated.

Figures 5-5a, b, and c represent the cases of no loss, finite loss, or negative finite loss in inventory. The negative finite loss or, rather, gain in inventory is sometimes observed in several tiered reservoirs or overburden collector zones when the gas from the deep horizon moves on to some shallower formation in finite quantity. The p/z vs. $Q_{metered}$ trace moves to the right in the event of loss and to the left in the event of gain. If the reservoir volume does not change, its slope remains constant.

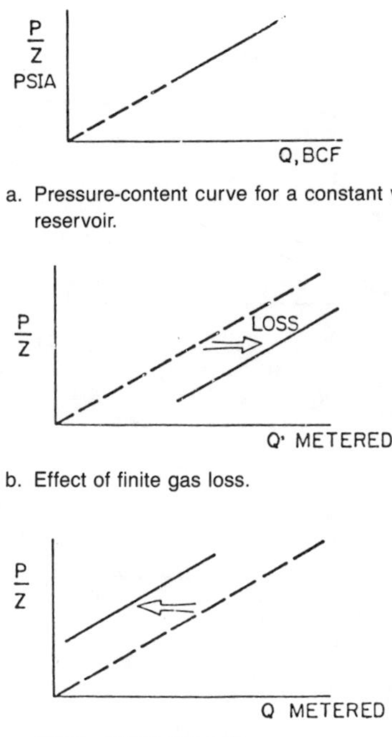

a. Pressure-content curve for a constant volume reservoir.

b. Effect of finite gas loss.

c. Effect of finite gas gain.

Figure 5-5. Pressure-content plots for constant volume storage reservoirs.

Figure 5-6[10] shows the pressure-content curve for a reservoir with no loss but subject to enlarged volume. The opposite is shown in Figure 5-7[10], a case illustrating again no loss but effect of shrunk reservoir volume. Figure 5-8[10] shows the combined effect of finite loss and shrunk reservoir volume.

Even in nearly-constant volume reservoirs, the behavior of p/z vs. Q trace does not strictly follow a reversible straight line path. It traces a loop called hysteresis loop as shown in Figure 5-9[10]. During each injection cycle, p/z vs Q trace stays above the straight line. During each withdrawal cycle, the trace stays below the ideal straight line. There are several reasons for this. The major reason is that the pressure surveys are, in general, not sufficiently long to permit measurement of truly equalized reservoir pressures.

The pressure data taken after a period of injection tend to manifest the pressure cone developed around the injection wells before it dissipates into an equilibriated pressure condition. The opposite occurs during the withdrawal season, as the pressures reflecting the withdrawal of storage gas develop deep sinks around the wells. In general, when the field is shut-in for pressure survey, there is not sufficient time for the pressures to build up to the equilibrium value. Accordingly, they consistently show p/z to read below the straight line representing equilibrium. The opposite occurs after injection.

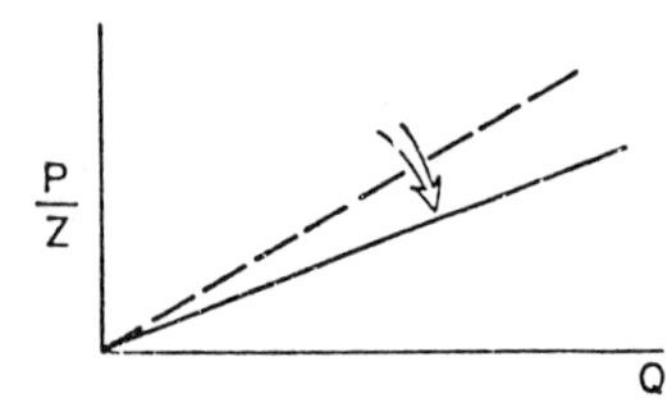

Figure 5-6. Pressure-content curve for a reservoir subject to volume increase.

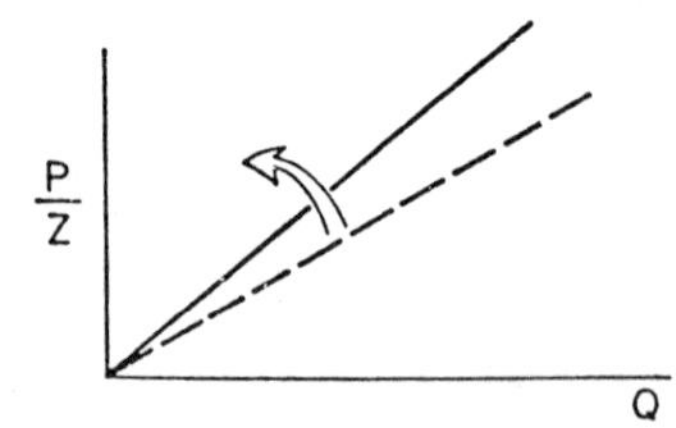

Figure 5-7. Pressure-content curve for a reservoir subject to volume decrease.

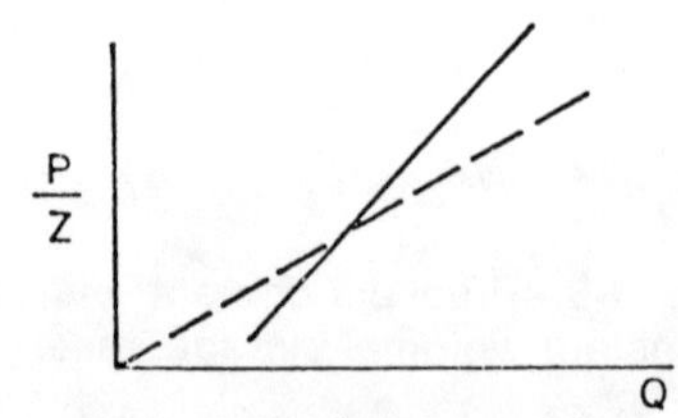

Figure 5-8. Pressure-content curve for a reservoir subject to volume decrease and finite loss.

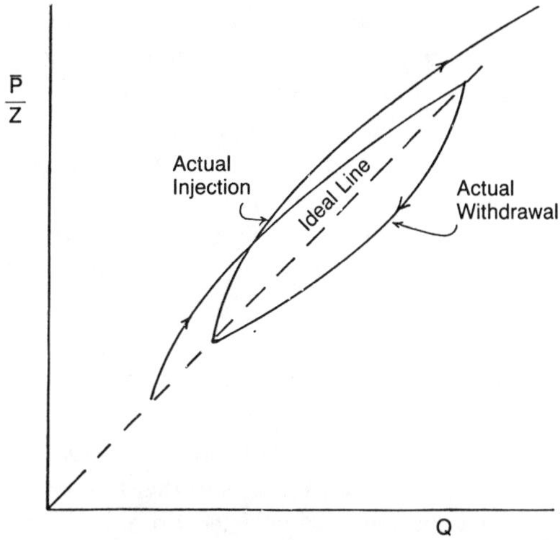

Figure 5-9. Pressure-content pattern on a volumetric storage reservoir showing hysteresis.

There are other effects, some of which are related to change in pore volume due to movement of water in and out of the gas bubble. Others are the results of errors in treatment of data, such as ignoring the necessity of volumetric averaging, using only surface data, or, in some cases, neglecting the effect of standing water leg at the bottom in calculating the bottom-hole pressures corresponding to measured surface pressures.

Pressure-Content Behavior Observed in Aquifer Storage

The pressure-content performance of an aquifer becomes more complicated due to the movement of water which takes place variably during the storage cycle. In aquifer storage, as the production resumes each winter in response to gas dispatch, it may be seen that sometimes little water movement takes place until later. The early portion of the withdrawal curve, therefore, appears nearly straight from upper right to lower left, as shown in Figure 5-10a. Late in the season, however, as more and more water influx occurs into the gas bubble, the pressure is raised due to shrinking gas-bubble volume. This causes the curve of p/z vs. content to tangentially curve to the left of the straight line. The early injection usually takes place against still incoming aquifer brine, giving a higher slope to p/z-content curve before resuming a linear trend from lower left to upper right as shown in Figure 5-10b. Near the end of the injection season, as more water is being pushed away from the gas storage bubble, the increase in reservoir pore volume causes the pressure curve to become less steep than shown in linear trend.

The extent of the injection season into late fall depends upon the weather. If the injection ceases sometime before the demand for gas materializes, or continues at a low rate, the movement of water away from the gas bubble slows down to a bare creep. Under these conditions, if and when the production of gas resumes, there is little if any water movement into the storage matrix. This results in the nearly linear decline shown in Figure 5-10a. On the other hand, if vigorous injection had continued until the last day of the injection season, the production operations, when resumed, would face a still outgoing aquifer. This particular operation causes a steep pressure decline early into withdrawal before resuming a linear trend some time later. This type of performance is shown as portion AB on the pressure-content curve of Figure 5-10c.

The near-parallel trend observed between corresponding injection or withdrawal branches of the p/z curve is indicative of the year-by-year growth of storage reservoirs being developed on aquifers.

Figure 5-11[10] depicts the various phases of a stable and mature aquifer storage performance. Branch A represents the ideal vertical pressure rise af-

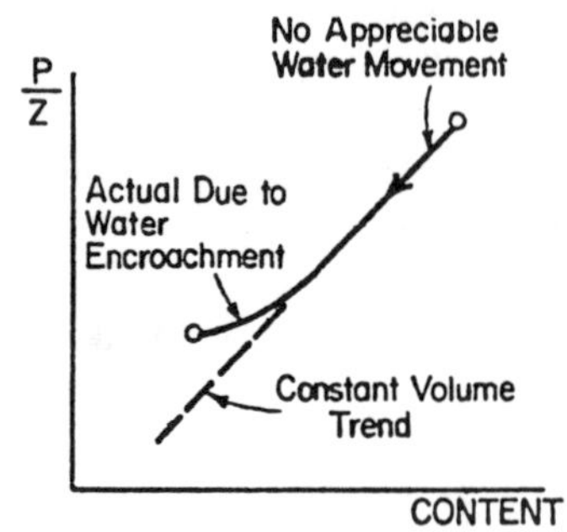

a. Production.

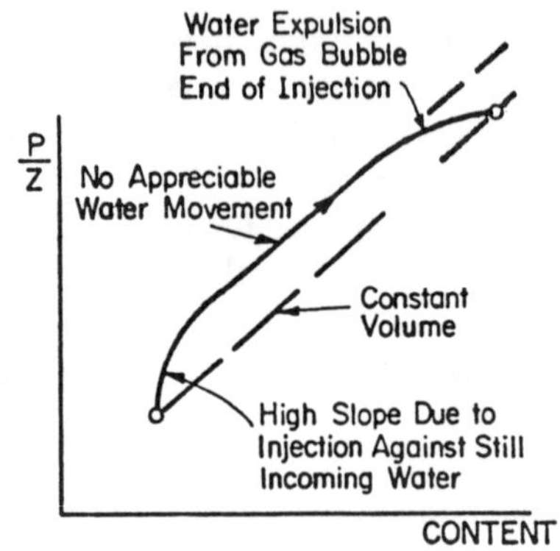

b. Injection.

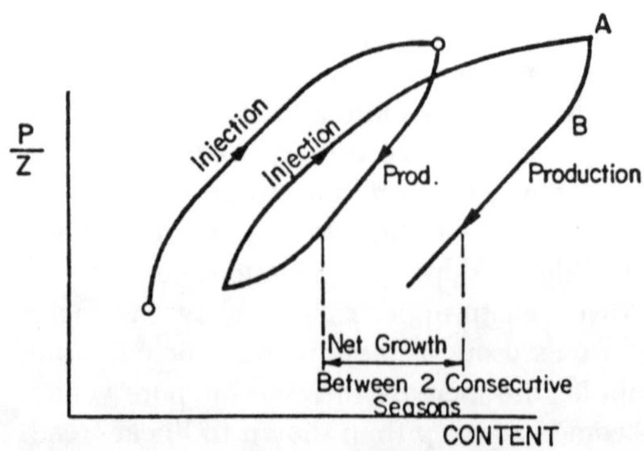

c. Composite pressure-content hysteresis.

Figure 5-10. Pressure-content relationship observed on aquifers.

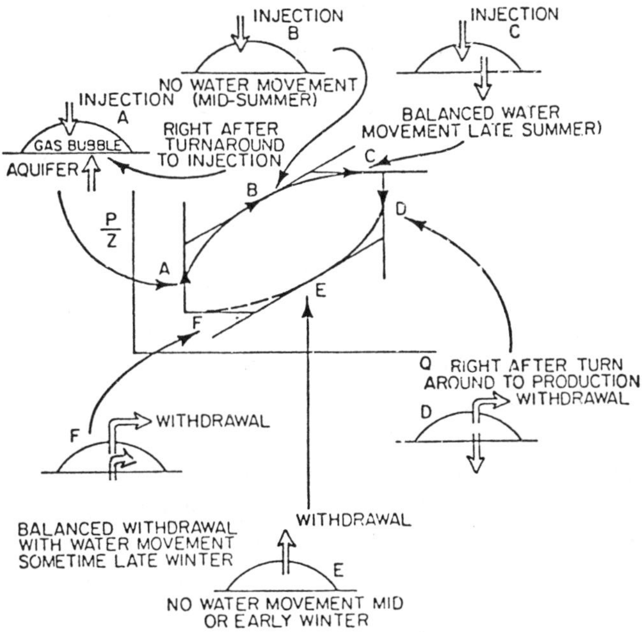

Figure 5-11. Various phases of a stable and mature aquifer storage.

ter turnaround as water is still invading, while the well is shut-in, and shortly after it is committed to injection against incoming water.

Branch B represents the nearly constant volume performance that occurs tangentially when water movement stops before turning around and starting to go away from the gas bubble.

Branch C is the performance observed at the end of summer, when the pore volume equivalent of injected gas balances the pore volume vacated by effluxing water. The inventory change does not then involve appreciable pressure change.

Branches D, E, and F represent diametrically opposite phenomena to A, B, and C upon withdrawal.

The area of the curved hexagon is the hysteresis loop for the aquifer. The area is observed to become large when the permeability is small. Figure 5-12 shows the pressure-content performance on an aquifer subject to both growth and leak.

Use of Several Key Indicator Wells

Whenever more than one key observation well is available and there is not appreciable water drive, use of the computer permits reasonably accu-

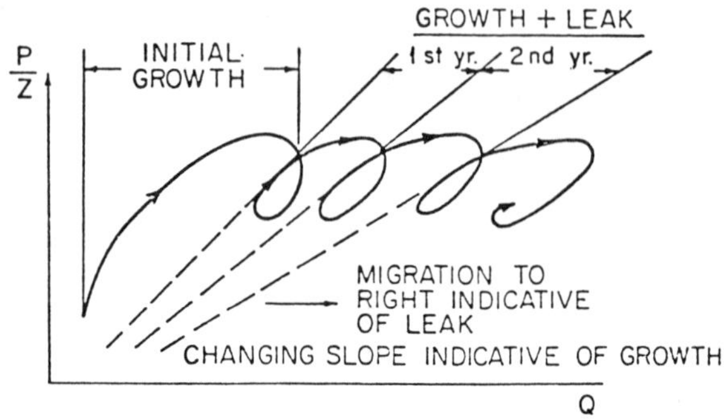

Figure 5-12. Pressure-content performance in an aquifer storage subject to growth and leak.

rate calculation of equalized reservoir pressure by determining a proper weighting factor for each individual pressure reading.

If the equalized reservoir pressure is referred to as $\bar{p}$ (bottom-hole, psia) and individual observation well pressure as p_i, one may formally set:

$$\frac{\bar{p}}{\bar{z}} = \sum_{i=1}^{N} \xi_i \frac{p_i}{z_i} \tag{5-5}$$

where ξ_i is a fractional quantity determining the extent to which the pressure in drainage area A_i affects the eventual equilibrium pressure $\bar{p}$ for the field.

Now, combining Equations 5-4 and 5-5:

$$\frac{\bar{p}}{\bar{z}} = \frac{RT_R}{V} \frac{Q\ 10^9}{379} \quad Q \text{ in billions} \tag{5-6}$$

$$\sum_{i=1}^{N} \xi_i \frac{p_i}{z_i} = \frac{RT_R}{V} \frac{Q \times 10^9}{379} \tag{5-7}$$

The coefficients ξ_i are equal to the ratio of pore volume associated with the drainage area of well "i" and the total pore volume in the storage horizon.

Equation 5-7 may be rewritten as:

$$Q = \sum_{i=1}^{N} \frac{379V}{RT\ 10^9} \xi_i \frac{p_i}{z_i} \tag{5-8}$$

If we let:

$$\beta_i = \frac{379V}{RT_R\ 10^9}\ \xi_i,$$

From the above:

$$Q = \sum_{i=1}^{N} \beta_i \frac{p_i}{z_i} \tag{5-9}$$

Having a number of observation wells available, providing p_i values for each well at a time when the present inventory is Q (z_i being directly calculable from p_i, gravity, or composition of gas and the reservoir temperature), Equation 5-9 may be regarded as a number of linear equations with N unknowns; i.e., i = 1, 2, 3, N. The number of equations is simply the number of times data points are available in p_i and Q's.

As an example: if five observation wells distributed throughout a storage field provide the pressure and inventories for twenty different times during, say, two years of operations, the equations look like the following:[10]

$$\beta_1 \frac{p_1^{(1)}}{z_1} + \beta_2 \frac{p_2^{(1)}}{z_2} + \ldots\ldots\ \beta_5 \frac{p_5^{(1)}}{z_5} = Q^{(1)}$$

$$\ldots\ldots\ldots\ldots\ldots\ldots\ldots\ldots\ldots = Q^{(2)}$$

$$\ldots\ldots\ldots\ldots\ldots\ldots\ldots\ldots\ldots$$

$$\ldots\ldots\ldots\ldots\ldots\ldots\ldots\ldots\ldots$$

$$\beta_1 \left(\frac{p_1}{z_1}\right)^{(20)} + \beta_2 \left(\frac{p_2}{z_2}\right)^{(20)} + \ldots\ldots\ \beta_5 \left(\frac{p_5}{z_5}\right)^{(20)} = Q^{(20)} \tag{5-10}$$

In this set of equations (5-10), the superscripts refer to the number of times that inventory pressure data are available. The subscripts refer to the number of observation wells available.

This set of equations is a system of 20 equations with five unknowns, in which $(p/z)_i^{(m)}$ and $Q^{(m)}$ are fed to the computer as pressure-content data points.

The system is solved using a linear regression technique on the computer which provides the best set of values β_1, β_2, ... β_5 necessary for the calculation of the equalized $(\bar{p}/\bar{z})$ for the reservoir. The sum $\left(\sum_{1}^{N} \beta_i \left(\frac{p}{z}\right)_i\right)$ may be regarded as the calculated inventory to be compared with Q, the actual metered inventory.

The plot of such calculated vs. actual metered inventory is shown in Figure 5-13. It is a plot of calculated vs. actual metered inventories for a dry storage field A where 25 data points were available on 13 observation well pressures from data over a number of years.

As an added feature, the computer program selects the particular combination of 13 observation wells which best correlates the data in terms of the linear equivalence of calculated vs. measured inventories.

In this particular case, three wells (numbers 2, 4, and 9) do much better at representing the reservoir pressure at any time than all thirteen wells combined or any one well in particular. The pressure-content hysteresis exhibited by well number 5 is shown to be spectacularly reduced to almost a straight line when data are used from wells 2, 4, and 9 in conjunction with their specific β factors. The fact that the calculated performance is about 2.6 billion to the right of and parallel to a 45° line of perfect agreement is an indication of finite loss due to blow-out on early primary production in this field. It has been shown[10] that properly weighted pressure data from several

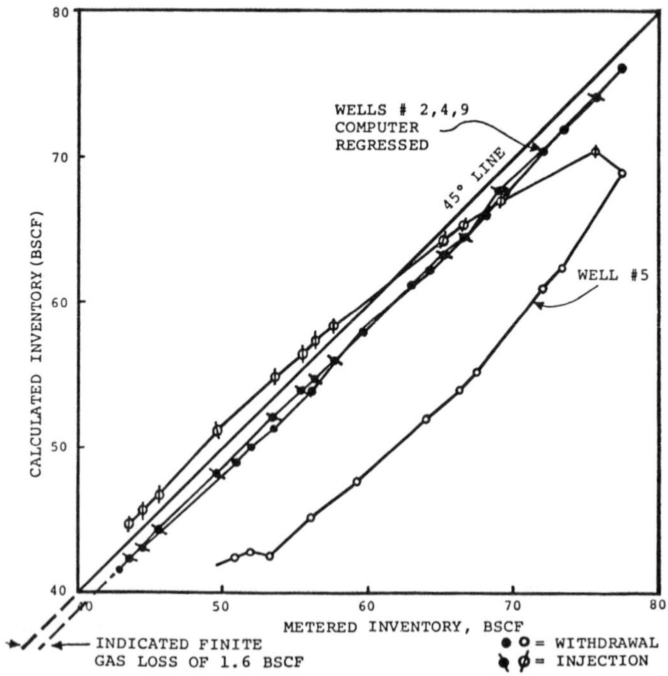

Figure 5-13. Effect of using several observation wells on pressure-content hysteresis.

wells provide a more realistic $\bar{p}/\bar{z}$ average for the field and reduce the extent of hysteresis manifested by pressure-content data from single observation wells.

Use of Edge Observation Wells in Inventory Verification

As an indirect means of inventory verification, observation of water levels from edge wells has been suggested in the past by Katz and Coats[7] and Katz and Tek.[8] Recent research on locating gas/water interface in storage has proposed mathematical models and special tests in which water levels recorded on edge observation wells are used to calculate distance and extent of gas fingering away from the storage bubble.[6]

METHODS FOR DETERMINING AND IDENTIFYING LEAKS

Underground storage reservoirs have sometimes been subject to migration and eventual loss of gas from the storage horizon. The amount of such gas lost is of interest to the industry for reasons related to safety, economy, and engineering. Unmetered gas escape from the storage environment could be due to many causes. It may be due to reservoir or caprock imperfection, mechanical imperfection, accidental blowout, pipeline leaks caused by corrosion, corrosion leaks in the wells, etc. Table 5-2 summarizes three major causes for unmetered loss of inventory. It is recognized that most of the losses or seepage in well heads, gathering systems, compressor plants, and pipelines are minor, subject to direct detection, and reparable or written-off. Unmetered loss of gas from underground occurs in two distinct areas:

1. Around the wellbores
2. In the reservoir away from the wellbores

Table 5-2
Causes and Areas for Unmetered Loss of Inventory

Wellheads	Wellbores	Reservoirs
Christmas Tree	Casing Collars	
Gathering System	(mechanical)	Fractures
Surface Equipment	Corrosion	Faults,
Compressor	(internal, external	Areas of Minimum
	or both)	Closure,
		Caprock thresholds
Pipeline	Cement Bond	Down Structure
	Leaks	
	Blowouts	Fingering

It is easier to detect, control, and repair leaks from wellbores. Several logging and surveying techniques are available and are effectively used in correcting wellbore problems. These include temperature surveys, noise logs, pressure testing of casing and tubing, cement bond surveys, corrosion logs, caliper surveys, and downhole photography. Remedial means involve cementing, liners, recompletion, corrosion control, and general equipment overhaul.

Reservoir leaks, on the other hand, are more difficult to detect and control. They may be due to wellbore-induced or incipient fractures, faults, saddles, caprock imperfections, diffusion, dissolution, and uncontrolled fingering. Remedial measures involve pressure control, locating collector zones, recycling, and monitoring. There have been just a few cases where the only remedy left was abandonment of the project for reason of safety and economy.

Use of tracers is sometimes helpful in determining the leak path, whether from the wellbores or across the caprock. Recently, isotope chemistry has been helpful in identifying storage gas which has migrated away.

Computer-oriented techniques related to history match studies are sometimes helpful in determining leaks from storage. Threshold leaks from caprocks can occur from areas with low threshold pressure after the extent of the gas bubble has grown sufficiently to reach the particular area. In this case, the history match on the computer becomes an aid in the timewise reconstruction of the leak event.

The research on inventory and migration at the University of Michigan has resulted in simulation of leaks from caprocks as distinguishable from those across wellbores.

If history of recorded inventory verification track (p/z vs. metered inventory) suggests time-dependent changes, it appears possible to determine:

1. Whether and/or when the reservoir is leaking
2. Whether the leak is across the caprock or from wellbores

Analytical methods to account for leaks have been developed in two modes.

Threshold Leaks across Caprocks or Reservoir Limits

If the storage reservoir is leaking at a location away from the wellbore, the uncontrolled movement and escape of gas may be across the threshold of the caprock or laterally across the reservoir limits. In either case, the flow of gas is from porous media to porous media.

In representing such a leak, the basic dry reservoir model is used as a starting point or benchmark for comparison.[9]

$$\sum_{i=1}^{N} \xi_i \frac{p_i}{z_i} = \frac{RT}{379V} Q \qquad (5\text{-}11)$$

Here "i" designates each observation well, whose number may be up to N, and Q is the inventory actually present in the gas storage bubble.

The actual inventory Q may be represented as the total metered inventory Q_m minus the total unmetered or lost inventory:

$$Q = Q_{metered} - Q_{lost} \qquad (5\text{-}12)$$

The cumulative amount of leak from the reservoir may be written as the time integral of the leak rate:

$$Q_{lost} = \int_{t_o}^{t} c_p \, (p^2 - p^2{}_o)^n dt \qquad (5\text{-}13)$$

where c_p = the performance coefficient and n = the slope of the back pressure equation describing the leak. Equation 5-13 simply implies that the rate of leak from the storage unit across the caprock or laterally into another adjoining porous media is a typical, empirical back pressure relationship through a certain performance coefficient cp and a slope term 'n.' In combining 5-12 and 5-13 with 5-11, one gets the following inventory pressure relation for the leaking reservoir:

$$\frac{p}{z} = \frac{RT}{379V} \left[Q_{metered} - \int_{t_o}^{t} c_p \, (p^2 - p^2{}_o)^n dt \right] \qquad (5\text{-}14)$$

If the leak is pressure-dependent, Equation 5-14 indicates that whenever the gas bubble pressure p exceeds p_o (the leak threshold), the trace of p/z vs. Q plot will cease to be linear and will trace a path which simulates a leaking storage performance. In selecting different values for cp and n, one can trace many different p/z vs. $Q_{metered}$ curves. This will provide a *type curve* which should be helpful in diagnosing the existence and nature of the leak. Equation 5-14 has been programmed and plotted on the computer for a simulated case assuming a reasonable set of values for c_p, p_o, and n. The result is shown in Figure 5-14.

For the case illustrated in Figure 5-14, with "n" assumed to be equal to 1 (Darcy Flow), for three years of injection/withdrawal operations with gas bubble pressure p exceeding p_o part of the time, the curves generated indicate distinct linear behavior on both injection and withdrawal periods as long as the pressure p is below p_o, the leak threshold pressure. By contrast,

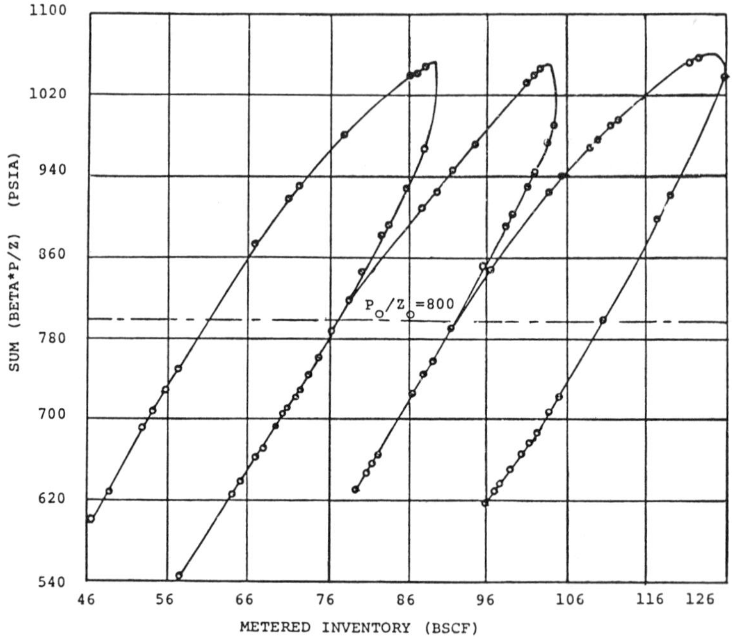

Figure 5-14. Simulation of leaks in caprocks.

when p exceeds p_o, the trace becomes a curve with a saw-tooth trend migrating consecutively to the right.

Figure 5-15 shows the effect of leak threshold p_o on p/z vs. metered inventory trace for various cases of leak as compared to the base case of infinite leak threshold pressure.

By generating several different sets of simulated leak type-curves for different n, p_o, and c_p values, one can not only diagnose a leak, but also determine the nature of it. If the method indicated above is used in conjunction with detecting leaks from aquifers, a special effort must be made to differentiate the effect of water movement and of the rate of growth of storage gas, since both would also tend to make the pressure-content trace nonlinear, as shown in Figures 5-14 and 5-15.

Leaks from Wellbores

If the leak is occurring through one or more wellbores, one or more corrosion pinholes, or imperfect casing collar joints, the nature of flow is from a void space (inside wellbore exposed to storage pressure) through an orifice-like passage to a low-pressure porous environment. The flow is compressible. The pressure upstream is nearly equal to the storage pressure. The pressure downstream depends upon the depth and the local subsurface brine gradient. In most circumstances of wellbore leak, the ratio of downstream pressure to upstream pressure is equal to or less than the *critical pressure ratio* for storage gas ($K = c_p/c_v = 1.3$).

Under these conditions, the rate of leak can be modeled and calculated from the critical flow equation, which results in a time-integrated unmetered loss quantity:

$$Q_{lost} = \int_o^t \left[\frac{Ap}{\rho} \left(\frac{g_c KM}{RT} \left(\frac{2}{K+1} \right)^{\frac{k+1}{k-1}} \right)^{1/2} \right] dt \tag{5-15}$$

For the case of a wellbore leak, the inventory-pressure relation becomes:

$$\frac{p}{z} = \frac{RT}{379V} 10^9 \left\{ Q_{metered} - \int_o^t \left[\frac{Ap}{\rho} \left(\frac{g_c KM}{R} \left(\frac{2}{k+1} \right)^{\frac{k+1}{k-1}} \right)^{1/2} \right] dt \right\} \tag{5-16}$$

In Equation 5-16, A is the total area through which the leak is occurring, T the temperature upstream of the leak, ρ the density at standard conditions, $K = c_p/c_v$, M the molecular weight of the gas, R the gas constant, and t is the time (all in English engineering units, book inventory in BCF).

Equation 5-16 has been programmed and solved for a number of injection withdrawal cycles simulating wellbore leak behavior for the reservoir and gas storage operation described previously. It may be noted that p/z vs. metered Q this time looks more or less like a succession of alternating slopes.

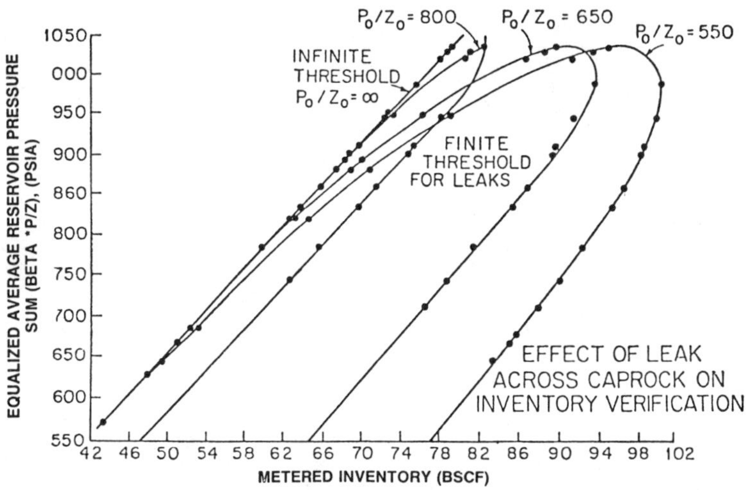

Figure 5-15. Effect of leak across caprock on inventory verification.

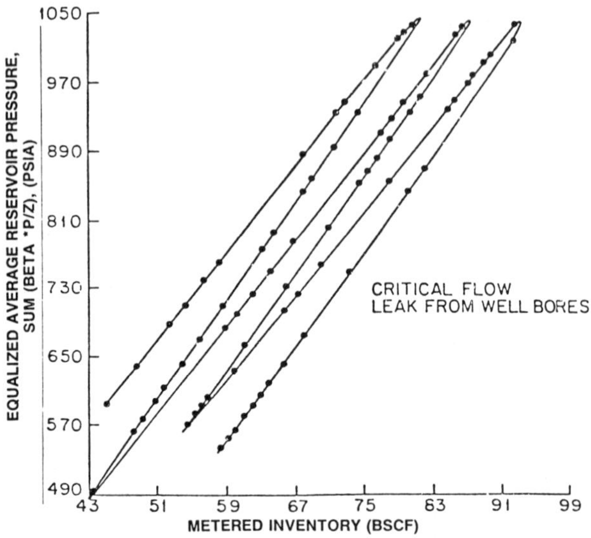

Figure 5-16. Simulation of leak from wellbore.

The case calculated and directly plotted as computer output is shown in Figure 5-16.

If the area of leak is deep enough, the criteria for critical flow might not be met. In those cases where the ratio of the downstream pressure to the upstream pressure is greater than the critical pressure ratio, the non-critical flow equation should be used, as given below:

$$Q_{\text{list N.C.}} = \int_0^t \frac{Ap}{\rho}\left[\frac{2g_cK}{(K-1)}\frac{M}{RT}\left(\frac{p_o}{p}\right)^{2/K}\left(1-\left(\frac{p_o}{p}\right)^{\frac{K-1}{K}}\right)\right]^{1/2} dt \tag{5-17}$$

This leads to the inventory-pressure relationship:

$$\frac{p}{z} = \frac{RT\,10^9}{379V}\left\{Q_m - \int_0^t \frac{Ap}{\rho}\left[\frac{2g_cK}{(k-1)\,RT}\left(\frac{p_o}{p}\right)^{2/K}\left(1-\left(\frac{p_o}{p}\right)^{\frac{K-1}{k}}\right)\right]^{1/2} dt\right\} \tag{5-18}$$

where ρ_o denotes pressure outside of the wellbore. All English engineering units, Q_m (book inventory in BCF). Several refinements are possible in developing diagnostic curves for wellbore leaks. The case of several wellbores vs. a single wellbore would be one. When the depth at which the leak is occurring is great enough to invalidate the critical flow equation, the rate of leak becomes a nonlinear function of the pressure ratio rather than an apparently linear function of upstream pressure. Further research in such factors could lead to determination of not only evidence of wellbore leak but of depth at which it may be occurring as well.

Pressure-time history and other physical properties used in the calculations have been obtained from typical injection withdrawal data on actual storage reservoirs.

References

1. Cierici, G. L., C. G. Ciucci, and G. Long. "Experimental Research on Gas Reservoirs Subjected to Water Drive," Section II, Paper 17 PD, Italy, World Petro. Congress, Tokyo, 1976.
2. Geffen, T. M., D. R. Parrish, G. W. Haynes, and R. A. Morse. "Efficiency of Gas Displacement from Porous Media by Liquid Flooding," *Trans. AIME*, 195, 29, 1952.
3. Gardner, G. H. F., J. Downie, and M. R. J. Wyllie. "Problems in the Recovery of Gas from Aquifers Used for Gas Storage," *J. of Inst. of Petro.* (Jan. 1962) 48, 457.
4. Neilson, R. "Non-Recoverable Gas in Underground Storage," AGA Project, 119–74, Pipeline Research Committee, Nov. 1978.
5. Tek, M. R., M. A. Ibrahim, and D. L. Katz. "Threshold Pressure in Gas Storage," Monograph, American Gas Association, Arlington, Virginia, 1970.
6. Udegbunam, E. O. "Migration of Natural Gas from Storage Reservoirs," Ph.D. Thesis, University of Michigan, 1983.
7. Katz, D. L. and K. H. Coats. *Underground Storage of Fluids*, Ulrich's Books, Inc., Ann Arbor, MI, 1968.
8. Katz, D. L. and M. R. Tek. "Overview on Underground Storage of Natural Gas," *J. of Petro. Tech.* (July 1981).
9. Tek, M. R. "Economic Prospects and Technical Problems in Underground Storage," Annual Technical Conf. of Canadian Gas Assoc., Vancouver, B.C., Nov. 1982.
10. Tek, M. R., E. O. Udegbunam, S. Schwartz, R. L. Lee, R. W. Logan, and J. S. Ahluwalia. "Inventory–Migration-Deliverability in Underground Storage," Monograph, American Gas Association, Arlington, Virginia, 1986.

6

MOVEMENT OF WATER IN CONTACT WITH NATURAL GAS

During the years 1960–1963, a research project on "Engineering Studies on Movement of Water in Contact with Natural Gas" was conducted at the University of Michigan under the direction of the author and Dr. D. L. Katz. Much of the material highlighted in this chapter relates to or was learned during those years and is included by permission of the American Gas Association. For the reader interested in the detailed background as well as procedures, the main reference is "Movement of Underground Water in Contact with Natural Gas," Donald L. Katz, M. Rasin Tek, K. H. Coats, M. L. Katz, S. C. Jones, and M. C. Miller, Monograph, American Gas Association, New York, 1963.

INTRODUCTION AND BACKGROUND

The environment of underground storage operations is either naturally occurring porous and permeable gas, oil or condensate reservoirs, or water-bearing formations having sufficiently high porosity and permeability. The naturally occurring gas reservoir or the original oil or condensate reservoir is frequently subject to movement of water in surrounding and/or underburden strata to some extent. It is only in the case where a given reservoir is a stratigraphic trap surrounded by shale or other material devoid of porosity and permeability that one does not encounter change in the pore volume occupied by gas.

Whether a producing reservoir is subject to continual pressure depletion due to removal of hydrocarbons from pore space, or a storage reservoir subject to cyclic changes in pressure in response to repeated injection/withdrawal cycles, the brine in surrounding areas or in the underburden often moves in and out of the reservoir pore space. The need to understand the movement of water was recognized more than 50 years ago by Shiltuis and Hurst[1] during the operation of East Texas Field.

In studies related to reserves and recoverability, the movement of water adjacent to oil and gas fields has been essential in material balance calculations. Problems in hydraulic seal around storage fields, the effect of water influx/efflux on storage pressures, performance analysis of storage reservoirs or depleted oil and gas fields, and design and development of aquifer storage became recognized during the early sixties.[2]

While the need to understand the nature of the movement of water was present during the development of producing fields, the difference in the operation of underground storage reservoirs indicated an even more critical need for detailed calculations of water movement. Early procedures in underground storage extended the techniques established in production to the storage fields. The prolific development of the underground storage industry since the thirties and forties, with so many reservoirs subject to different extents of water drive, justified development of certain practices and applications unique to storage. This is mainly because storage, in some respects, is so different from production. First of all, the entire top gas content of a storage reservoir is in general produced in, say, 60 to 130 days. By contrast, a producing gas field may take 15 to 30 years to produce its recoverable content.

The rate of pressure decline in a storage field, toward the end of the withdrawal season, may be as high as 10 to 20 psi/day. The spacing on a producing field may be of the order of 160 to 640 acres, while on storage it may be as low as 20 acres.

The need for monitoring against uncontrolled movement of gas justifies the drilling of several observation wells in aquifers near the storage horizon. This provides unique data of a quantity and quality seldom encountered in production operations.

Finally, the practice of *overpressure* or *delta-pressure* in underground storage almost invariably results in expansion of pore volume occupied by gas through efflux of water from the storage horizon. The extent, nature, and rate of such efflux is important in development of the storage reservoir.

AQUIFERS IN STORAGE

An *aquifer* is a water-filled underground zone extending many miles in lateral distances either as a blanket zone or a thick layer. The aquifer rock is sufficiently porous and permeable so that it involves large quantities of brine mobile to some extent. Aquifer is the generic name given to both the fluid in the pore space and the solid structural rock environment around it. Some aquifers are replenished by surface waters where they outcrop. Others are in the form of stratigraphic traps and, as such, are isolated from surface waters. Figure 6-1 shows a typical regional cross section where water flows from a high- to a low-level outcrop.[3]

Aquifers provide the energy to produce oil and gas through expansion of water into the reservoir pore volume. A celebrated example is the Woodbine aquifer adjacent to East Texas Field (see Figure 6-2).[4]

In the midwest, prominent aquifers near market storage areas are St. Peter, Galesville, and Mt. Simon. These are blanket sands extending over Illinois and eastern Iowa. Table 6-1 shows examples of aquifers encountered in production and storage.

The Marshall aquifer in Michigan, of Mississippian age, is contiguous to several storage fields developed in Stray Sands. The relationship between the Stray Sand and Marshall aquifers is shown in Figure 6-3. The discovery pressure prevailing in an aquifer usually corresponds to a gradient indicated by the density of brine indigenous to a particular region. In general, this gradient corresponds to the hydrostatic pressure prevailing in the area. On a broad regional basis, the static level of the water in the well may just coincide with the surface level or be located at some distance from the surface. Occasionally, the pressure in the aquifer will be such that the water could stand at a level higher than the well head. Called an artesian condition, this is shown in Figure 6-1.

Some aquifers are so large in areal extent that they justify the assumption of infinite size in calculations. Others are small and confined. When they are adjacent to several fields from which oil and/or gas are produced, they become subject to expansion of water with some slight depressurization. Sometimes production of water from small limited aquifers over long periods of time results in a small decrease of pressure. This depressurization rarely exceeds the order of a few psi per year.

In order to inject gas into an otherwise virgin aquifer, one must exceed the discovery pressure of the aquifer, which maintains a static head of water all

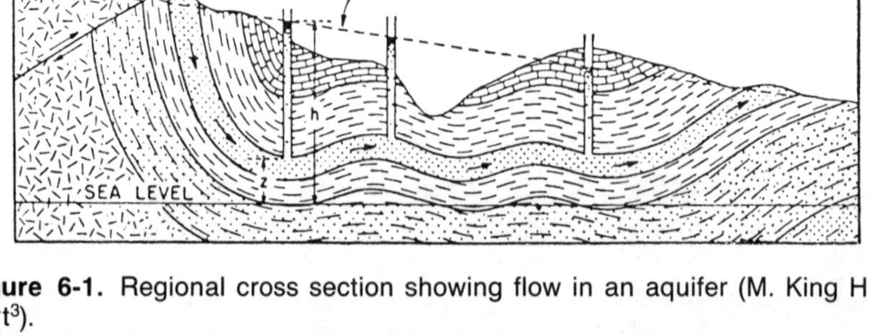

Figure 6-1. Regional cross section showing flow in an aquifer (M. King Hubbert[3]).

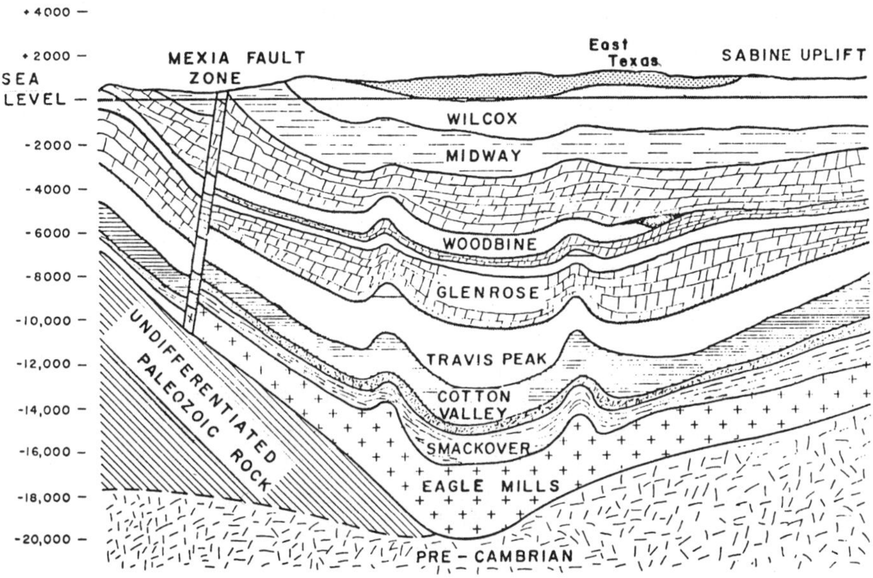

Figure 6-2. East-West cross section of East Texas Basin showing the Woodbine Aquifer.

Table 6-1
Examples of Aquifers in Production and Storage

Name of Aquifer	Location	Nature of Use	Remarks
Woodbine	Texas	East Texas Oil Field	Extends to 100 miles
Embar Ellenberger	West Texas	Several oil fields	Very large
St. Peter	Illinois, Iowa	Storage	
Galesville	Illinois, Iowa	Herscher Storage	
Mt. Simon	Illinois, Iowa	Manlove Storage Herscher Storage Troy Grave Storage Redfield Storage	Very thick sand
Marshall	Michigan	Several Stray Sand reservoirs	
Thaynes	Wyoming	Leroy Storage	
Dakota J. Sand	Nebraska Colorado	Huntsman Storage Several oil fields	
Beynes	France	Storage	
Domengine Sand	California	Los Medanos Storage	

the way near or in excess of the elevation of the well site. This practice is referred to as *overpressuring.*

When an aquifer pressure is increased in an area immediately underneath a gas bubble, a transient pressure condition gets established in the aquifer in the vicinity of the gas bubble. This is shown in Figure 6-4.

When gas is injected to raise the static pressure in the aquifer by Δp, water is compressed beneath the gas bubble and begins to move away through transient pressurization of the water in the sand. The pressure gradients at times t_1, t_2, t_3 . . . keep propagating toward the far reaches of the aquifer, as shown in Figure 6-4.

The volumetric variation experienced by water under the influence of pressure change is measured in terms of compressibility of the water. The compressibility (c) of water is equal to volume change per-unit-volume per-pressure-change at formation temperature. This is given by

$$c = -\frac{1}{V}\left(\frac{\partial V}{\partial p}\right)_T \tag{6-1}$$

For water, $c = 3 \times 10^{-6}$ volume/volume × psi.

The compressibility of water varies with the amount of mineral content and dissolved gases.

When water is compressed in a porous environment, the combined compressibility of the water and rock must be used in calculations. Under these

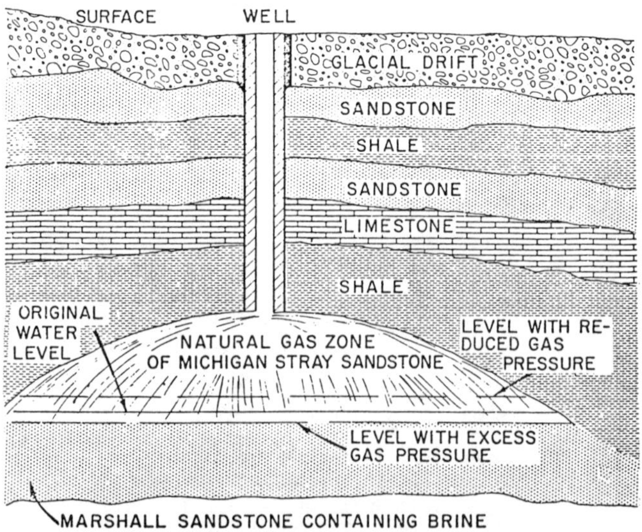

Figure 6-3. Michigan Stray sand and Marshall Aquifer.

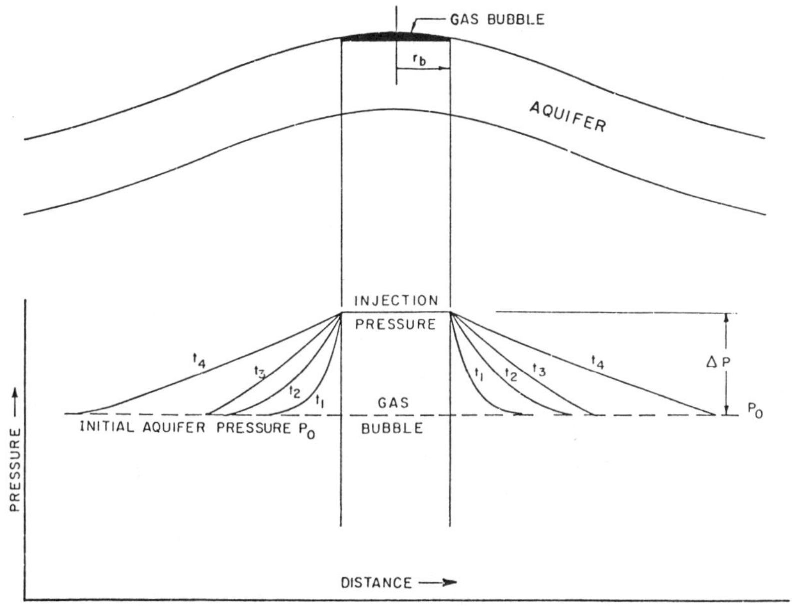

Figure 6-4. Pressure-time distribution in an aquifer adjacent to a gas storage bubble.[2]

conditions, in general, the compressibility becomes about twice that of pure water.[5,6]

FLOW EQUATIONS AND GEOMETRIC MODELS

The theory of flow through porous media given in Chapter 3 provides the foundation upon which equations and procedures were developed for calculation of movement of water in contact with natural gas.

The partial differential equation describing the movement of the brine in the aquifer surrounding the storage gas is given by:

$$\nabla^2 p = \frac{\phi \mu c}{k} \frac{\partial p}{\partial t} \tag{6-2}$$

The above partial differential equation, along with appropriate initial and boundary conditions, has been solved for the following geometric models:

- Radial
- Linear

- Thick sand
- Hemispherical
- Elliptical

For each of the models given above, Reference 2 gives the basic solution for either constant terminal pressure or constant terminal rate boundary conditions. These solutions are summarized in Table 6-2.

Table 6-2
Summary of Flow Equations for Water Movement
Constant Terminal Rate Case, $p = p_o - k_r e_w p_t$
Constant Terminal Pressure Case, $W_e = k_p(p_o - p)Q_t$

Parameter	Radial Model	Linear Model	Thick Sand Model	Hemispherical Model
t_D	$\frac{.00633\ kt}{\phi\mu\ cr_t^2}$	$\frac{.00633\ kt}{\phi\mu\ cL^2}$ $\frac{.00633\ kt}{\phi\mu\ c}$	$\frac{.00633\ kt}{\phi\mu\ cr_b^2}$	$\frac{.00633\ kt}{\phi\mu\ cr_t^2}$
k_r	$\frac{25.15}{kh}$	Infinite- $178\ \mu/Ak$ Finite- $158\ \mu L/Ak$	$\frac{25.15}{kh}$	$\frac{50.2}{kr_b}\mu$
k_p	$6.283\ \phi\ cr_b^2h$	Infinite- $1.13\ \phi\ cA$ Finite- $\phi\ cAL$		$3.14\ r_b^3\phi c$
p_t	Tabulated (Ref. 2)	Infinite- $\sqrt{\theta}$ Finite-Fig.	Appendix 1	$1 - e^{t_D}$ $[1 - erf(t_D)]$
Q_t	Tabulated (Ref. 2)	Infinite- Finite		$t_D + 2\sqrt{t_D/3.14}$

where e_w = constant water influx or efflux rate, ft^3/day
p = pressure at gas water interface, psia
A = cross-sectional area, ft^2
c = compressibility, (vol)/(vol)(psi)
k = permeability, millidarcies
L = length of linear finite aquifer, ft
μ = viscosity, centipoise
ϕ = porosity, fraction
r_b = radius gas bubble, ft
t = time, days
θ = transformed time, ft^2

Figure 6-5 shows the radial flow model depicting the assumed geometry between the gas bubble and the surrounding aquifer.

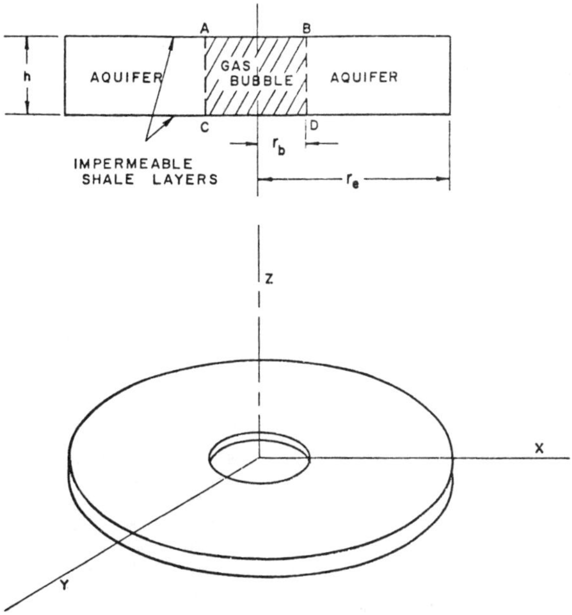

Figure 6-5. Idealized radial flow model showing the assumed geometry between the gas bubble and aquifer.

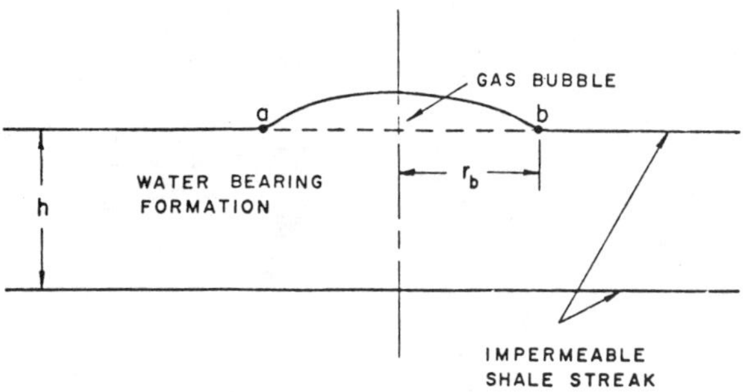

Figure 6-6. The thick sand flow model.[2]

Figure 6-6 shows the thick sand flow model, where the flow in the aquifer has a vertical component as well near the gas bubble. This is the case where one would have discernible bottom water-drive along with edge water-drive, and the dimensionless pressure function for constant terminal rate becomes dependent on vertical to horizontal permeability ratio.

Figure 6-7 shows the hemispherical flow model where the assumed position of gas/water interface becomes a spherical surface of radius equal to the

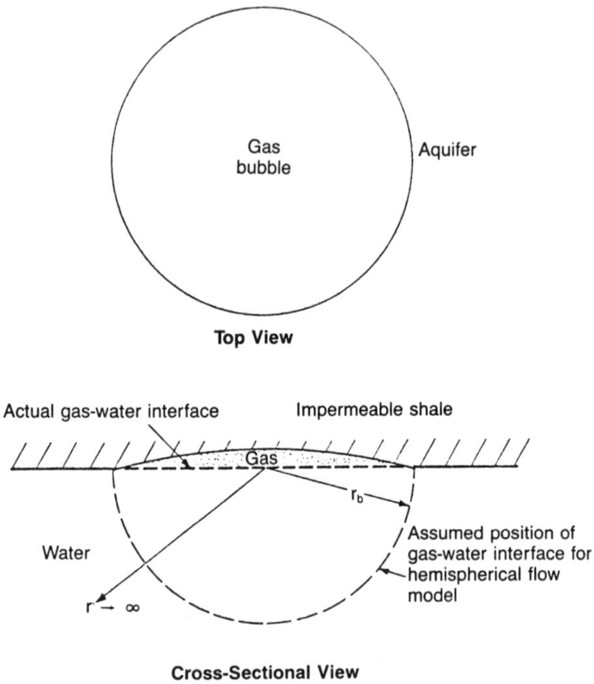

Figure 6-7. The hemispherical flow model.[2]

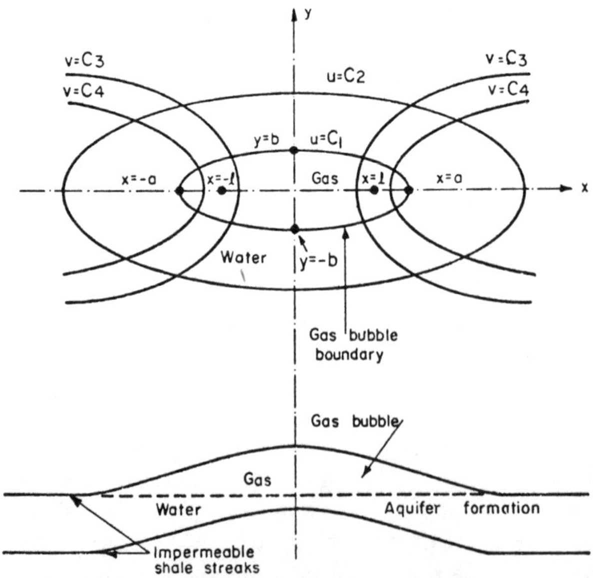

Figure 6-8. The elliptical flow model.[2]

gas-bubble radius. Figure 6-8 shows the elliptical flow model. Figure 6-9 shows a comparison between dimensionless pressure function (for constant terminal rate) for various models. Figure 6-10 shows a comparison of dimensionless water inlux (for constant terminal pressure) for various geometric models.

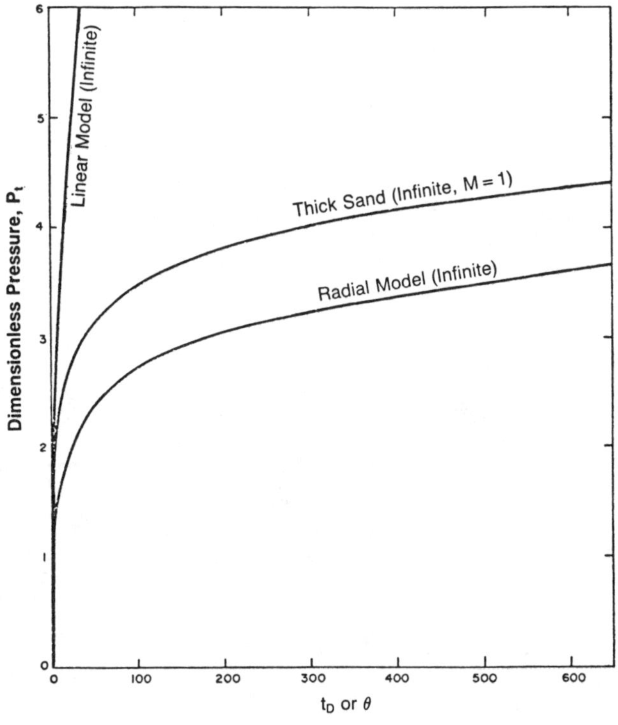

Figure 6-9. A comparison of dimensionless pressure functions Pt for various models.[2]

The Resistance Function Model

The *Resistance Function Model* is a generalized method for calculating the performance of the aquifer from field data. Because the various geometric models discussed earlier all involve restricting assumptions and idealized configurations, performance predicted by the models often differs substantially from the real-world data collected from storage operations. When such data are available, a dimensionless resistance function may be developed by trial and error, matching the observed results with the performance calculated from the so-called resistance function. One such method, developed and published in Reference 2, is a modification of the *resistance function* concept formulated by Hutchinson and Sikora.[7] A treatment using both analog and digital computers is also given in the literature by Hicks, Webber, and Ledbetter.[8]

An example of a resistance function curve is shown in Figure 6-11. It can be seen that during early times, when the aquifer behaves as if of infinite extent, and for which not many data are available, the resistance function curve z(P_t or Q_t) is approximated using geometric models. Later, when data reflecting aquifer boundaries become available, one develops the z-curve from field data subject to certain constraints and smoothing. Once developed, the z-curve permits calculation of either the pressures (in constant terminal rate) or rates (in constant terminal pressure).

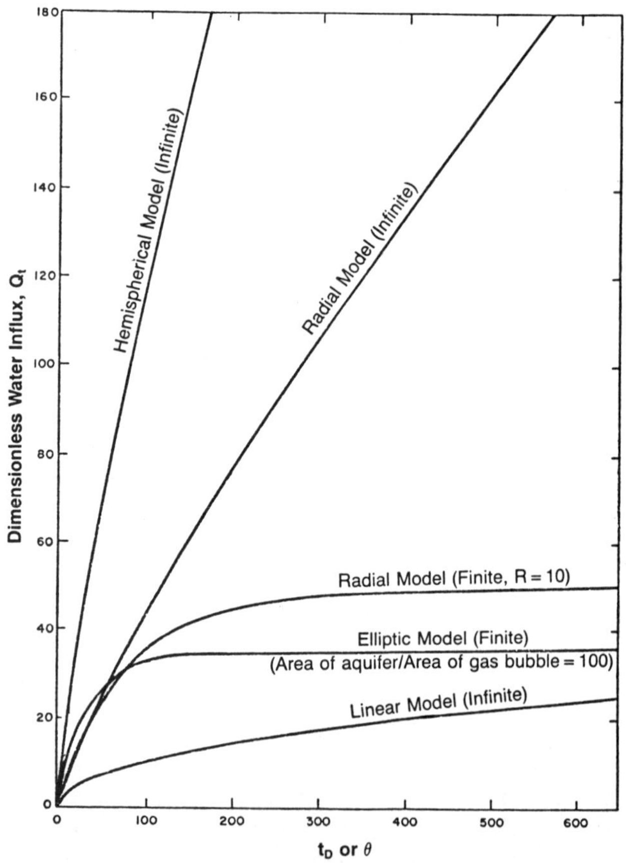

Figure 6-10. Comparison of dimensionless water influx Qt for various geometric models.[2]

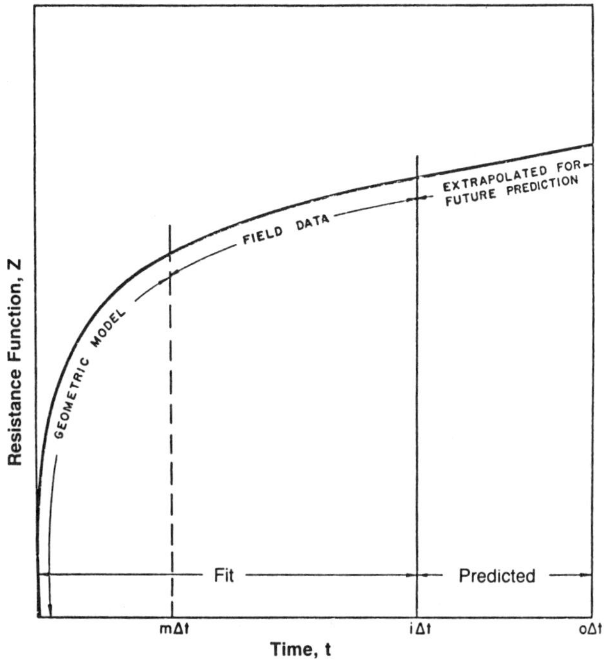

Figure 6-11. The resistance function curve.[2]

THE PRINCIPLE OF SUPERPOSITION

The fundamental solutions to the flow equation for water through porous media are given for boundary conditions that do not change with time. In aquifer storage problems, we therefore have direct solutions for cases when the water flowrate in or out of the gas bubble or the gas bubble pressure is held constant. In actual operation of storage reservoirs, both the pressure and the rate across the boundaries are subject to continual change. Under these conditions, the basic solutions must be modified to account for the time-dependent boundary conditions. This is done by invoking the *principle of superposition*. The principle, first enunciated by Duhamel[9] and sometimes mathematically referred to as the convolution integral, amounts to substituting for the time-variant boundary value a succession of step functions and then systematically adding the effect of each step with proper time duration to form a general solution. An example of the principle is given in Figure 6-12. It can be seen that, once the declining pressure curve is represented by a series of step declines, p_1, p_2, etc., the composite effect on cumulative water influx is developed by superposing the effect of each pressure drop.

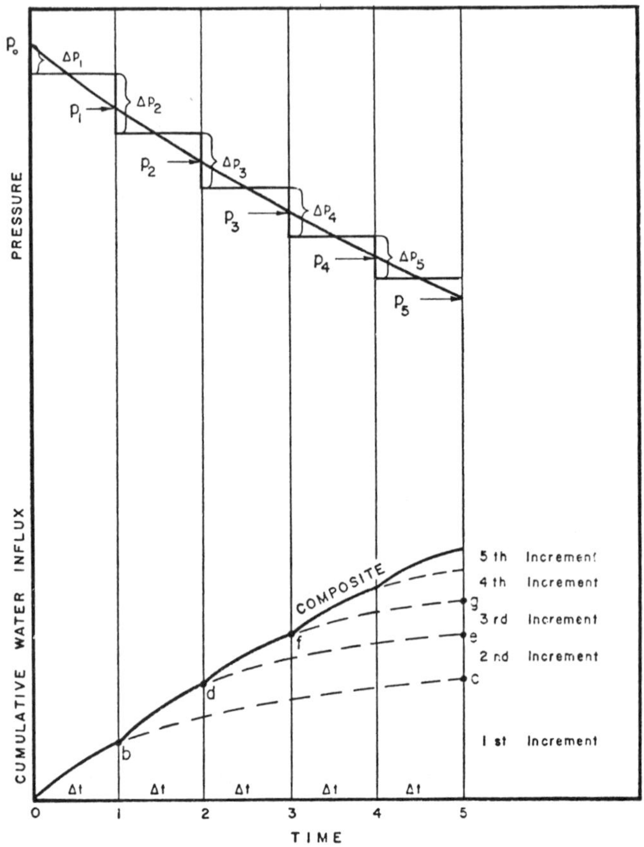

Figure 6-12. Cumulative water influx for varying pressures in aquifer storage.[2]

Figure 6-13 shows the cumulative water movement in response to a cyclically varying boundary pressure. Similar superposition may be made for computed cumulative pressures resulting from step changes in boundary rates. In theory and practice, the superposition principle is shown to be exactly correct for systems described by linear partial differential equations. When the equations are quasi-linear (such as the ones describing the flow of natural gas), the superposition principle still applies, but only as a reasonable approximation. In the rigid mathematical sense, the superposition equation is derived in integral form and called the *convolution integral.* A summation of a finite number of terms is given here in the form of series because of convenience for digital computation.

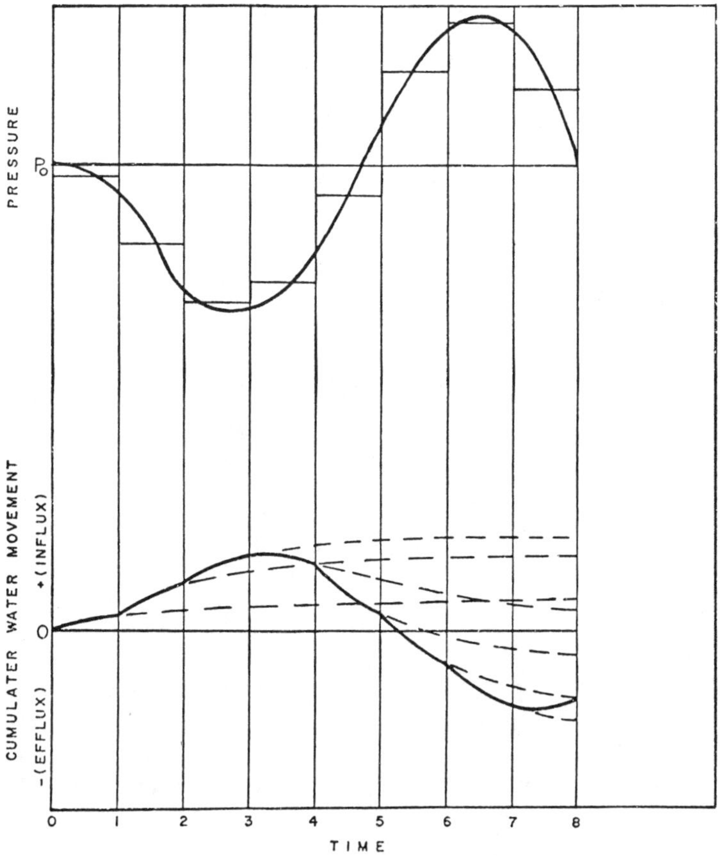

Figure 6-13. Cumulative water influx by superposition of cyclically varying boundary pressure.

Variable Terminal Rate by Superposition

For water movement, the equation giving the boundary, or gas bubble, pressure as a function of time for constant flowrate (i.e., the rate of water across the boundary) is given[2] as:

$$p_o - p = k_r \, e_W \, p_t \tag{6-3}$$

where p = Pressure at gas/water interface, psia
p_o = Initial, uniform aquifer pressure, psia
e_W = Constant terminal rate of water into the gas bubble, ft^3/day

p_t = Dimensionless pressure drop from tables or graphs[2]
k_r = Constant for given model

The particular values for k_r, t_D, and P_t may be computed and obtained from the information summarized in Table 2.

When the water influx or efflux rate is not constant, Equation 6-3 is modified by using the superposition principle, as follows:

$$p_o - p_n = k_r \sum_{j=1}^{n} e_{w_{n-j+1}} p_{t_j} \tag{6-4}$$

where j = counter in the running sum
$e_{w_{n-j+1}}$ = average water rate during the particular time interval, ft^3/day

The nth time interval is defined by:

$$t_n = n\Delta t \tag{6-5}$$

$$\Delta p_{t_j} = p_{t_j} - p_{t_{j-1}} \tag{6-6}$$

$$t_j = j\Delta t \tag{6-7}$$

By expanding the terms and rearranging Equation 6-4:

$$p_o - p_n = k_r \sum_{j=0}^{j=n-1} e_{w_j} p_{t(n-j)} \tag{6-8}$$

where $e_{w_j} = e_{w_{j+1}} - e_{w_j}$ (6-9)

and $p_{t(n-j)} = p_t$ evaluated at $t_D = (n - j)\Delta t_D$ (6-10)

Variable Terminal Pressure

The general solution for constant terminal pressure is given by:

$$W_e = k_p(p_o - p)Q_t \tag{6-11}$$

where W_e = cumulative water influx, ft^3

Again, the particular values for K_p and t_D may be computed from Table 6-2.

The superposition principle applied to the pressure case results in the following equation:

$$w_{e_n} = k_p \sum_{j=0}^{n-1} \Delta p_j \, Q_{t(n-j)} \tag{6-12}$$

where $Q_{t(n-j)} = Q_t$ evaluated at $t_D = (n - j)\Delta t_D$ (6-13)

$$\Delta p_j = \frac{p_{j-1} - p_{j+1}}{2} \; j \geqq 1 \tag{6-14}$$

and $\Delta p_o = \dfrac{p_o - p_1}{2}$

Example 6-1

Problem. A natural-gas producing reservoir with initial discovery pressure $p_o = 700$ psia is being depleted according to a specific schedule given below. The aquifer surrounding the field may be assumed infinitely large.

It is desired to calculate the water influx.

The following data are available:

$c = 7 \times 10^{-6}$ vol/(vol)(psi), effective compressibility
$k = 100$ mdcy
$h = 30$ ft
$r_b = 5{,}000$ ft
$\phi = 0.15$
$\mu = 1$ cp

Flow geometry is radial and horizontal.

The pressure schedule is given in Table 6-3.

Table 6-3
Pressure Schedule for Example 6-1

Time, months	Pressure, psia
0	700
3	680
6	645
9	610
12	570

Solution. By superposition principle:

$$W_{e_n} = 6.283\phi c r_b{}^2 h \sum_{j=0}^{n-1} \Delta p_j Q_{t(n-j)} \qquad (6\text{-}15)$$

where W_{e_n} is the cumulative water influx corresponding to time $t = n\Delta t$.
Now, using a time increment $\Delta t = 3$ months:

$$\Delta p_0 = \frac{p_0 - p_1}{2} = \frac{700 - 680}{2} = 10 \text{ psi}$$

$$\Delta p_1 = \frac{p_0 - p_2}{2} = \frac{700 - 645}{2} = 27.5 \text{ psi}$$

$$\Delta p_2 = \frac{p_1 - p_3}{2} = \frac{680 - 610}{2} = 35 \text{ psi}$$

$$\Delta p_3 = \frac{p_2 - p_4}{2} = \frac{645 - 570}{2} = 37.5 \text{ psi}$$

$$t_D = \frac{0.00633 \text{ kt}}{\phi \mu c r_b{}^2} = \frac{0.00633 \times 100 \times 90}{0.15 \times 1 \times 7 \times 10^{-6} \times (5{,}000)^2} = 2.17$$

From appendix of Reference 2:

$$Q_{t_1} = 2.575 \text{ at } t_D = 2.17$$

$$Q_{t_2} = 4.113$$

$$Q_{t_3} = 5.448$$

$$Q_{t_4} = 6.692$$

Now, for n = 1:

$$W_{e_1} = 4{,}950 \sum_{j=0}^{j=0} \Delta p_j Q_{t(n-j)} = 4{,}950 \times 10 \times 2{,}575 = 127{,}500 \text{ ft}^3$$

n = 2:

$$W_{e_2} = 4{,}950 \sum_{j=0}^{j=1} \Delta p_j Q_{t(n-j)} = 4{,}950\ (\Delta p_o Q_2 + \Delta p_1 Q_1)$$

$$= 4{,}950\ [10(4.113) = 27.5\ (2.575)]$$

$$= 555{,}000\ \text{ft}^3$$

n = 3:

$$W_{e_3} = 4{,}950\ [\Delta p_0 Q_{t_3} + \Delta p_1 Q_{t_2} + \Delta p_2 Q_{t_1}]$$

$$= 4{,}950\ [10(5.448) + 27.5(4.113) + 35(2.575)]$$

$$= 1{,}230{,}000\ \text{ft}^3$$

n = 4:

$$W_{e_4} = 4{,}950\ [\Delta p_0 Q_{t_4} + \Delta p_1 Q_{t_3} + \Delta p_2 Q_{t_2} + \Delta p_3 Q_{t_1}]$$

$$= 4{,}950\ [10(6.692) + 27.5(5.448) + 35(4.113) + 37.5(2.575)]$$

$$= 2{,}250{,}000\ \text{ft}^3$$

APPLICATION OF WATER MOVEMENT TO DEVELOPMENT OF AQUIFER STORAGE FIELDS

Engineering development of projects storing gas in aquifers involves calculation of the movement of water in contact with natural gas. The evaluation of a potential storage structure involves water-movement calculations based on geometric models. Pump tests may be made prior to gas injection to evaluate the caprock and determine *in situ* permeability and compressibility.

In Situ Permeability and Compressibility from Pump Tests

Laboratory tests on cores are normally made to determine the matrix permeability of aquifer zones. However, the formation may have a different effective permeability from that obtained by averaging core data. Fractures not included in test cores will influence the *in situ* permeability as will other non-uniformities. Pumping tests on a well may be used to obtain the *in situ* permeability. Coincident with this pumping, pressure observations on an adjacent well completed in the aquifer will provide data for computation of the *in situ* compressibility of the water-sand system. Figure 6-14 illustrates the location of wells which may be used in a pump test.

In any pump test, the preferred method is to pump at a constant rate. When only a single well is available, the water level in the pumping well is needed to obtain the *in situ* permeability. When there is an observation well in addition to the pumping well, both k and c can be obtained without

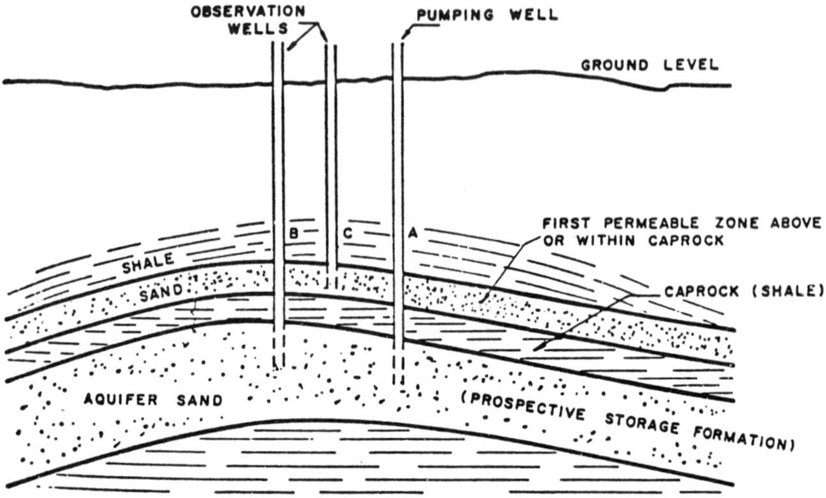

Figure 6-14. Pumping of water wells to obtain *in situ* permeability and compressibility as well as cap rock leakage.

knowledge of water level in the pumping well. Table 6-4 lists some of the combinations of data which can be used. The table gives information required to compute k and c from various combinations of data.

Table 6-4
Variety of Tests for In Situ Permeability and Compressibility

Calculations presented as:	Figure 6-14 *Well A* (with pump and water level recorder)	Figure 6-14 *Well B* (with water level recorder)	Computations which can be made
Case I	Pump well, shut-in well, measure rate of pressure build-up	Not used	Unsteady state computation of k
Case II	Pump well, recorder not used	Observe level for drawdown and for build-up	Unsteady state computation of k, c
Case III	Pump well, measure rate of pressure drawdown	Observe levels during drawdown	Pseudo steady state computation of k in addition to those of Case I and Case II

Case 1: Pump Test on a Single Well, Drawdown and Build-up Pressure Observed on Pumping Well Only. The *in situ* permeability and compressibility of an aquifer can be calculated from the drawdown test on a

well pumping at a constant rate. The *point source* solution given by Horner in field units is:

$$p = p_o + \frac{70.6\ q\mu}{kh}\ Ei\left[\frac{-\mu\phi c r_w{}^2}{4(0.00633)kt}\right] \qquad (6\text{-}16)$$

where p_o = initial aquifer pressure, psia
r_w = well radius, ft
t = time, days
p = pressure at the wellbore at time t, psia
ϕ = formation porosity, fraction
k = permeability, millidarcies
μ = fluid viscosity, centipoise
c = effective compressibility, vol/(vol)(psi)
q = water pumping rate, Bbls/day

The exponential integral function (Ei) is defined as:

$$-Ei(-x) \int_x^{\infty} \frac{e^{-y}}{y}\, dy \qquad \text{(D-38, Appendix D)}$$

It may be represented (46) by the series:

$$-Ei(-x) = \ell nx + .5772 - x + \frac{x^2}{(2)(2!)} - \frac{x^3}{(3)(3!)} + \ldots \frac{(-1)nx^n}{(n)(n!)} \qquad \text{(D-43)}$$

If the value of x is small; i.e., less than 0.01, then Ei(− x) can be approximated by:

$$-Ei(-x) = \ell nx + .5772 \qquad \text{(D-44)}$$

Substitution of Equation (D-44) in Equation 6-16 yields:

$$p = p_o + \frac{70.6\ q\mu}{kh}\left\{\ell\, n\ \frac{\mu\phi c r_w{}^2}{4(0.00633)kt} + .5772\right\} \qquad (6\text{-}17)$$

$$\text{or: } p = p_o + \frac{162.6\ q\mu}{kh}\left\{\log_{10} \frac{\mu\phi c r_w{}^2}{4(0.00633)kt} + .2508\right\} \qquad (6\text{-}18)$$

$$\text{or: } p = p_o + \frac{162.6\ q\mu}{kh}\left\{\log_{10} \frac{\mu\phi c r_w{}^2}{4(0.00633)kt} - 0.3513\right\} \qquad (6\text{-}19)$$

If the well is pumped at a constant rate, then a plot of p vs. log t will approach a straight line. The slope of this line, pounds per-square-inch per-cycle, is equal to $-162.6\ q\mu/kh$. If q, μ, and h are known, the *in situ* permeability can be calculated from the slope. While it may appear that *in situ* compressibility, c, may also be calculated from these equations, the calculation is sensitive to errors in measurement of the radius of the wellbore and, as such, is believed impractical.

Unfortunately, skin effect at the well can give an erroneous slope on the drawdown curve. Therefore, the pressure build-up curve is usually used to obtain a more accurate value for the permeability. If the total pumping time is designated by t_o, and the shut-in time by Δt, the well pressure after shut-in is given by superimposing two solutions in the form of Equation 6-19:

$$p = p_o + \frac{162.6\ q\mu}{kh}\left[\log_{10}\frac{\mu\phi c r_w^2}{0.00633\ K(t_o + \Delta t)} - 0.3513\right]$$

$$- \frac{162.6\ q\mu}{kh}\left\{\log_{10}\left(\frac{\mu\phi c r_w^2}{0.00633\ k\Delta t}\right) - 0.3513\right\} \tag{6-20}$$

which simplifies to:

$$p = p_o - \frac{162.6\ q\mu}{kh}\log_{10}\left(\frac{t_o + \Delta t}{\Delta t}\right) \tag{6-21}$$

A plot of pressure vs. $\log_{10}(t_o + \Delta t)/\Delta t$ will approach a straight line after the afterflow and skin effects of the well damp out. The *in situ* permeability can be calculated from this slope expressed in pounds per-square-inch per-cycle. The slope is equal to $-162.6\ q\mu/kh$.

Case 2: Pump Test on a Single Well, Drawdown, and Build-up Pressure Observed at an Observation Well Only. The *in situ* permeability and compressibility of an aquifer can be calculated from the drawdown and build-up pressures at an observation well. In Equation 6-17, r_w is replaced by r, the distance between the pumping well and the observation well. The pressure of the observation well is given by:

$$p = p_o + \frac{162.6\ q\mu}{kh}\left\{\log_{10}\left(\frac{\mu\phi c r^2}{0.00633\ kt}\right) - 0.3513\right\} \tag{6-22}$$

where p_o is the initial pressure of the observation well. There is no change in the equation describing the pressure build-up, Equation 6-21. The formation *in situ* permeability can be calculated from the same procedure as outlined in Case 1*.

An effective compressibility value may be determined by using Equation 6-22. If the observation well pressure is plotted as $(p - p_o)$ versus $\log_{10}t$, then the slope of the straight line portion of this curve is $-162.6\ q/kh$ and the intercept of this straight line (value of p at $\log_{10} t = 0$ or $t = 1$) is:

$$I = \frac{162.6\ q\mu}{kh}\left\{\log_{10}\left(\frac{\mu\phi cr^2}{0.00633\ k}\right) - 0.3513\right\} \tag{6-23}$$

from which c can be calculated. Alternatively, any point $(\Delta p, \log t)$ on the straight-line portion of the curve may be used in the following equation to calculate c:

$$p - p_o = \Delta p = -\frac{162.6\ q\mu}{kh}\log_{10}t + \frac{162.6\ q\mu}{kh}\log_{10}\left(\frac{\mu\phi cr^2}{.00633\ k} - 0.3513\right) \tag{6-24}$$

Case 3: Pump Test on a Single Well, Drawdown and Build-up Pressures Observed at the Pumping Well and One Observation Well. The quasi-steady-state formula can be used at sufficiently large times. The pseudo-steady-state is reached when the rate of pressure decline, dp/dt, is constant and the same at both wells. Now, using Darcy's Law:

$$q = \frac{kA}{\mu}\frac{dp}{dr} \tag{6-25}$$

where q is the waterflow rate into the producing well. Rearranging Equation 6-25 and integrating between two wells:

$$\int_{p_1}^{p_2} dp = \int_{r_w}^{r} \frac{q\mu}{kA}\,dr = \int_{r_w}^{r} \frac{q\mu}{k(2\pi r)h}\,dr \tag{6-26}$$

or:

$$p_2 - p_1 = \frac{q\mu}{2\pi kh}\,\ell n\left(\frac{r}{r_w}\right) \tag{6-27}$$

* In this particular case, the magnitude of dimensionless time must be verified to be of such magnitude to justify logarithmic approximation of the Ei function.

and hence:

$$k = \frac{q\mu \, \ell n \left(\frac{r}{r_w}\right)}{2\pi h(p_2 - p_1)} \tag{6-28}$$

Equation 6-28 expressed in field units is:

$$k = \frac{325 \, q\mu \, \log_{10} \left(\frac{r}{r_w}\right)}{h(p_2 - p_1)} \tag{6-29}$$

which can be used to calculate the *in situ* permeability. The distance r_w is the radius of the pumping well and r is the distance between the pumping well and the observation well.

Example 6-2

Problem. An aquifer is being considered for gas storage. The sand occurs at a depth of 1,250 feet and is considered a suitable structure. The average physical properties of the sand determined from core data area: porosity = 14.5%, permeability = 168 millidarcies, and sand thickness = 164 feet.

A well in the aquifer was pumped at 1,370 barrels per day and the pressure was observed at both the pumping well and a neighboring well 500 ft away during pressure drawdown and build-up. Predict the *in situ* permeability of the sand and the effective compressibility of the aquifer. The pressure build-up data for the pumping well are given in Table 6-5 and pressure drawdown data for the observation well are given in Table 6-6.

Solution

A. Determination of the *in situ* permeability using the pressure build-up data on the pumping well.

In the following equation:

$$p = p_o - \frac{162.6 \, q\mu}{kh} \log_{10} \left(\frac{t_o + \Delta t}{\Delta t}\right) \tag{6-21}$$

substitute $\Delta p = p - p_o$ to obtain:

$$\Delta p = -\frac{162.6 \, q\mu}{kh} \log_{10} \left(\frac{t_o + \Delta t}{\Delta t}\right) \tag{6-30}$$

From the build-up data, determine $(t_o + \Delta t)/\Delta t$ and Δp for each data point.

Table 6-5
Pressure Build-up Data—Pumping Well

Time Since Pumping ceased Hours, t	$\frac{t_o + \Delta t}{\Delta t}$	Pressure p*, Feet of water	Δp Feet of water	psi
0.05	72.05/.05 = 1441	1060	−25	−10.8
0.1	72.1/.1 = 721	1066	−19	− 8.25
0.167	72.167/.167 = 433	1062	−23	− 9.95
0.33	72.33/.33 = 219	1064	−21	− 9.09
1	73/1 = 73	1065	−20	− 8.67
2	74/2 = 37	1068	−17	− 7.35
3	75/3 = 25	1070	−15	− 6.50
4	76/4 = 19.75	1074	−12	− 5.20
9	81/9 = 9	1075	−11	− 4.76
55	127/55 = 2.31	1080.5	−10	− 4.33
70	142/70 = 2.03	1081	− 4.5	− 1.95
118	190/118 = 1.61	1082.2	− 2.8	− 1.21
190	262/190 = 1.38	1082.3	− 2.7	− 1.17

$t_o = 72$ hours = 3 days, pumping time

* Initial pressure prior to pumping = 1085 feet of water
Pumping rate = 1370 barrels water per day, 40 gpm.

Table 6-6
Pressure Drawdown—Observation Well

Pumping time, hours	Pressure, feet of water	p, feet of water	p, psi
0	1085	0	0
0.33	1084	0	0
0.75	1083.84	− 0.16	−0.0693
1.33	1083.05	− 0.95	−0.0411
2.0	1082.02	− 1.98	−0.855
3.0	1080.73	− 3.27	−1.41
4.0	1079.68	− 4.32	−1.87
5.0	1078.26	− 5.74	−2.44
8.0	1077.17	− 6.83	−2.96
10.0	1076.60	− 7.40	−3.20
20.0	1074.55	− 9.45	−4.08
40.0	1072.28	−11.72	−5.08
60.0	1070.53	−13.47	−5.32
72.0	1070.03	−13.97	−6.04

A plot of p vs. $\ell n(t_o + \Delta t/\Delta t)$ is shown in Figure 6-15. The slope is −3.55 pounds per-square-inch per-cycle. Hence:

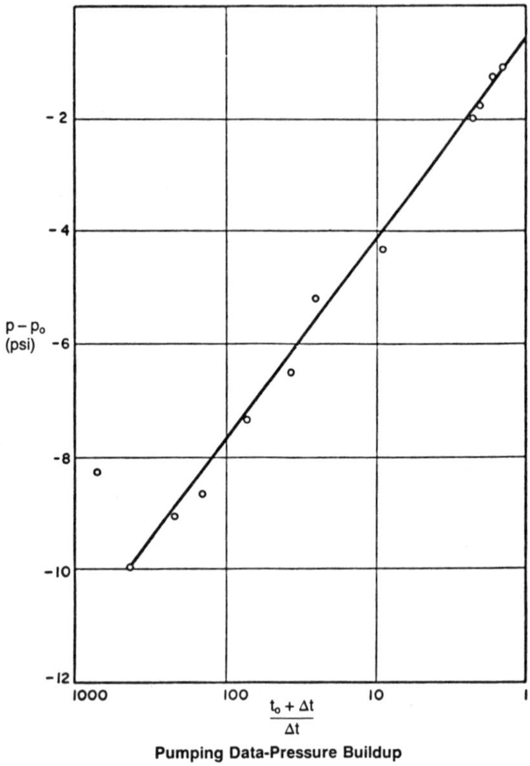

Figure 6-15. Pressure build-up curve for example problem.

$$q\mu\left(-\frac{162.6}{kh}\right) = -3.55$$

or:

$$k = \frac{(162.6)(1{,}370)(1)}{(3.55)(164)} = 383 \text{ md}$$

B. Determination of the *in situ* permeability and compressibility using the drawdown data on the observation well.

Substituting the pressure drop $p = p - p_o$ in Equation 6-24 yields:

$$p = \frac{162.6\, q\mu}{kh}\left[\log_{10}\frac{cr^2}{0.00633\, kt} - 0.3513\right] \tag{6-31}$$

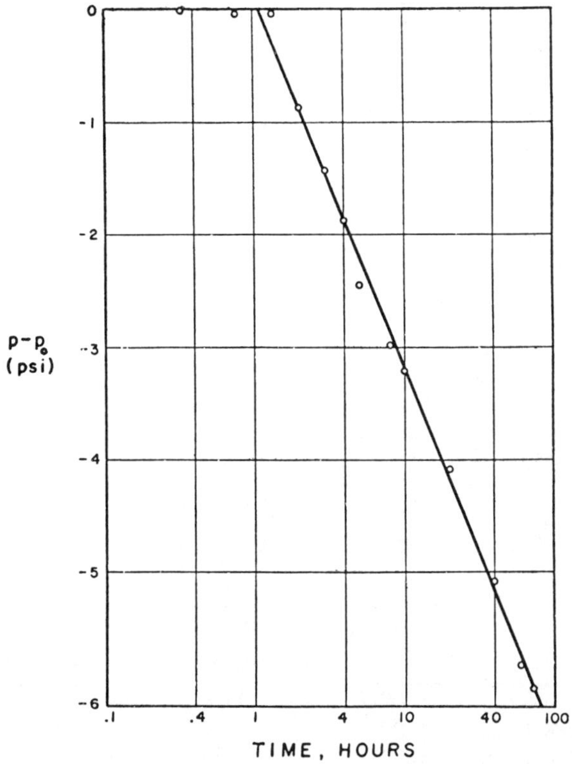

Figure 6-16. Drawdown curve for example problem.

From the data, calculate the pressure drop for each time step:

A plot of p vs. $\log_{10}t$ is shown in Figure 6-16. The slope is − 3.31 pounds per-square-inch per-cycle. Hence:

$$q\mu\left(-\frac{162.6}{kh}\right) = -3.31 \text{ pounds per-square-inch per-cycle}$$

or:

$$k = \frac{(162.6)(1{,}370)(1)}{(3.31)(164)} = 410 \text{ md}$$

For a truly homogeneous formation, the two k values calculated above should be identical. In case they differ, the second k value determined from the observation well pressure plot should be used in the following compressibility calculations.

Calculate the compressibility at t = 72 hours (3 days), $\Delta p = -6.04$ lb/in.^{2} From Equation 6-31:

$$\log_{10}\left(\frac{\mu\phi c r^2}{0.00633\ kt}\right) = \frac{\Delta p}{\left(\dfrac{162.6\ q\mu}{kh}\right)} + 0.3513$$

$$= \log_{10}\left(\frac{\mu\phi c r^2}{0.00633\ kt}\right) = \frac{-6.04}{3.31} + 0.3513 = -1.4734$$

$$\frac{\mu\phi c r^2}{0.00633\ kt} = 0.0339$$

$$c = \frac{(0.0339)(0.00633)(410)(3)}{(1)(.145)(500)(500)} = 7.3 \times 10^{-6}(\text{vol})/(\text{vol})(\text{psi})$$

Pressure Observations Across the Caprock

For an observation well above the caprock zone to be effective in taking observations over periods of days, it should be completed in a zone of greater permeability than the caprock. The observation of a pressure drawdown for a well complete in the first permeable zone above the caprock, while a well in the proposed storage zone is pumped, is considered to be the best confirmation that leakage is taking place between the two zones.

Early Development of Storage Zones in Aquifers

When an aquifer storage field is ready for gas injection, one would expect to have at least two storage wells completed on top of the structure with water standing in these wells somewhere near the surface. How fast should gas be injected initially? Water moves relatively slowly, and it will take time for gas to displace the water into the formation. It is recommended that gas pressure be applied to the well head in 100 lb/in.^{2} increments with observation of the recession of the water level in the wellbore. The gas injection pressure for depths of 1,200 ft or more could well be 100 to 150 lb/in.^{2} above the initial aquifer pressure. Sudden increases in wellhead pressure with a full water column would increase the bottom hole pressure almost instantaneously, running the risk of fracturing the reservoir rock. It may be two or three days before the formation will begin to take gas at a rate equal to some 50% of its ultimate capacity.

Once gas injection has started, the neighboring observation well should show a pressure response in a matter of hours and is likely to show the presence of gas in a few days. The initial wafer of gas may be very thin, especially for layered systems. This means that during initial gas injection, gas could travel relatively great distances. This gas layer may travel down-structure as much as 100 ft before the effects of buoyancy arrest the downward movement.

Figure 6-17 gives a view as to the mechanism by which gas enters the water-bearing rock. The instability of gas displacing water can be very great. Gas will preferentially enter the zones with the highest permeability. Once gas starts moving into a zone, it will continue to flow, because the pressure is transmitted more easily through this gas layer than through the water layers due to the differences in viscosity of water and gas. Figure 6-17 also shows how the gas/water interface may be conceived to develop. Water retained in the layers between those containing gas will drain down through the gas layers by a slow process. Eventually the gas bubble will develop a "bottom," but during high rates of injection, even in later years, the instability effects may create a jagged interface between the gas bubble and the water zone.

Although only one well is shown in Figure 6-17, simultaneous injection may occur in a group of wells. The practice is often followed by injecting

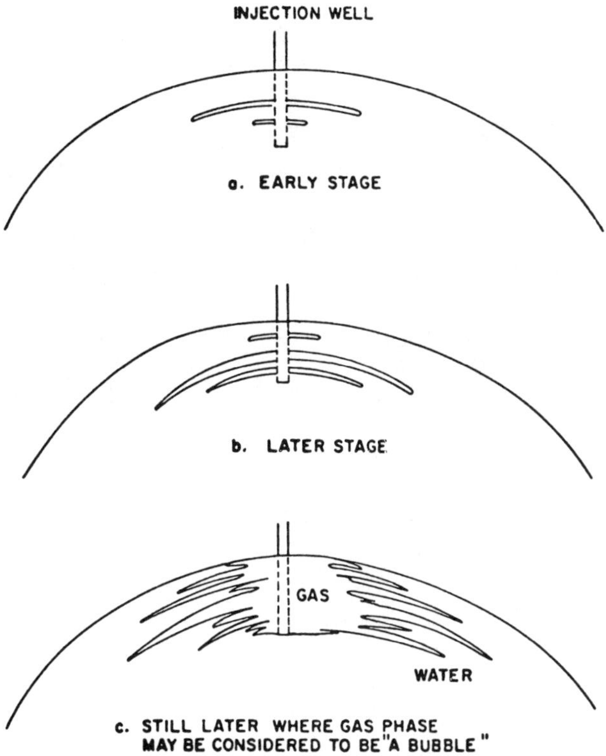

Figure 6-17. Development of gas bubble in an aquifer.

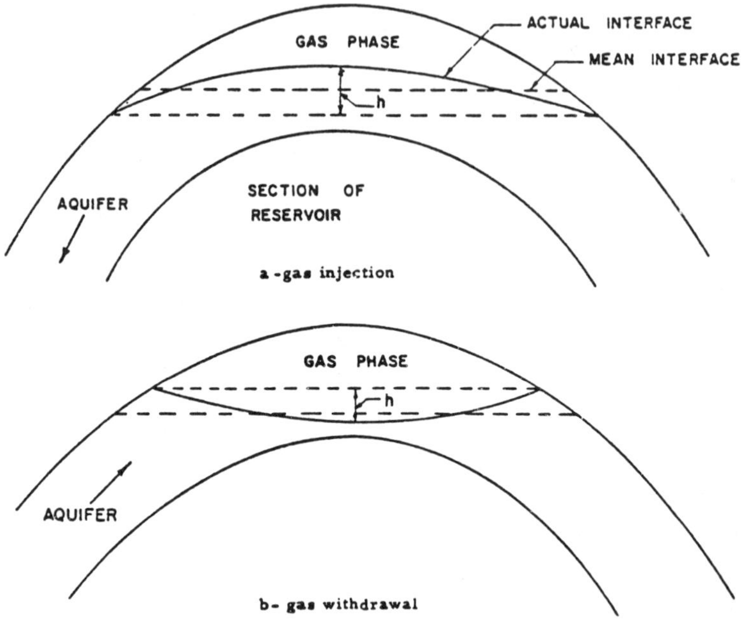

Figure 6-18. Illustration of bubble bottom shape for aquifer.

into the second, third, and fourth wells, etc., only after the gas zone has reached the well in question. This, in effect, assures that only one gas bubble is being created. It is expected that a series of separate bubbles simultaneously created around a group of wells will eventually coalesce. However, in the early days, one can conceive of some undue interference with water drainage patterns by having several gas zones. In a matter of one to three years, the water in the sand at the "ceiling" of the reservoir will have drained down to its residual content and the gas bubble will correspond to the structure of the caprock.

Figure 6-18 shows a sketch of an aquifer bubble after full coalescence has taken place and much of the water has drained out. In effect, the gas bubble is now similar to a water-drive gas field. When water is being pushed out, the gas water contact is concave downward (see Figure 6-18), since some head of water is required to move the water from the center of the bubble to its periphery. The shape is concave upward (see Figure 6-18) upon gas withdrawal.

References

1. Schiltuis, G. G. and W. Hurst. "Variations in Reservoir Pressures in East Texas Field," *Pet. Trans. AIME* (1935) 114, 164.

2. Katz, D. L., M. R. Tek, K. H. Coats, M. L. Katz, S. C. Jones, and M. C. Miller. "Movement of Underground Water in Contact with Natural Gas," Monograph, American Gas Association, New York (1963).
3. Hubbert, M. K. "Entrapment of Petroleum under Hydrodynamic Conditions," *Bull. AAPG* (1953) 37, 1954.
4. Bell, J. S. and J. M. Shepherd. "Pressure Behavior in the Woodbine Sand," *Trans. AIME* (1951) 192, 19.
5. Fatt, I. "Pore Volume Compressibility of Sandstone Reservoir Rocks," *Trans. AIME* (1958) 213, 362.
6. Hall, H. N. "Compressibility of Reservoir Rocks," *Trans. AIME* (1953) 198, 309.
7. Hutchinson, T. S. and V. J. Sikora. "A Generalized Water Drive Analysis," *Pet. Trans. AIME* (1959) 216, 169.
8. Hicks, A. L., A. G. Weber, and R. L. Ledbetter. "Computing Techniques for Water Drive Reservoirs," *Trans. AIME* (1959) 216, 400.
9. Churchill, R. V. *Modern Operational Mathematics in Engineering*, McGraw Hill, 1944.

7

THRESHOLD PRESSURE IN UNDERGROUND STORAGE

During the years 1967–1970, a comprehensive research program was conducted at the University of Michigan on "Study of Threshold Pressures for Gas Displacing Water from Caprock of Underground Storage Reservoirs" under the sponsorship of the Pipeline Research Committee of AGA. This program resulted in publication of the monograph entitled "Threshold Pressure in Gas Storage" in 1970[1] authored by M. A. Ibrahim, M. R. Tek, and D. L. Katz. The purpose of this chapter is to document, highlight, and summarize the findings published in the monograph with permission from the American Gas Association.

The first recorded experiment in underground storage was the successful injection of natural gas into a depleted gas field in 1915 in Welland County, Ontario, Canada. The gas injected during the summer was withdrawn the following winter to meet peak shaving requirements.

The next storage project was in the U.S. at Zoar Field, south of Buffalo, New York in 1916. The third recorded underground storage operation was in Menifee Field, Kentucky in 1919 by Central Kentucky Natural Gas Company. By 1931, nine storage pools were in operation in six different states with total storage capacity of 18 BCF. In 1930, the Bow Island Field in Alberta, Canada was used to store gas from Turner Valley Field for conservation purposes.

The continuing success and phenomenal growth of underground storage in the U.S. resulted in over 7 trillion ft^3 of gas in storage presently, with some 4.89 TCF recoverable gas, as of 1979. Underground storage is practiced in some 380 pools in 26 states. World-wide, underground storage is a significant sector of the energy industry in Great Britain, The Netherlands,

France, Italy and Germany. In other countries such as Yugoslavia, Iraq, Spain, Australia, and New Zealand, significant developments are now taking place. All of these developments critically rely on caprock integrity to prevent upward migration of storage gas. In the early days, almost all of the gas in storage was in depleted producing fields. Each reservoir was recognized as being blessed with a caprock which, after all, had held the gas over geological eons. During those years, no one questioned the integrity of caprock because discovery pressures were never exceeded.

Underground storage reservoirs operate between two pressure limits. At the end of the injection period, when the working inventory is restored to its value at the start of withdrawal, the pressure is maximum. On the other hand, when the full working inventory is withdrawn and sent to market, the storage pressures become equal to a minimum value. This occurs at the turnaround to injection season. How far down one can reduce the storage pressure depends upon the ability to compress the gas so it can be sent to the pipeline for transmission to market. The obverse question—how far up can one go in storage pressures or inventory?—is a function of the reservoir, related to the depth in the form of a ratio called the *pressure gradient.* For instance, the bottom-hole discovery pressure divided by the average depth of the reservoir is called the *discovery gradient.* If the water that saturates the pore space in the crust of the earth is pure, the pressure gradient would be 0.433 psi/ft. If, on the other hand, the pore space is full of brine (heavier than pure water) the gradient would be, say, 0.53 psi/ft or higher. Such fluid pressure gradients encountered in discovery of reservoirs are shown in Figure 7-1 as pressure-depth curves having slopes corresponding to brine densities.

Figure 7-2 shows the range of pressure gradients encountered in underground storage technology. The limiting curve to the right corresponds to formation lifting pressure gradient of 1 psi/ft. This is a value one must safely stay away from in storage because it corresponds to the composite density and weight of the overburden. If and when that gradient is reached or exceeded in reservoirs having depths less than 5,000 ft, the formation will part and horizontal fractures will develop. When the pressure gradient corresponding to the maximum operating pressure of a storage field exceeds the discovery gradient for that particular field, the practice is called *overpressuring* or *Delta-pressuring.* Delta-pressure is a necessity in aquifer storage. It is safely practiced with due caution in storage in depleted gas or oil reservoirs. When such a reservoir is pressured beyond its discovery gradient, however, one must not exceed the threshold pressure corresponding to the caprock in the overburden. Accordingly, the threshold pressure is that pressure which will cause the gas to penetrate into and move continuously across the caprock. It is either directly measured or estimated by engineering correlations developed and documented in Reference 1.

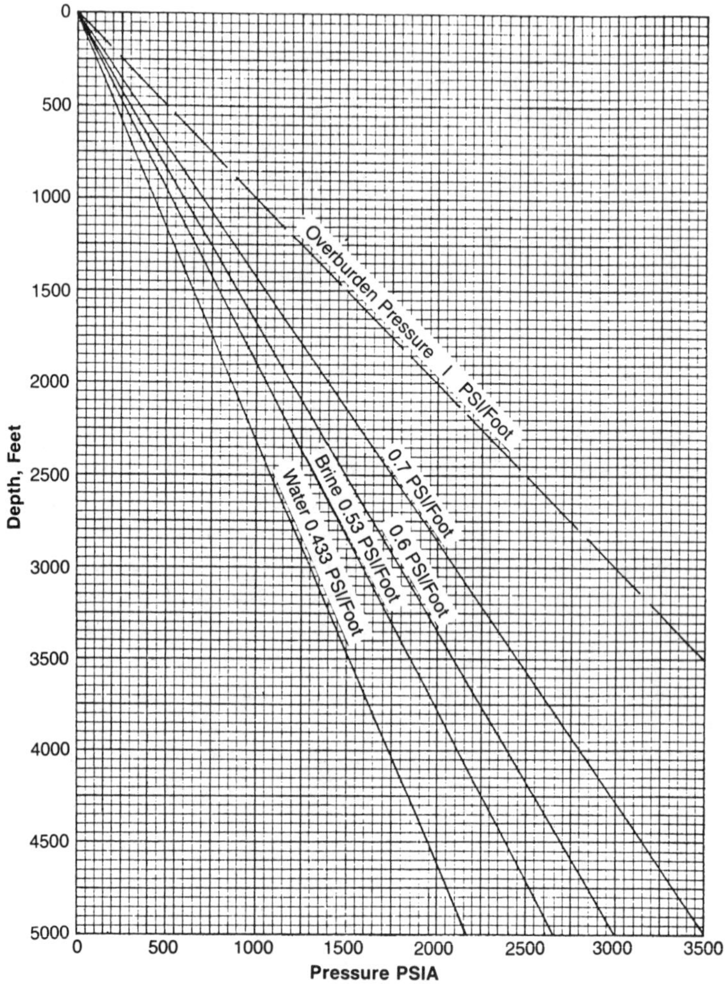

Figure 7-1. Fluid pressure gradients in the earth.[1]

CAPROCKS AND THRESHOLD DISPLACEMENT

A schematic view of a caprock cross section is given in Figure 7-3 (a and b). The term caprock means it overlies the gas storage reservoir. In the overburden, one usually finds other permeable zones. The caprock may be shale, or carbonate as dolomite, limestone, or anhydrite. The shales and most carbonates have some measurable permeability.

The gas in the sand at A does not enter the shale even though under higher pressure. The gas does not drive the water out of the fine pores be-

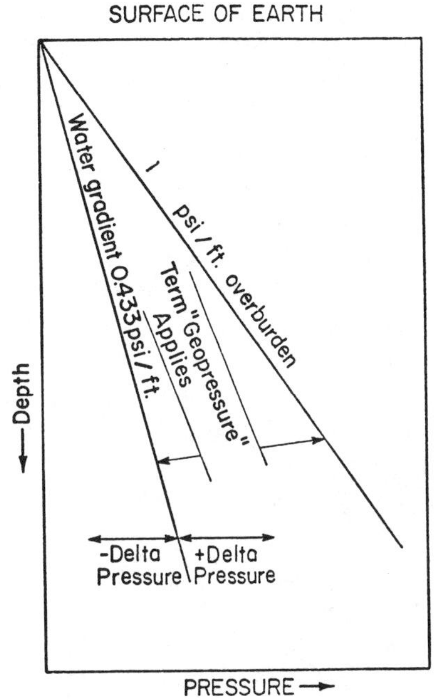

Figure 7-2. Range pressure gradients in storage.[1]

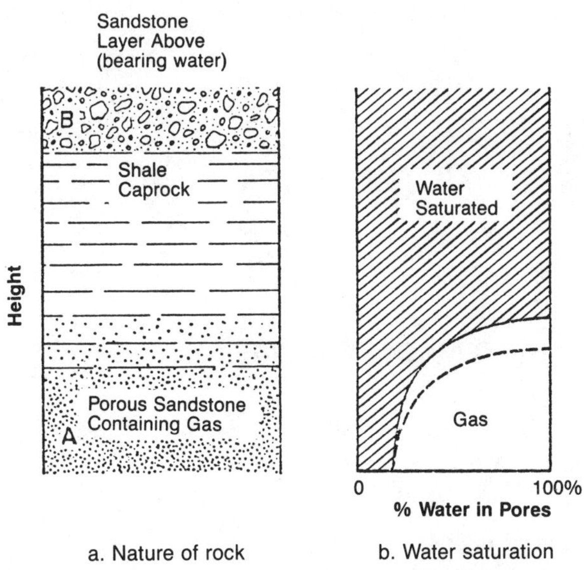

Figure 7-3. Nature of the gas-water contact in the caprock.[1]

cause surface forces of water wetting the solid rock hold the gas back. So, the water saturation takes on the distribution shown in Figure 7-3b. At some point there is an interface between the storage gas and the water-filled pores of the caprock. Depending upon the pressure gradient, the saturation may vary as shown by solid and dotted curves.

To simulate this behavior of caprock, the threshold pressure test was developed using cores. Figure 7-4 shows how the core, saturated with water, is held by a rubber sleeve. With gas under pressure below the core and water at atmospheric pressure above the core, the surface forces in the caprock specimen confine the gas. The test consists of increasing the gas pressure until the gas overcomes surface forces and displaces water from the caprock, flowing at the top of the core. The purpose of the test is to determine the pressure differential required for gas to displace water from caprock cores.

The concept of displacement pressure for gases to displace water from porous solids has been known for many years.

Figure 7-4. Diagram of high pressure core holder for threshold displacement tests.[1]

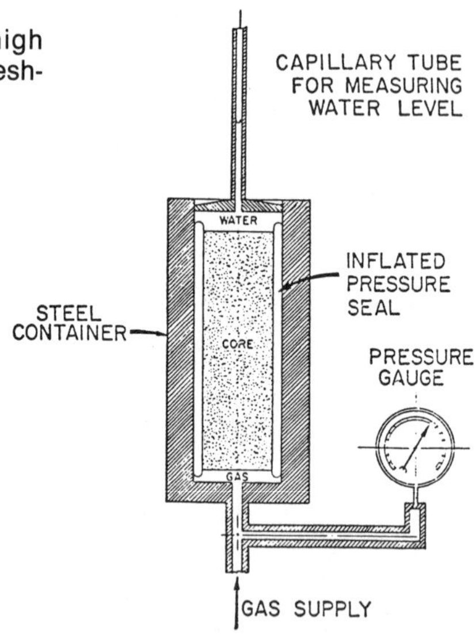

CAPROCKS IN GAS STORAGE

Understanding the performance of underground storage reservoirs involves intimate knowledge of several factors which affect the characteristics of the reservoir. These include:

1. Properties of reservoir solids
2. Geometric and structural features
3. Characteristics of reservoir fluids

4. Thermodynamic conditions prevailing (the temperature and pressure)
5. Operating conditions imposed upon the reservoir (withdrawal, injection rates, and pressure schedules)

Properties and variables in some of these factors are better understood than others. The caprock in the overburden of the storage reservoir will be discussed next as it relates to its main function containment of storage inventory against migration.

Geological Background

The majority of the storage fields, including aquifers and depleted oil or gas reservoirs, are located within or adjacent to large structural basins into which thousands of cubic feet of sediments were deposited during geologic times.

The clastic sediments were derived from adjacent land masses and transported by various agencies—usually wind and water—into their depositional environment. The carbonates, on the other hand, generally represent *in situ* chemical precipitates.

During the compaction and aging process, some of the strata took on the characteristics to become caprocks rather than reservoirs containing hydrocarbons. During geologic times, sandstone layers generally retained permeability so that the fluids would flow through them readily. Carbonates, on the other hand, were subjected to solution and precipitation processes from percolating waters. Carbonate strata could become either caprocks with low porosity and permeability or oil and gas reservoirs with higher porosity and permeability.

During geologic times, in addition to compaction and deformation processes, it is known that tectonic forces often created discontinuities in the form of faults and fractures.

In considering the geology of a given reservoir, the areal or basin characteristics become part of the underlying formation. In a basin, there may be trends of increasing salinity of water as the strata dips down. The porosity may decrease with deeper burial.

Although there is a spectrum of porosities and permeabilities for underground rock layers, a distinction can be made between the strata which are potential caprocks and those which have accumulated hydrocarbons. Table 7-1 gives typical porosity and permeability ranges for the rock matrices for caprocks and producing reservoirs.

Nature and Evaluation of Caprocks

The upward seepage of hydrocarbons by buoyancy through the sedimentary rocks is occasionally prevented by blanket-type rocks constituting the

Table 7-1
Ranges of Porosity and Permeability of Hydrocarbon Bearing Sands and Caprocks

Porosity, percent		Permeability, millidarcies
Oil	6–35	0.1 to 4,000
Gas	4–35	0.01 to 7,000
Caprocks	1–12	zero to 0.0001

upper boundaries of reservoirs. Such rocks, which effectively interdict the net vertical migration of gas or oil, are called *caprocks.* Shales, limestones, and anhydrites are often discovered as natural caprocks to many oil- and gas-bearing strata.

The rock comprising the storage container for natural gas obviously must have more porosity than the overlying caprock. Likewise, the transmissibility of fluids through such a rock must be many times greater than that of the caprock. Essentially all the core studies made by the petroleum industry over the past four decades were concerned with the reservoir material, not the caprock. The advent of aquifer storage and the practice of overpressuring introduced the need for close examination of caprock cores.

There are three primary areas of concern in studying a potential caprock for a storage reservoir. The first is its geometric shape. Figure 7-5 shows the basic elements of an underground storage reservoir, including the inverted saucer shape to confine the gas within a specified area. The closure, the vertical distance from the top of the storage container to the highest point at

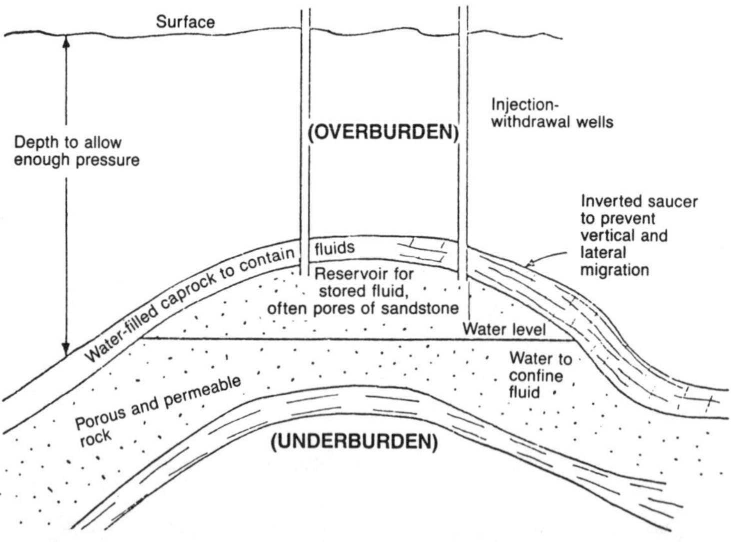

Figure 7-5. Basic elements of underground storage reservoirs.

which gas may leave the container, is of primary interest as one measure of the caprock shape. The shape is determined by geologic mapping based on data obtained from wells.

The second concern has to do with the nature of the caprock matrix. The permeability, porosity, and threshold displacement pressure should be measured on core specimens. The selection of core samples and observations made while inspecting well cores of caprocks are important considerations.

The third concern for caprocks is with regard to their continuity and their freedom from anomalies. The extrapolation of information on cores to the entire caprock confining the gas can be made with assurance only after evaluating the continuity of the strata upon which measurements were made.

A distinction should be made between converted oil and gas reservoirs used for gas storage, and aquifers, i.e., man-made gas storage reservoirs. For the converted storage reservoir, the accumulation process for oil and gas over geologic time has proven that the above three desired characteristics of the reservoir were present. When using high delta pressures in such reservoirs, one would often verify that the caprock has appropriate threshold displacement pressures. Likewise, should pressure cycles designed for the operation of storage reservoir cause the gas bubble to increase in size and expose new caprock to gas below it, then the shape, continuity, and anomalies become important.

Heterogeneities of significance may and do occur in gas storage reservoirs and in their associated caprocks, some of which are of sufficient magnitude to produce significant variations in permeability, both horizontal and vertical.

Many storage reservoirs have caprock material overlying them which is from twenty to several hundred feet thick. It behooves the observer to ascertain that within such a thick caprock there is a continuous layer of sufficiently low permeability and high threshold pressure so as to provide the retention capability desired. A thin shale, 5 ft in thickness, would be ample to retain gas if it were continuous. Discussions of caprock thickness often have more to do with implications of continuity and ability of cementing operations to seal the wellbore than of the gas percolation through the rock layer itself.

The importance of the sampling procedure of the caprock material for analysis cannot be overemphasized. Facies changes which represent discrete lateral and vertical variations in lithologies must be recognized, if the resulting core analyses are to be properly interpreted.

PRESSURE GRADIENTS RELATED TO CAPROCKS

The capacity of caprocks to hold the gas beneath is related not only to their tight porous structure, but, more importantly, to the water seal provided by their 100% water saturation. Accordingly, it becomes desirable as well as necessary to understand the hydrostatics of pressure distributions related to caprocks.

The gradient of fluid pressures prevailing in the pore space in the subsurface are related to depth and the specific gravity of the fluid column. The pore pressure existing in the crust of the earth may be evidenced by the height to which water will stand in a well drilled and completed to a water-bearing sand as shown in Figure 7-6. The height to which water stands in well W above point A provides what is called a *piezometric pressure level* at depth D.

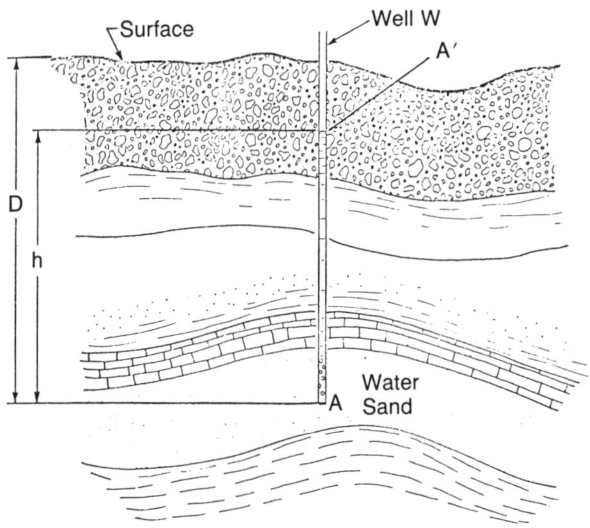

Figure 7-6. Cross-section relating the depth to hydrostatic column of native water.

Observing that the well is opened to atmosphere at top and to the water-bearing sand formation through the perforations at the bottom, the actual pressure prevailing at point A may be computed from simple hydrostatic considerations as follows:

$$p_A = p_{A'} + \rho_w \frac{g}{g_c} h \tag{7-1}$$

where $p_{A'} \cong p_{a'}$ atmospheric pressure and, accordingly:

$$p_A = p_a + \rho_w \frac{g}{g_c} h \tag{7-2}$$

where p_A = pressure at point A, pounds-force per square feet abs.
p_a = atmospheric pressure, lbs/ft^2 abs.
ρ_w = density of water, pounds-mass/ft^3
g = acceleration of gravity = 32.17 ft/sec^2
h = height of water column, ft

$$g_c = \text{conversion factor} = 32.17 \left(\frac{lb_m}{lb_f}\right)\left(\frac{ft}{sec^2}\right)$$

If it may be assumed that the liquid column is pure water having the density of 62.4 pounds-mass/cu ft, then Equation 7-2 may be simplified to give directly the static pressure gradient in psi per foot:

$$p_A - p_a = \frac{62.4}{144} \times \frac{32.17}{32.17} h \tag{7-3}$$

$$\Delta p = 0.433\ h,\ psi \tag{7-4}$$

Equation 7-4 permits direct conversion of height above datum h to gage pressure prevailing at the datum.

If the water is brackish and has certain salinity, then the pressure gradient, $\Delta p/\Delta h$, may be equal to values such as 0.45 psi/ft or 0.52 psi/ft for densities of 1.04 or 1.2 gm/cc.

Figure 7-1 shows various such fluid pressure gradients encountered in earth strata. In this figure, the uppermost curve is the gradient due to the weight or density of the overburden itself.

Now, going back to Figure 7-6, it may be remarked that if the standing leg of water h exceeds the total depth D of the well, the water would cause positive gage pressure at the well head and may be produced naturally as an artesian well.

If the well were completed in a tight shale, it would take a very long time for it to fill up with water, the production being limited by the very low permeability of the shale.

HYDRAULICS OF THE WELLBORE ENVIRONMENT

Any natural gas reservoir at discovery has a step change in pressure within the caprock due to a capillary retention phenomenon. Figure 7-7 shows the pressure gradients for such a natural gas field. Water in the caprock is subject to the same hydraulic gradient AB as the water underlying the gas reservoir. The pressure gradient prevailing in the gas reservoir is that of a gas column as shown by curve BC. These two pressure curves, when

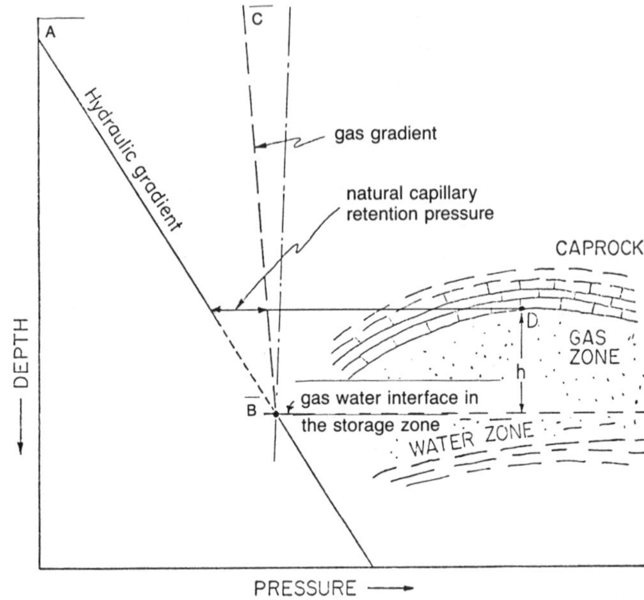

Figure 7-7. Illustration of natural capillary retention pressure.[1]

viewed at the level of the gas-water interface within the caprock (D), will indicate a step change in pressure between the gas phase and the water phase called the Natural Capillary Retention Pressure.[1] Its value is computed as follows:

$$P_{NCR} = \frac{h \times (\rho_w - \rho_g)g}{144\ g_c} \tag{7-5}$$

where P_{NCR} = natural capillary retention pressure, psi
h = height of gas column, caprock to gas/water contact, ft
ρ_w = density of water column, lb_m/ft^3
ρ_g = density of gas column, lb_m/ft^3
g/g_c = (lb_f/lb_m)

It has been known that capillary phenomena were responsible for retaining oil and gas in underground reservoirs. The low permeability alone would not confine hydrocarbons, which would continue to percolate across the caprocks due to their buoyancy. Thus, the capillary retention by the caprock pore is responsible for confining natural gas at discovery pressures. The use of delta pressures implies an increase in the natural capillary retention pressure at the caprock.

Figure 7-8 relates the pressure gradients above, through, and below the caprock to the formations and fluids encountered at various levels. The cross section (A) depicts a section of the storage reservoir with point A, the original water level, near the surface; point B, the top of the caprock, presumably in contact with a porous and permeable zone above; and point C the bottom of the caprock adjacent to transition from caprock to gas storage sand below. Point D may be considered the gas/water interface at some point in the development of the storage reservoir. Point E is a point in the water-bearing storage zone at some greater depth (the maximum possible depth the delta pressure will displace water).

Under normal conditions, before gas injection, one would find a relatively smooth hydraulic pressure gradient to prevail from near the surface down to the storage zone. This pressure gradient often corresponds to that of relatively pure water, say 0.44 psi per foot of depth. Accordingly, at point B, 2,050 ft from the surface or 2,000 ft below original water level, the pressure would be 880 psig or about 895 psia. Likewise, the pressure at point C at the

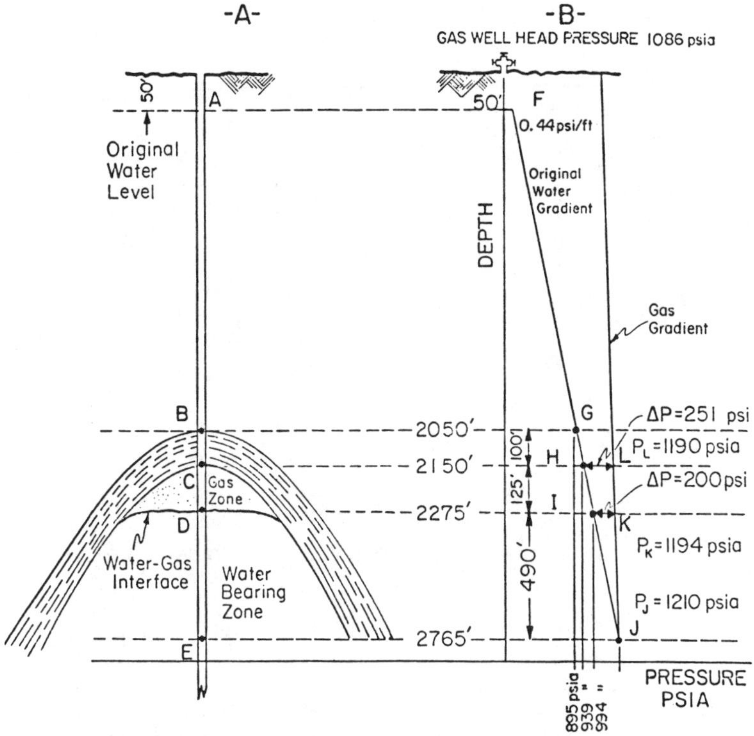

Figure 7-8. Pressure gradients in aquifer storage.

bottom of the caprock, 100 ft deeper, would be 44 psi more or 939 psia.* Point D, at 125 ft deeper, would be subject to another increase in pressure of 55 psi, making it 994 psia as shown in part B to the right depicting the gradients.

Now let us consider what happens to the pressures when the pressure at point D at the gas/water interface is increased by 200 psi due to injection of gas displacing the water. In the particular case illustrated here, Figure 7-8, the point K, that which corresponds to point D, is shown at a pressure 200 psi higher than the prevailing hydrostatic shown as point I. The actual pressure distribution within the caprock is shown in Figure 7-9 as the curve

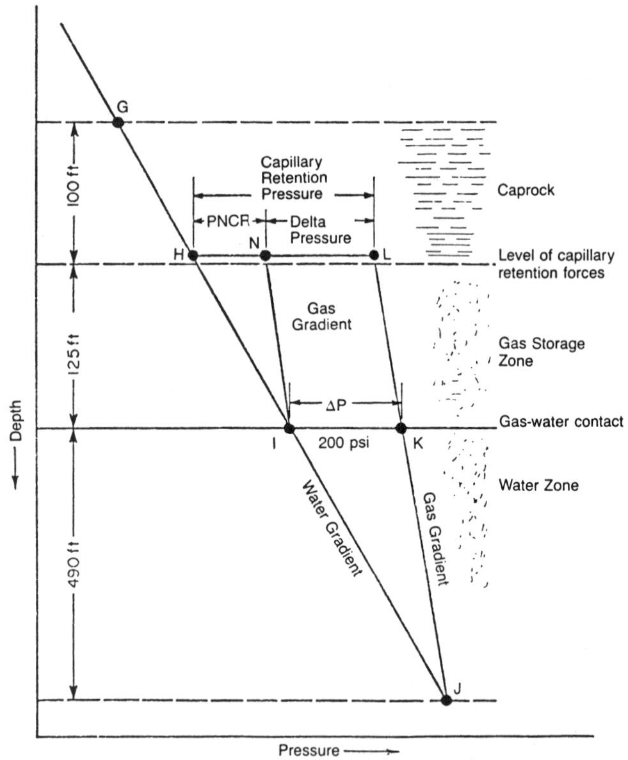

Figure 7-9. Pressure gradient in caprock with delta pressure.[1]

* In actual field practice, one may encounter 5, 10, or 20 psi difference between pressures corresponding to normal hydraulic gradient and the actual pressures prevailing in various individual zones. Sometimes, these pressure differences may be caused by water production when one is close to villages or other water production wells. Seasonal or climactic changes for water pressures, difference in water movement in different zones, salinity of water, other factors may also account for differences in the pressures mentioned above.

GHNL. The portion HNL of this curve represents that transition between water pressures at the top of the caprock to the water pressure just at the gas/water contact.

CONCEPTS RELATED TO PHYSICS OF THRESHOLD DISPLACEMENT

The threshold displacement phenomenon involves the displacement of a wetting phase from the pore space by a non-wetting phase, the process starting with a porous medium fully saturated by water. In true engineering perspective, the threshold displacement became a focal point of interest as a result of the successful practice of operating depleted oil or gas reservoirs in gas storage service at pressures exceeding their discovery values and using delta pressures in aquifer storage. The fact that oil or gas has indeed accumulated through geologic times indicates that the water seal in the caprock is effective at least for pressures in the oil- or gas-bearing formation up to and at least equal to the discovery values. Natural capillary retention pressure has received little attention but it is generally known that surface effects retained our oil and gas reserves rather than the low permeability itself.

In order to be precise and clear, one may refer to the actual *in situ* condition prevailing at the interface between the gas of the porous and permeable formation and the native water saturating the tiny pores of the tight and much less porous caprock. If the actual pressure level is then considered, the threshold pressure carries the connotation of threshold pressure level. On the other hand, if one merely considers the difference between the gas pressure and the prevailing hydrostatic pressure in the caprock when the threshold displacement occurs, then the threshold pressure term implies the connotation of threshold displacement pressure differential to be designated in units of psi rather than psia or psig. Hereafter, the term threshold pressure refers to the displacement pressure differential.

The position, shape, and location of gas/water interface depends upon the forces that act on both wetting and non-wetting phase fluids. These forces are due to gravity (or buoyancy), viscous friction in the case of flow, and pressure difference due to capillarity. The capillary forces which control the interface are of the same nature as the forces which cause the liquid to rise in a capillary tube. The capillary rise, in fact, is often used as the basis for laboratory measurement of the surface tension.

Figure 7-10 shows the relationship between the pressure discontinuity across the capillary and the gradient due to the density of the liquid. The sketch at the left labeled A shows how the internal forces within the water layer wetting the wall of the glass pull the water up through the height of capillary rise until gravity forces on the water column balance the surface forces. If above the meniscus, in the capillary tube, a gas pressure equal to

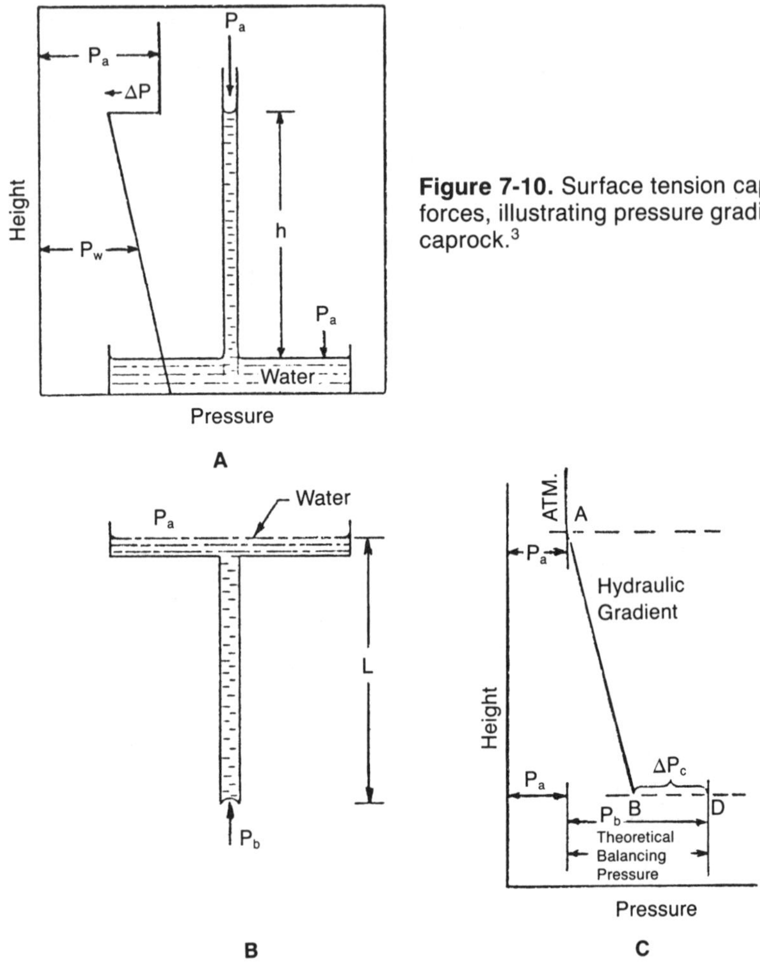

Figure 7-10. Surface tension capillary forces, illustrating pressure gradient in caprock.[3]

P_a (atmospheric pressure) plus ΔP (due to either capillarity or water head) is applied, the meniscus would be depressed down to the main water level. The sketch B shows an inverted capillary. In this case, not only capillarity, but the gravity as well, pulls the water downward from the main container. In this case, the pressure required to push the meniscus back would be the sum of the hydraulic equivalent of AB plus the capillary discontinuity BD as shown on sketch C.

Concept of Capillary Pressure (Single Capillary Tube)

When a single meniscus separating two phases, one wetting the solid and the other non-wetting, are considered (again referring to Figure 7-10), the pressure discontinuity ΔP_c or BD, across the curved meniscus, is called capillary pressure. The equation defining the capillary pressure is given as:

$$P_c = P_{nw} - P_w \tag{7-6}$$

where P_c = capillary pressure
P_{nw} = pressure of the non-wetting phase fluid at the level of meniscus
P_w = pressure of the wetting phase fluid at the corresponding point

A general equation is given frequently in the literature relating the above defined capillary pressure to the principal radii of curvature of the meniscus:

$$P_c = \sigma\ (1/R_1 + 1/R_2) \tag{7-7}$$

where σ = surface tension of interfacial tension between wetting and non-wetting phases
R_1 and R_2 = the principal radii of curvature of the meniscus.

It must be noted that the force units in P_c must be consistent with the force unit included in the surface tension term; i.e., if the pressures are to be expressed in pounds-force per square foot, then σ must be given in units of pounds-force per foot. In the CGS system, σ is in dynes/cm and pressures in dynes/cm^2.

As mentioned earlier, the rise of water in a capillary may be related by a simple force balance or by considerations of fluid statics to radius of capillary surface, forces due to surface tension (or interfacial tension), and the contact angle of meniscus with the capillary wall, Figure 7-11.

$$2\pi\ r\ \sigma \cos\theta = \pi\ r^2\ h\ \rho\ g/g_c \tag{7-8}$$

where r = radius of capillary, ft
σ = surface tension, pounds-force/ft
θ = contact angle, radians
h = height of capillary rise, ft
ρ = density of water, pound mass/ft^3
g = acceleration of gravity = 32.17 ft/sec^2
g_c = conversion factor = 32.17 (lb_m/lb_f) ft/sec^2

Now in the above analysis, if the density of both wetting and non-wetting fluids are considered, starting with the pressure prevailing at reference da-

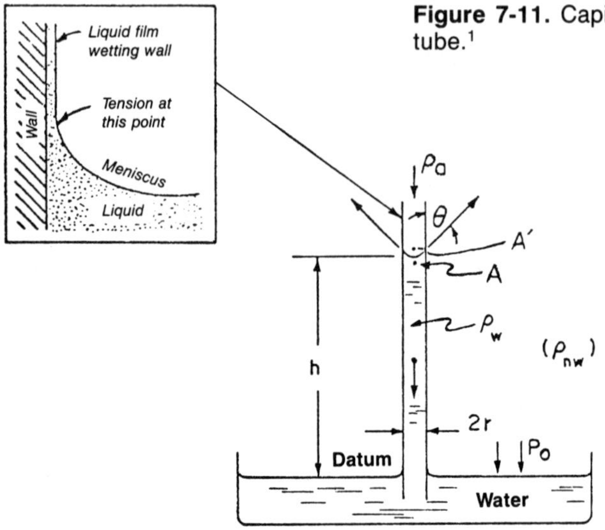

Figure 7-11. Capillary rise in a small tube.[1]

tum, the wetting and non-wetting phase pressures may respectively be evaluated at points A and A′, as follows:

$$P_{A'} = P_a - \rho_{nw}\, g/g_c\, h \tag{7-9}$$

$$P_A = P_a - \rho_w\, g/g_c\, h \tag{7-10}$$

Now, subtracting Equation 7-10 from Equation 7-9

$$P_{A'} - P_A = (\rho_w - \rho_{nw})\, g/g_c\, h \tag{7-11}$$

But:

$$P_{A'} - P_A = P_c \tag{7-12}$$

by the definition of capillary pressure. Accordingly:

$$P_c = (\rho_w - \rho_{nw})\, g/g_c\, h \tag{7-13}$$

Equation 7-13 relates the capillary pressure to the capillary rise and the density difference between the wetting and non-wetting phases.

For the case of a simple experiment with single capillary tube involving air and water, $\rho_{nw} = \rho_{air} << \rho_{water}$, then:

$$P_c \cong \rho_w\, g/g_c\, h \tag{7-14}$$

and substituting this in Equation 7-8:

$$2\pi \text{ r } \sigma \cos \theta = \pi \text{r}^2 \text{ P}_c \tag{7-15}$$

which in turn gives:

$$\text{P}_c = 2\sigma \cos \theta / \text{r} \tag{7-16}$$

Equation 7-16 indicates that in a single, simple capillary, the capillary pressure depends upon surface (or interfacial) tension, contact angle (a measure of the degree of wettability), and the radius of capillary. In most work involving water with glass or rock, cos θ is taken as a unity, since complete wetting is assumed.

Capillary Pressure in Porous Media

If the container of Figure 7-11 had several capillaries of different radii immersed in it, the water would rise to different heights at different capillaries depending upon their radii, Figure 7-12. Now if one views a porous medium, a complex network of capillary pores interconnected in various directions, intersecting one another, and each of variable radius, it becomes clear that at any height considered there may exist a capillary such that the inferface between wetting and non-wetting phases may be present. Accordingly, then at each point considered, one may think of the value of the capillary pressure as related to height above datum, densities of wetting and non-wetting fluids, surface (or interfacial) tension, and pore radius (or radius of capillary).

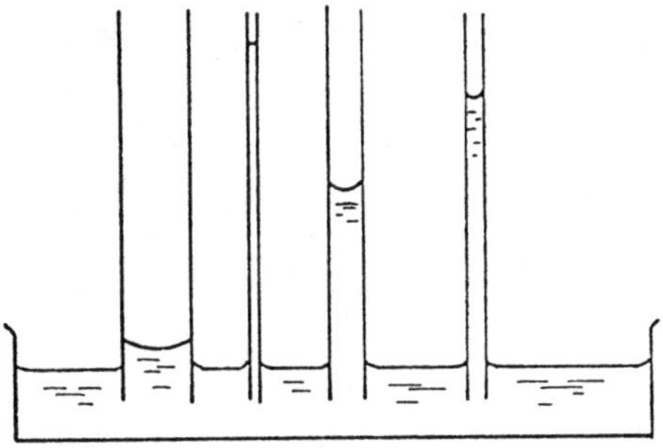

Figure 7-12. Capillary rise in tubes of different radii.[1]

Now consider a vertical tube containing sand standing in contact with water, as shown in Figure 7-13. One can observe that water would rise above A by capillarity to certain saturation distribution with height if the sand pack or core was bone dry at the start of the process. On the other hand, if the sand pack or core was fully saturated to begin with, then the saturation distribution above A (level of 100% water) would result from water draining down under the influence of gravity and against the resistance of capillarity. This process is called *drainage*. The opposite process, the one starting with the porous environment bone dry, is called *imbibition*. The capillary pressure curves corresponding to drainage and imbibition are not the same. The phenomena related to irreversible paths followed during drainage and imbibition is called *capillary hysteresis*.

Underground storage reservoirs developed on aquifers or the ones on depleted fields subject to "bottom" or "edge" water drive undergo repeated cycles of imbibition and drainage as the water moves in and out of the matrix occupied by the gas bubble. The zones affected become subject to two-phase flow where permeabilities become effectively reduced to relative permeability values and where some gas becomes entrapped at residual and immobile equilibrium saturation values.

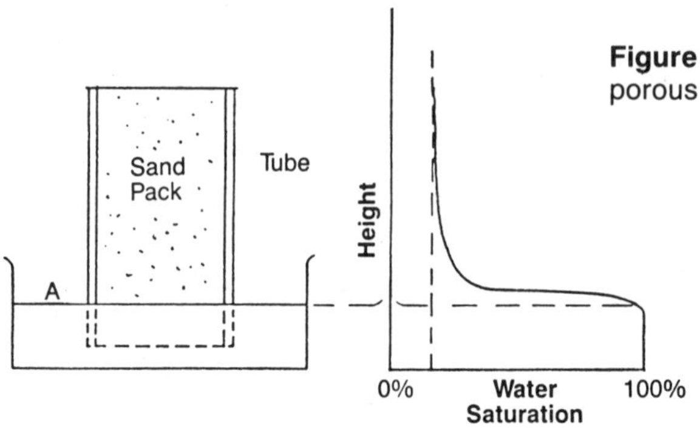

Figure 7-13. Capillary rise in a porous medium.[1]

GRADING OF CAPROCKS BY THRESHOLD DISPLACEMENT PRESSURES

All caprocks having integrity to retain the storage gas are fully saturated with water. If and when the threshold displacement condition is reached, the intrusion of gas represents drainage of gas from the caprock at $S_w = 100$ percent point.

Figure 7-14 shows a comparison between the drainage capillary pressure curves for different porous rocks. The intercepts of the various curves corresponding to sandstones, carbonates, shales, etc. at $S_w = 1.0$ axis represents the threshold displacement pressure for that particular caprock. The various curves are denoted by ranges of permeability showing, at least qualitatively, the wide corresponding variation in threshold displacement pressures. The point corresponding to anhydrites of no porosity and no permeability may

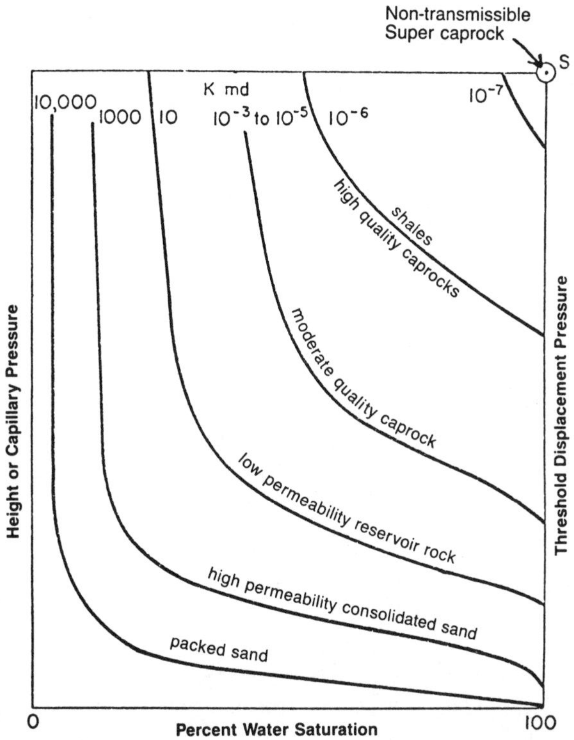

Figure 7-14. Variations in capillary pressure behavior for various rocks.[1]

be viewed as the limiting trend of drainage curve, a vertical straight line above S. The point is labeled *super caprock* because it represents the highest threshold pressure encountered.

CORRELATIONS FOR ESTIMATING THRESHOLD PRESSURES

Literature data, as well as new experimental measurements, were reported in the doctoral research by L. Kent Thomas.[4,5] Figure 7-15 shows his data, as well as those of Wyllie and Rose,[4] and Rose and Bruce.[4] The correlation over four logarithmic cycles is considered to be reasonably good.

Another correlation, purely empirical, suggested by Thomas on the basis of least squares fit of data was:

$$P_T = 7.37\ (1/k)^{0.43} \qquad (7\text{-}17)$$

where k is the permeability in millidarcies and P_T is the threshold pressure in psi.

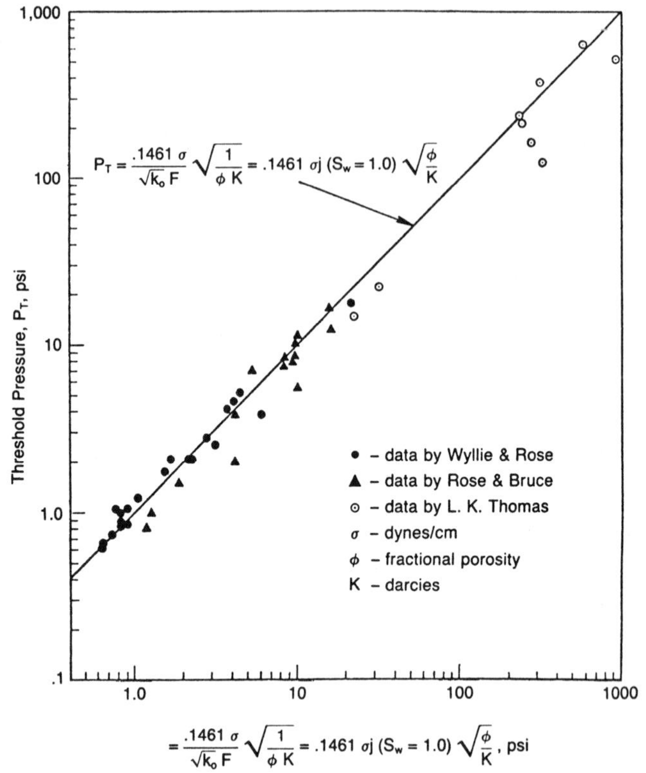

Figure 7-15. Correlation of threshold pressure with other properties of porous media (Thomas).[4,5]

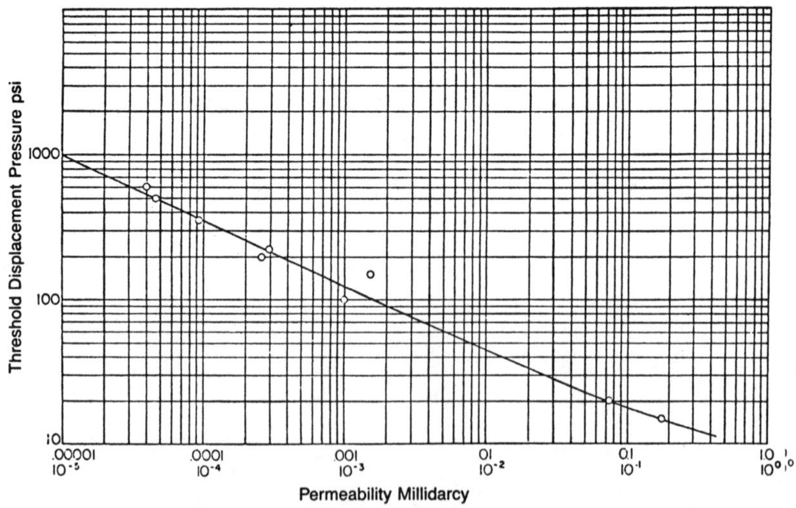

Figure 7-16. Threshold displacement pressure as a function of permeability.[1]

This correlation and the data upon which it was based are shown in Figure 7-16.[1]

References

1. Ibrahim, M. A., M. R. Tek, and D. L. Katz. "Threshold Pressure in Gas Storage," Monograph, American Gas Association, Inc., Arlington, Virginia, 1970.
2. Katz, D. L. "Aquifer Behavior in Gas Production," Oil and Gas Compact Bulletin (1960) XIX, No. 1, p. 14.
3. Katz, D. L., M. R. Tek, and M. A. Ibrahim. "Threshold Displacement Pressure in Gas Storage," Proc. of AGA Transmission Conference, April 13–14, 1970.
4. Thomas, L. K. "Threshold Pressure in Storage," Ph.D. Thesis, University of Michigan, 1968.
5. Thomas, L. K., D. L. Katz, and M. R. Tek. "Threshold Pressure Phenomena in Porous Media," *SPE Journal* (1968) Vol. 8, No. 2, p. 174.

8

HANDLING LIQUIDS IN NATURAL GAS PIPELINES

Episodes of condensate problems in natural gas pipeline systems in the early 1970s prompted industry to investigate two-phase flow in natural gas systems. The research at the University of Michigan, started in May 1971 under the auspices of the Michigan Gas Association, was enlarged and subsequently supported by the Pipeline Research Committee of the American Gas Association for a three year period starting in January of 1973. A monograph, "Retrograde Condensation in Natural Gas Pipelines," was completed in 1975.[1] The bulk of the research on the project was devoted to understanding the phase and chemical composition of the hydrocarbon systems and the ability to predict the quantity of condensate that would form under given pipeline conditions.

An auxiliary portion of the study involved the handling of hydrocarbon condensate when it did form. It was shown that the condensate formed was usually low volume percent liquid in transmission pipelines, but that the concentration of the liquid could increase many-fold en route to the market. It is believed that most of the physical problems with condensates interfering with metering, regulation, and compression were due to an accumulation of liquid from several sources by *route selectivity.* Methods for removing liquids from pipeline systems involve drips or separators in current practice.

One of the results of the research program was the use of extended gas analysis beyond the hexanes and heavier out to about C_{12}. Recent studies on fields producing retrograde condensate fluids have further extended the range of higher components to C_{28}.[2] Recent work based on thermodynamic equations of state such as the Peng Robinson Equation, implemented on the computers, have shown that one can reliably predict the dew point and liquid drop out using extended compositions "pseudoized" into major groups.[3]

Table 8-1 shows the equilibrium composition of a natural gas system in a pipeline with liquid and vapor in equilibrium at 615 psia and 40°F. This composition was used to predict the density and molecular weight of the phases and the surface tension of the hydrocarbon system reported.

Table 8-1
Composition and Properties of Typical Condensate in Pipeline
Natural gas condensate in pipeline at 615 psia, 40°F

Constituents	Gas phase	Liquid Phase
C_1	.96045	.1837
C_2	.03034	.0410
C_3	.00554	.0261
iC_4	.00115	.0142
nC_4	.00109	.0207
iC_5	.00047	.0233
nC_5	.00036	.0198
C_6	.00035	.0631
C_7	.00017	.0980
C_8	.00006	.1450
C_9	.000016	.1466
C_{10}	.000007	.1157
C_{11}	.000002	.0600
$C_{12}+$	.000001	.0439
Density g/ml	0.0356	0.6894
Mol. wt	17.5	94.6
Surface tension		12.5 dyne/cm

FLOW REGIMES IN TWO-PHASE FLOW

The distribution of vapor and liquid phases in the pipeline is quantitatively measured in terms of *liquid-hold-up*, i.e., volume fraction of the pipe occupied by the liquid phase. The nature of distribution of each phase sharing the pipe volume defines the *flow regime* prevailing at that point. Figure 8-1 shows the flow regimes occurring in horizontal natural gas pipelines due to the presence of retrograde condensate. For extremely low gas velocities, normally well below design flowrates, liquid and gas may separate almost completely in stratified flow. With higher flowrates, but still a small fraction of the total line capacity, small droplets of liquid may enter the vapor phase and the pipe wall becomes wetted with a wavy liquid film. At high velocity representing flow equal to substantial volumes, possibly over the range of 60% to 100% of the design flow capacity, an annular mist condition prevails. Here small droplets of liquid are entrained throughout

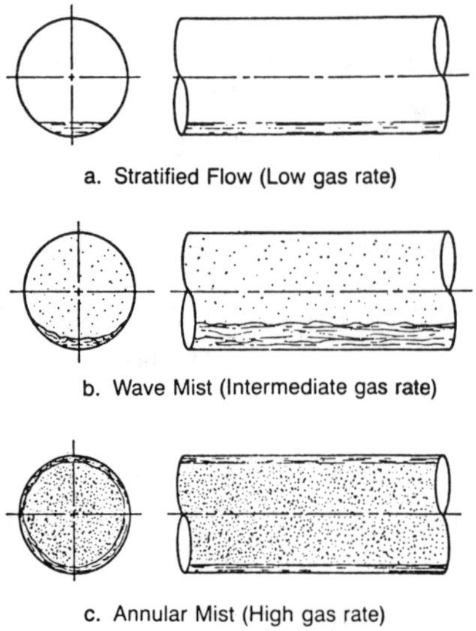

Figure 8-1. Flow regimes commonly occuring in horizontal natural gas pipelines.[1]

the turbulent gas phase, moving in random directions. Many of these droplets hit the wall of the pipe and coalesce with the wavy film of liquid moving along the wall. The dispersed phase in the gas stream is occasionally replenished by droplet formation from the crest of the waves on the ripples along the wall.

Figure 8-2 is a flow regime map developed for horizontal two-phase flow. At increased liquid concentrations, sometimes the pipe geometry becomes subject to intermittent slugs where single-phase liquid results from bridging of the liquid phase. This regime, where no particular phase is continuous, is called *slug flow regime*.

A survey was made of the pipelines by contacting Pipeline Research Committee members for their experiences. The results of the survey, given in detail in the monograph,[1] showed that the quantity of condensate collected from a pipeline system varied from essentially no liquid to up to 112 gals/million ft^3 of gas. Phase studies on the behavior of natural gas streams entering pipelines showed that dropping the pressure on the gas and dropping the temperature over the range normally occurring in pipeline systems, would give yields of the order of 2 to 10 or 15 gals/million ft^3 of gas. The procedure for prediction of the flow regime is given in detail in the monograph.[1]

To assist in understanding the volume percent liquid present in a pipeline when the yields of liquid are in the range of the survey, Figure 8-3 was prepared, typical of pipeline conditions. A liquid content of 1 gal/million over the range of pressure from 300 to 1,000 lb represents 0.0002 to 0.0012 volume percent liquid or 2 to 12 parts per million by volume liquid hydrocarbon. A content of 100 gals/million ft^3 under pipeline conditions represents 0.003 to no more than 0.12 volume percent liquid. Figure 8-2 shows that for this quantity of liquid present one may expect it to be in the annular mist flow regime. It should be appreciated that under certain circumstances, to be discussed later, the ratio of liquid to gas may increase by all of the liquid flowing in a stream accompanying a fraction of the gas when the two streams divide at a tee. In those cases, one may enter the transition region or even stratified flow at a very low gas velocity. Before going into a description of condensate removal, gas streams containing small quantities of liquid will be followed as they enter the branching on the road to the market to distribution systems.

During the course of this research, L. Oranje[4] published a paper on condensate behavior in gas pipelines in which he described the route selectivity exhibited by the liquid phase whenever the flow line encountered a lateral taking off a side stream. Visual observations at low pressure were made, as well as quantitative measurements of liquid following gas under pipeline conditions. Figure 8-4 shows typical results in which the part of the gas flow going into the branch is compared to the part of the liquid flowing into the branch. A 45° line has been drawn on the original figure. It may be seen that at low gas flows through the branch, the liquid tends to follow the side stream preferentially. As a gross

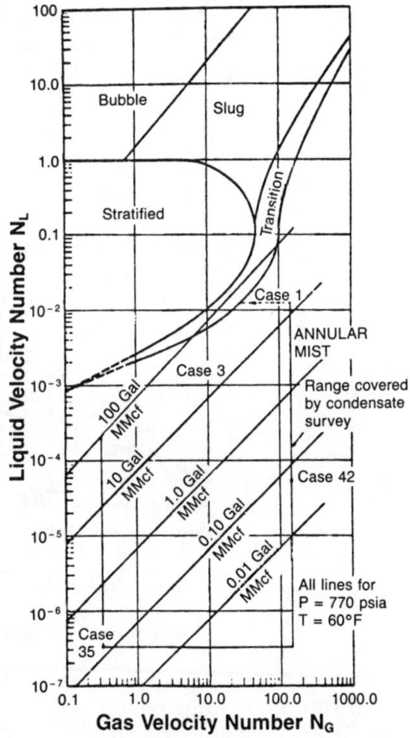

Figure 8-2. Flow regime map for horizontal two-phase flow.[1]

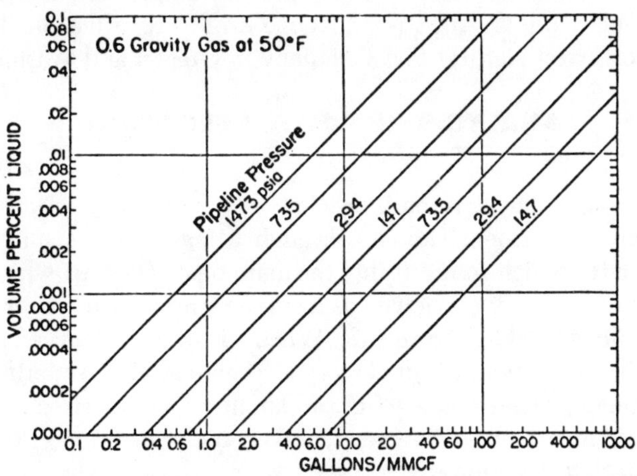

Figure 8-3. Volume percent liquid corresponding to gallons per MMCF 0.6 gravity gas, 50°F.[1]

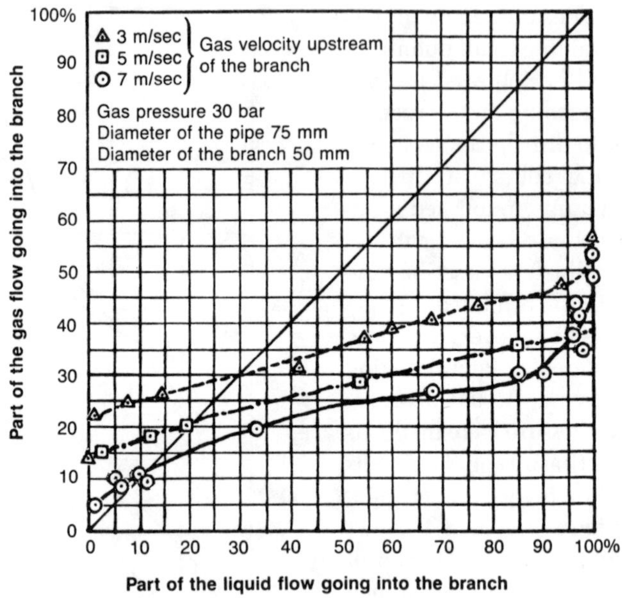

Figure 8-4. Data reported by Oranje.[4]

rule of thumb, Oranje concluded, "in case less than 20% of the gas stream enters the branch line, the liquid condensate will follow the straight line and not enter into the branch line. In case more than 35% of the gas stream enters into a branch line, nearly all of the liquid condensate will enter into the branch." This work was carried out at 435 psia pressure in a right angle 3-in. tee. Following Oranje's publication, Amies[5] reported on the experiences of Northwest Natural Gas Company in Oregon and Washington.

ANALYSIS OF ROUTE SELECTION

The following analysis is made of what creates the route selectivity of the liquid when the effluent flowing through a main pipe is split into two streams by a tee which connects into the main pipe. The part which is taken off and the lateral may be viewed as a curved stream near the branch connection, as illustrated in Figure 8-5a. When a flow is in a curved pipe, as in Figure 8-5b, the section a-b undergoes a secondary flow superimposed on the main stream. The secondary flow apparently becomes effective in carrying the liquid into the branch when relatively high rates of gas flow are taking place in that direction.

The data by Oranje, on percent gas into the lateral, have been plotted against the Reynolds number at the main pipe with parameters of percent liquid in the side-stream take-off (Figure 8-6). The lines of constant percent liquid take-off all exhibit definite minimums around the Reynolds number of 1 million for the main approach stream. It would also appear that, at extremely high Reynolds numbers, there would be an almost homogeneous mixture of extremely fine droplets of liquid in the flowing stream moving at exactly the same velocity as the gas. Such pseudo-homogeneous flow would cause the liquid phase to split at exactly the same ratio of gas stream split off into the tee. The various regions encountered in Figure 8-6, along with corresponding effects and mechanisms, are discussed in detail in the monograph.[1]

Although all geometries have not been explored, it is clear that secondary flow can cause liquid enrichment in a flowing stream which is subject to

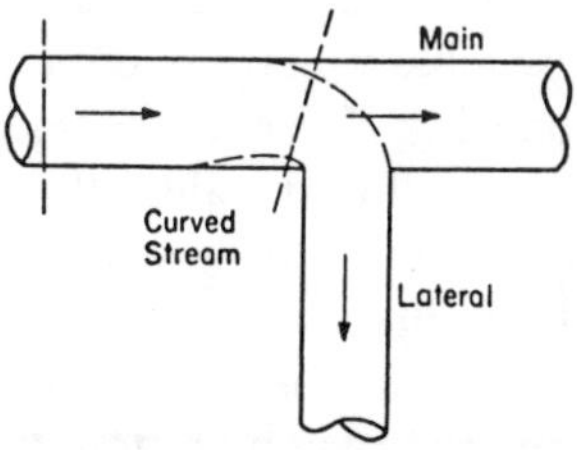

Figure 8-5a. Curves stream representation of side outlet flow.

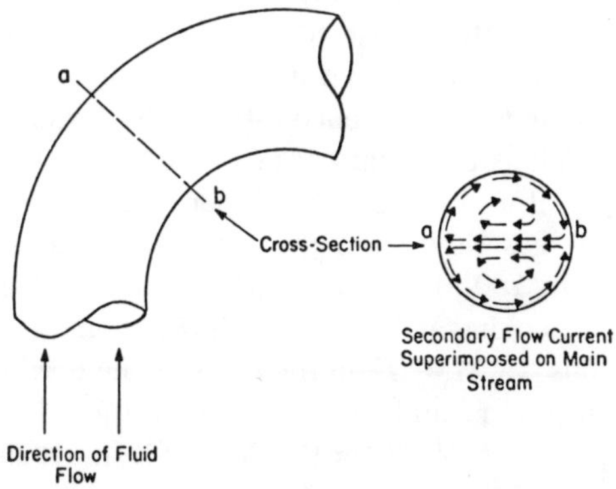

Figure 8-5b. Secondary flow in a curved pipe.

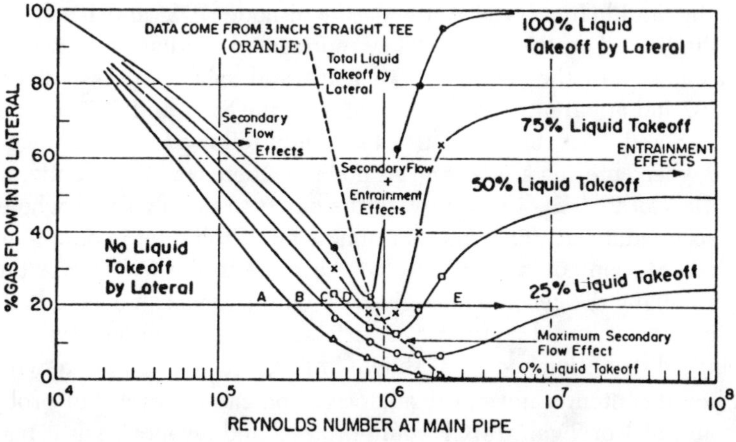

Figure 8-6. Generalized chart for route selectivity based on Oranje's data.[1]

branching where the branch stream carries some 30% or more of the gas in the lateral. In this manner the content of the stream can build up several times above the 5 to 15 gallons of condensate per million ft^3 originally formed in the main transmission system.

CASE STUDIES IN DISTRIBUTION SYSTEMS

Two case studies are given in the monograph, one of which is reproduced here. Figure 8-7 shows the distribution system of the Michigan Consolidated Company, which had liquid accumulation in a particular station: the Chalmers Regulator Station. Pipeline gas, which had been stored in the Belle River Mills storage field, is mixed with regular pipeline gas to enter the Lynch Regulator Station at the entrance to Figure 8-7. The stream branches, as shown in the flow diagram of Figure 8-7. In February 1975, condensate was encountered at the Chalmers Regulator Station with as much as 1,534 gallons of condensate collected in one day. The gas was flowing at a rate of 200 MCF per hour. Table 8-2 shows the liquid collected and the gas flow from the Lynch Regulator system entering the network for some 200 MCF leaving the side arm of the Chalmers Station.

Looking at Table 8-2, 1,534 gallons per day of condensate becomes 11 gallons per MMCF on the basis of the total gas leaving the Lynch Station.. On the other hand, the quantity of liquid leaving the Chalmers Station is some 319 gal/MMCF for the 200 MCF/hr. It should be appreciated that the 11.7 gal/MMCF entering the Lynch Station could have been the result of a branching prior to arrival there since the condensate formed from the storage field could well have been only 5 or 6 gal/MMCF.

As one analyzes the flow through Figure 8-7, all but nodes 7, 10, and 11 have less than 10% side-flow of gas and hence should have removed little if any of the liquid. Inquiry as to the nature of nodes 7, 8, and 10 reveal the flow as indicated on the inserts. They represented a situation in which the stream enters into the side outlet of a tee and splits into two diverging branches, and apparently the liquid followed the 84% and 89% of the gas on nodes 8 and 7 for this geometry. The liquid followed 25% of the gas at node 10 when entering a side arm. At node 11, there was a conventional tee with a flow of 27.7% of gas through the outlet to the regulator where the liquid accumulated. It appears that almost all of the liquid coming out of the Lynch Station could have selected the route to the Chalmers Station. Thus, following the general rules given by Oranje, one can conclude that a small amount of liquid content can enrich itself rather quickly when in a branching distribution system as shown in Figure 8-7. It is not known how much liquid content a normal regulator station can tolerate, but probably an amount of 1 or 2 gal/MMCF could proceed and revaporize as it reaches the low pressure systems prior to use.

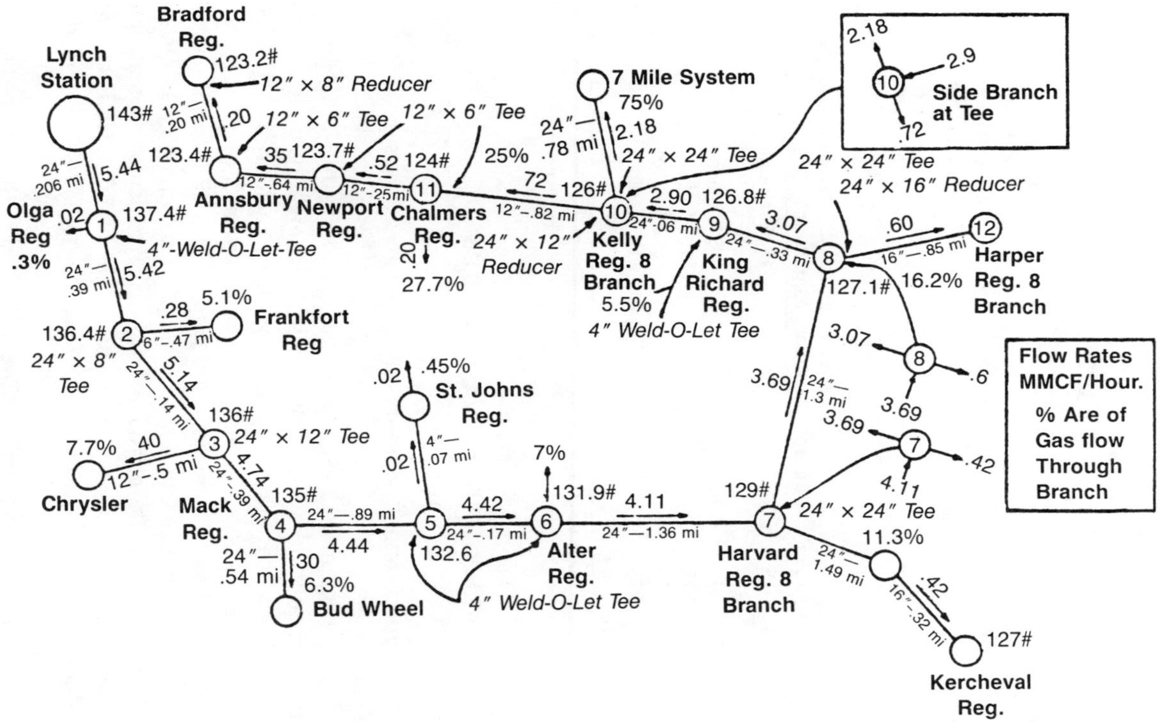

Figure 8-7. Representation of distribution from Lynch Station to Chalmers Station, Michigan Consolidated Gas Company.[1]

Table 8-2
Condensate Collection and Conditions in Waveney System, Chalmers Regulator Station[1]

Temperature Average			
Date	**Day**	**Atmospheric Temp. (°F)**	**Average Flow**
Feb. 2, 1975	Sunday	29°	5.18 MMcf/hr
Feb. 3, 1975	Monday	30°	5.44 MMcf/hr
Feb. 4, 1975	Tuesday	33°	5.23 MMcf/hr
Feb. 5, 1975	Wednesday	31°	5.11 MMcf/hr
Feb. 6, 1975	Thursday	22°	5.83 MMcf/hr
Feb. 7, 1975	Friday	21°	5.72 MMcf/hr
Feb. 8, 1975	Saturday	15°	5.00 MMcf/hr

Liquids Pumped from Chalmers Regulator	
Feb. 4, 1975	1534 Gallons
Feb. 5, 1975	76 Gallons
Feb. 6, 1975	57 Gallons
Feb. 7, 1975	1168 Gallons
Feb. 8, 1975	19 Gallons
Feb. 9, 1975	107 Gallons
Feb. 10, 1975	20 Gallons
Flowing Gas Temperature 32°F	

LIQUID REMOVAL DEVICES

When condensation cannot be prevented, it becomes necessary to put in liquid removal devices.

Pipeline drips are common devices used as shown in Figure 8-8. Knowing that the liquid normally follows the wall, Figure 8-8b might be preferable to 8-8a, which simply reduces the pipeline flow velocity. Such drips need to be placed in initial designs where liquid might accumulate after pressure reduction to the 300 lb level and where flowrates vary such as in distribution systems. The route selection of any liquid formed to accumulate, as at the Chalmers Station, should be considered for any particular system. Of course, if all the liquid had been removed while at the 11 gal/MMCF level at the Lynch Station, then there would have been no problem at any of the substation regulators.

It should be realized that certain lines are little used in the summer and a small amount of liquid may enter such lines and lie in the low sections. When cold weather comes, gas is flowed through the lines and sweeps the liquid toward the market either with or without concentration by route se-

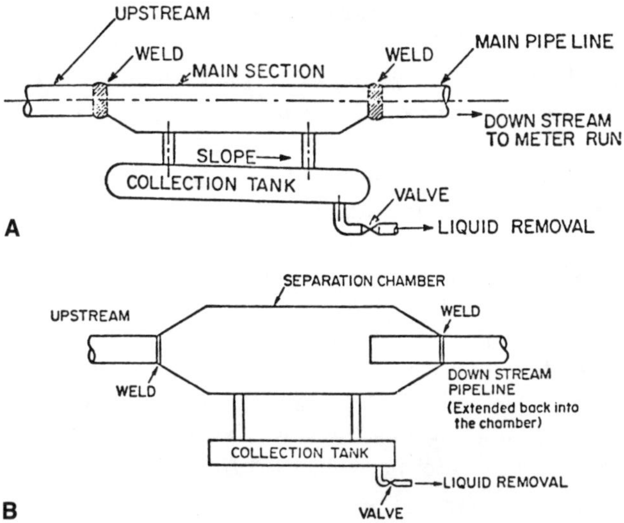

Figure 8-8. Pipeline drips, a. typical, and b. concentric, with wall liquid removal.

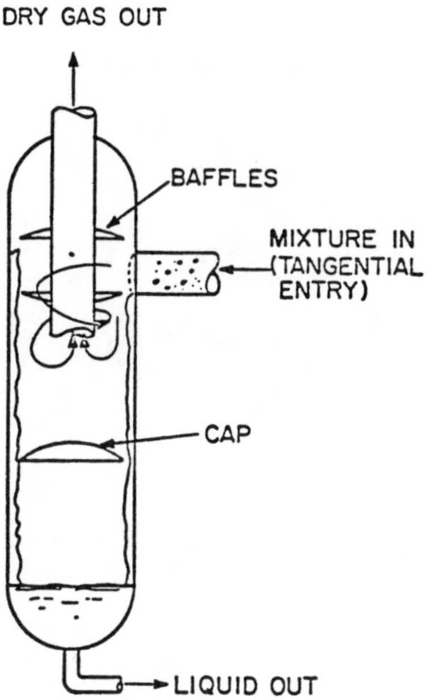

Figure 8-9. Spin separator (Courtesy Porta-Test).[1]

lection. In those cases where fairly large quantities of liquid come on occasion, line separators are useful, as shown in Figure 8-9. Such devices can handle large quantities of liquid, especially those which might come on a surge during start-up where liquid has accumulated during periods prior to operation.

Lubricating oil is commonly lost into the pipeline systems and is thought by some to be a primary contributor to condensate formation. Condensates obtained in the pilot unit or from drips were analyzed by distillation. Only the last fraction not distilled would have contained lubricating oil and it was 3% to 6% for most liquids. In one drip liquid, lubricating oil was identified, but it is not believed to be responsible for most of the condensates collected during that study.

References

1. Bergman, D. F., M. R. Tek, and D. L. Katz. "Retrograde Condensation in Natural Gas Pipelines," Monograph, American Gas Association, Arlington, Virginia, 1975.
2. Dixon, A. G. "Hydrocarbon Fluid Evaluation Part 1—Hydrocarbon Components," Proceedings, 8th Doha Technical Symposium (Gas Industry), Doha, Qatar, Nov. 26–28, 1984.
3. Coats, K. H. "Simulation of Gas Condensate Reservoir Performance," SPE paper number 10512, 1982.
4. Oranje, L. "Condensate Behavior in Gas Pipelines is Predictable," *Oil and Gas J.*, July 2 (1973).
5. Amies, W. T. "Method of Determining the Amount of Compressor Oil in a Pipeline by Sampling," Pacific Gas Association, Distribution Conference, San Diego, California, Feb. 19, 1974.

9

NUMERICAL SIMULATION OF FLOWS AND PRESSURES IN UNDERGROUND STORAGE

Computer simulation of natural gas flow through porous media became a useful and practical reality during the early fifties.[1-4] This was a result of the need for better understanding of gas reservoir behavior in the face of analytical difficulties in an attempt to solve the gas flow equations which had many nonlinearities. During those early years of computer technology and developing numerical methods, simulation—even though specific and numerical—appeared to be the only link between idealized theory and real-world practice. The effort continued through the early sixties while much new well test data[5,6] and studies directed to inertial, turbulence, non-Darcy effects[6-9] were underway. During the late sixties, computing capabilities at the University of Michigan were much enhanced by the pioneering work of Professors Wilkes and Carnahan.[10] The period also coincided with three major research projects sponsored by the American Gas Association in:

1. Movement of Water in Contact with Natural Gas.[11]
2. New Concepts in Underground Storage.[12]
3. Threshold Pressure in Gas Storage.[13]

Of the three monographs listed above, the one on new concepts included the first simulation of leaks from underground storage, developed by J. O. Wilkes. His work led to continuing efforts by others such as K. H. Coats, T. L. Gould, and S. P. Singh. A storage simulator called P.E.S.O.S. (Programs for Engineering Simulation of Storage) was developed between 1970 and 1973. Project PESOS paved the way for many simulators now marketed by several companies. During the following years and into the eighties when each year an underground storage course was offered at the University of Michigan to those from the industry, Professor Wilkes traditionally presented the main elements involved in simulation of underground storage reservoirs. The following is an abridged version of the outline of his presentation reproduced here with his permission.

FUNDAMENTAL EQUATIONS

The basic equations involved in simulation of storage are:

1. Unsteady-state mass balance (continuity)
2. Darcy's Law* (constitutive equations)
3. Equation of state (related to gas density)

The basic volume element considered in audit of gas fluxes in and out is shown in Figure 9-1. The simulator is essentially two-dimensional (x, y), but allows for variable h, the net formation thickness in the z direction.

Unsteady-state mass balance gives:

$$-\{\partial/\partial x\ (\rho uh) + \partial/\partial y\ (\rho vh)\} + mh = \phi h\ \partial\rho/\partial t \tag{9-1}$$

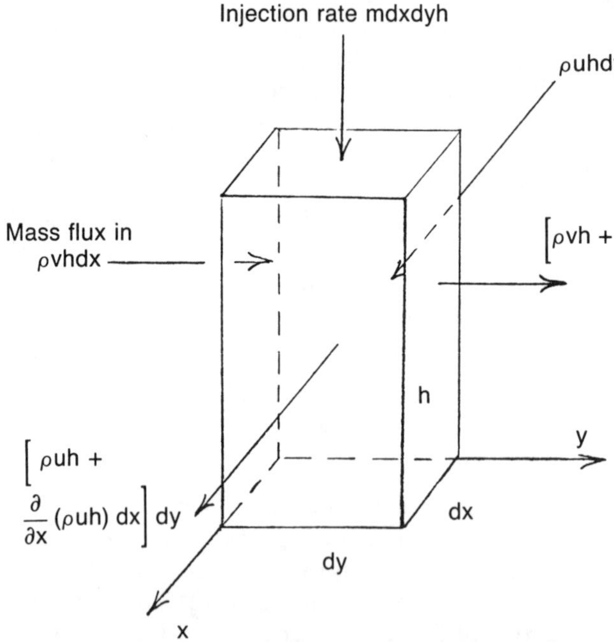

Figure 9-1. Mass fluxes in and out of a volume element.

* Later simulators developed at the University use a generalized form to include inertial and turbulent effects.

where ϕ = porosity
ρ = density
h = reservoir thickness
u,v = superficial velocities in x and y directions
m = mass injection rate per unit volume

Darcy's Law.

$$u = -k/\mu \, \partial p/\partial x \tag{9-2}$$

$$v = -k/\mu \, \partial p/\partial y \tag{9-3}$$

where k = permeability of the porous medium
μ = viscosity of the gas

Gas Density

$$\rho = Mp/zRT \tag{9-4}$$

where M = molecular weight
z = compressibility factor
R = gas constant
T = absolute temperature of formation

Combining Equations 9-1 through 9-4 and dividing through by M and R:

$$\frac{\partial}{\partial x}\frac{p}{zT}\frac{hk}{\mu}\frac{\partial p}{\partial x} + \frac{\partial}{\partial y}\frac{p}{zT}\frac{hk}{\mu}\frac{\partial p}{\partial y} + mh = \phi h\frac{\partial}{\partial t}\left(\frac{p}{zT}\right) \tag{9-5}$$

Next, one defines the *real gas potential* (also known as *pseudo-pressure*) $\Phi = \Phi(p)$ as:

$$\Phi = \int_{p_r}^{p} \frac{p}{z\mu} dp \tag{9-6}$$

where p_r = some reference pressure

$z = z(p)$

$\mu = \mu(p)$

By substituting Φ for p as shown in Chapter 3, the differential equation becomes:

$$\frac{\partial}{\partial x}\left(hk\frac{\partial \Phi}{\partial x}\right) + \frac{\partial}{\partial y}\left(hk\frac{\partial \Phi}{\partial y}\right) + \frac{m\ hRT}{M} = \phi h \frac{\partial (p/z)}{\partial t} \tag{9-7}$$

Now note:

$$\frac{\partial (p/z)}{\partial t} = \frac{\partial (p/z)}{\partial \Phi}\frac{\partial \Phi}{\partial t} \qquad \text{(chain rule)} \tag{9-8}$$

where

$$\frac{\partial (p/z)}{\partial \Phi} = \gamma \tag{9-9}$$

Now recall:

$$\rho_s = \frac{M\ p_s}{R\ T_s} \tag{9-10}$$

and

$$m = \rho_s\ q_s \qquad \text{(where s denotes standard condition)} \tag{9-11}$$

leads to:

$$\frac{\partial}{\partial x}\left(hk\frac{\partial \Phi}{\partial x}\right) + \frac{\partial}{\partial y}\left(hk\frac{\partial \Phi}{\partial y}\right) + \frac{q_s T p_s}{T_s} h = \phi h \gamma \frac{\partial \Phi}{\partial t} \tag{9-12}$$

Since $\Phi = \Phi(p)$, $z = z(p)$ then $p = p(\Phi)$ and $z = z(\Phi)$. Therefore,

$$\gamma = \frac{\partial (p/z)}{\partial \Phi}$$

becomes a known, calculable function of the pseudo-pressure, i.e., $\gamma = \gamma(\Phi)$.

SPECIFIC DESCRIPTION OF THE SIMULATION PROBLEM

Figures 9-2 and 9-3 depict a vertical section and plane view of the reservoir with blocks, wells, and surface gathering system. In the horizontal section, the storage reservoir is considered to be surrounded by blocks saturated with water, with no permeability to gas. The island in the middle depicts a tight section where the permeability to gas is again considered to be zero.

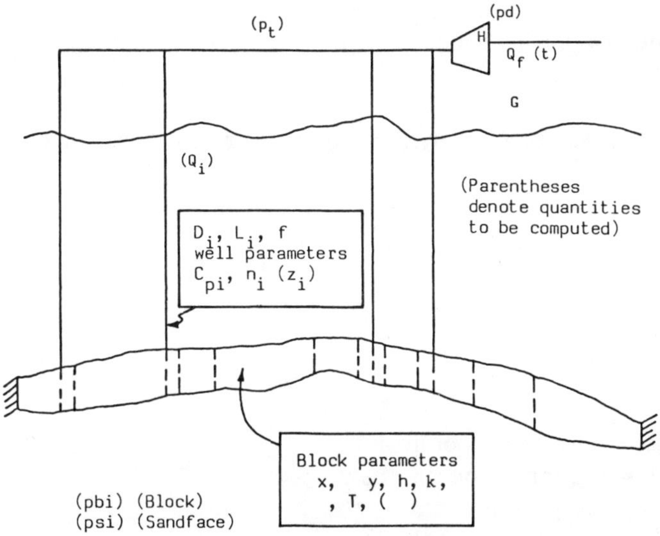

Figure 9-2. Vertical section through reservoir.

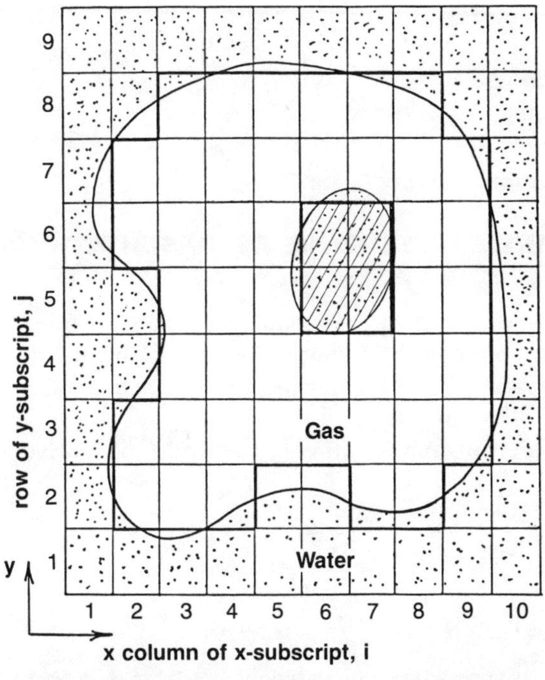

Figure 9-3. Horizontal section through reservoir.

The object of the simulation problem is as follows: Given the injection/production schedule $Q_f = Q_f(t)$, compute all pressures and well flowrates as functions of time.

Review of Governing Equations for Simulation

$$\frac{\partial}{\partial x}\left(hk\frac{\partial \Phi}{\partial x}\right) + \frac{\partial}{\partial y}\left(hk\frac{\partial \Phi}{\partial y}\right) + q_s\frac{Tp_s}{T_s}h = \phi h\gamma\frac{\partial \Phi}{\partial t} \tag{9-12}$$

$$\Phi = \int_{p_r}^{p} \frac{p}{z\mu}dp \tag{9-6}$$

where x,y = horizontal coordinates
h = net formation thickness
k = permeability
T_s = standard temperature
p_s = standard pressure
T = absolute formation temperature
q_s = injection or production rate, SCF per ft^3 of reservoir per unit time
ϕ = porosity

$$\gamma = \frac{d(p/z)}{d(\Phi)}$$

z = compressibility factor for gas

FINITE DIFFERENCE REPRESENTATIONS

The object is to calculate the pseudo-pressure and, following that, the pressure at every point at every time of interest. Figure 9-4 depicts the finite difference representation of the unknown potential Φ in both x and y directions as counted by integers i and j.

Central-difference approximation of the first Φ derivative is given by:

$$\frac{\partial \Phi}{\partial x} = \frac{\Phi_{i+1/2,j} - \Phi_{i-1/2,j}}{\Delta x} = \frac{\Delta_x \Phi_{i,j}}{\Delta x} \tag{9-13}$$

$$\text{where } \Delta_x \Phi_{i,j} = \Phi_{i+1/2,j} - \Phi_{i-1/2,j} \tag{9-14}$$

Finite-difference approximation at block (i,j), as shown in Figure 9-4.

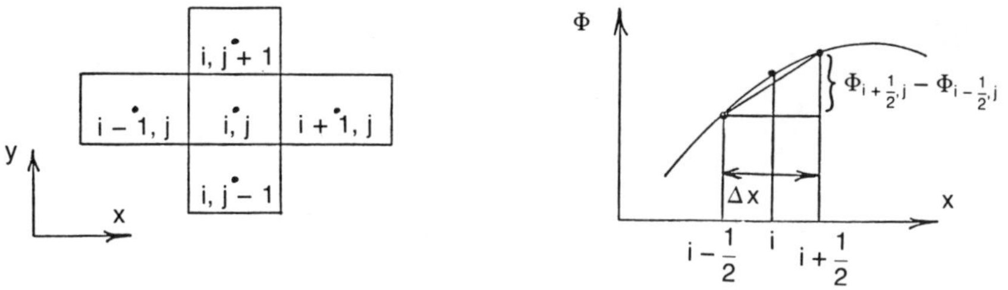

Figure 9-4. Finite-difference approximations to derivatives.

Note first, for example:

$$\frac{\partial}{\partial x}\left(hk\frac{\partial \Phi}{\partial x}\right) = \frac{1}{\Delta x}\Delta_x\left[(hk)_{i,j}\ \frac{\Delta_x \Phi_{i,j}}{\Delta x}\right] \tag{9-15}$$

Define transmissibilities:

$$a^x = hk\ \frac{\Delta y}{\Delta x}\ \frac{T_s}{T_{p_s}},\quad a^y = hk\ \frac{\Delta x}{\Delta y}\ \frac{T_s}{T_{p_s}} \tag{9-16}$$

Then:

$$\Delta_x(a^x{}_{i,j}\ \Delta_x\Phi_{i,j}) + \Delta_y(a^y{}_{i,j}\ \Delta_y\Phi_{i,j})$$

$$+\underbrace{\Delta x\ \Delta y(hq_s)_{i,j}}_{\substack{Q_{i,j}\\ \text{Volumetric}\\ \text{injection}\\ \text{rate to}\\ \text{block (i,j)}}} = \underbrace{\Delta x\ \Delta y(\phi h\gamma)_{i,j}}_{\substack{\nu_{i,j}\\ \text{constant}\\ \text{for}\\ \text{block (i,j)}}}\ \frac{\Delta \Phi_{i,j}}{\Delta t} \tag{9-17}$$

Time Table for Updating the Pseudo-Pressure

	Level	Value	Pseudo-Pressure
Time increasing	n + 1	(n + 1)Δt	Φ_{ij}^{n+1}
	n	nΔt	Φ_{ij}^{n}

Explicit Representation

All gas potentials are at the beginning of the time step, except in:

$$\frac{\Delta\Phi_{ij}}{\Delta t} = \frac{\Phi_{ij}^{n+1} - \Phi_{ij}^{n}}{\Delta t} \tag{9-18}$$

Written out in full, the FDA[†] becomes

$$a^x_{i+1/2,j}(\Phi^n_{i+1,j} - \Phi^n_{i,j}) - a^x_{i-1/2,j}(\Phi^n_{i,j} - \Phi^n_{i-1,j}) + a^y_{i,j+1/2}(\Phi^n_{i,j+1} - \Phi^n_{i,j}) - a^y_{i,j-1/2}(\Phi^n_{i,j} - \Phi^n_{i,j-1}) + Q_{i,j} = \nu_{ij}\gamma_{ij}\frac{\Phi^{n+1}_{ij} - \Phi^n_{ij}}{\Delta t} \tag{9-19}$$

Everything is known except Φ_{ij}^{n+1}, the pseudo pressure for block (i,j) at the end of the time step, which is then calculated. The process is then repeated for all blocks.

Implicit Alternating Direction (IAD) Method (for improved stability)

Time Levels	Potentials
$(n + 1)\Delta t$	Φ_{ij}^{n+1}
$(n + 1/2)\Delta t$	$\Phi_{ij}^{n+1/2} = \Phi^*_{ij}$
$n\Delta t$	Φ^n_{ij}

Proceed from n to n + 1 in two half-steps.

First $\Delta t/2$ (Implicit in the x direction):

$$a^x_{i+1/2,j}(\Phi^*_{i+1,j} - \Phi^*_{i,j}) - a^x_{i-1/2,j}(\Phi^*_{i,j} - \Phi^*_{i-1,j}) + a^y_{i,j+1/2}(\Phi^n_{i,j+1} - \Phi^n_{i,j}) - a^y_{i,j-1/2}(\Phi^n_{i,j} - \Phi^n_{i,j-1}) + Q_{i,j} = \nu_{ij}\gamma_{ij}\frac{\Phi^*_{ij} - \Phi^n_{ij}}{\Delta t/2}, \tag{9-20}$$

which can be rewritten as:

$$a_i\Phi^*_{i-1,j} + b_i\Phi^*_{i,j} + c_i\Phi^*_{i+1,j} = d_i \tag{9-21}$$

the ith member of a *Tridiagonal* system of simultaneous linear equations. Here, the coefficients a_i, b_i, c_i, and d_i are all known.

† Finite Difference Approximation.

This last equation is applied at every block i; imin, imin + 1, ..., imax in a typical row of blocks extending in the x direction. The horizontal configuration is shown in Figure 9-5.

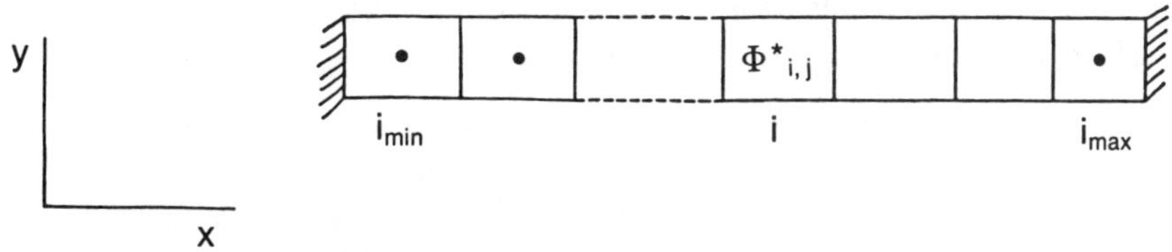

Figure 9-5. Horizontal configuration of blocks in implicit FDM scheme.

This resulting system of tridiagonal equations can be readily solved by a special purpose algorithm. The process is repeated for all rows of blocks in the x direction, at which stage all the Φ^*_{ij} at the intermediate time level are known.

Second Δt/2 (Implicit in the y direction):

$$a^x_{i+1/2,j}\,(\Phi^*_{i+1,j} - \Phi^*_{i,j}) - a^x_{i-1/2,j}\,(\Phi^*_{i,j} - \Phi^*_{i-1,j})$$

$$+\ a^y_{i,j+1/2}\,(\Phi^{n+1}_{i,j+1} - \Phi^{n+1}_{i,j}) - a^y_{i,j-1/2}\,(\Phi^{n+1}_{i,j} - \Phi^{n+1}_{i,j-1})$$

$$+\ Q_{i,j} = \nu_{ij}\,\gamma_{ij}\,\frac{\Phi^{n+1}_{i,j} - \Phi^*_{i,j}}{(\Delta t)/2}, \qquad (9\text{-}22)$$

which can be rewritten as:

$$a_j\Phi^{n+1}_{i,j-1} + b_j\Phi^{n+1}_{i,j} + c_j\,\Phi^{n+1}_{i,j+1} = d_j \qquad (9\text{-}23)$$

This is applied at every block j; jmin-, ..., jmax in a typical column of blocks extending in the y direction, again giving a tridiagonal system.

Then repeat for all columns of blocks in the y direction.

The vertical configuration is shown in Figure 9-6.

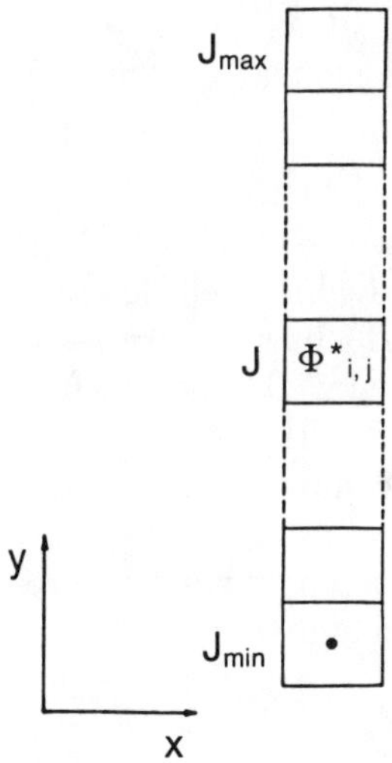

Figure 9-6. Vertical configuration of blocks in implicit FDM scheme.

COMPUTATION OF INDIVIDUAL WELL FLOWS

Figure 9-7 is a schematic representation of a number of wells (n_w) each connected vertically from bottom block average matrix points (b_1, b_2, ...b_i) to flowing bottom-hole sand face points (s_1, s_2, ...s_i) and connected at the surface to a common gathering system pressure p_t.

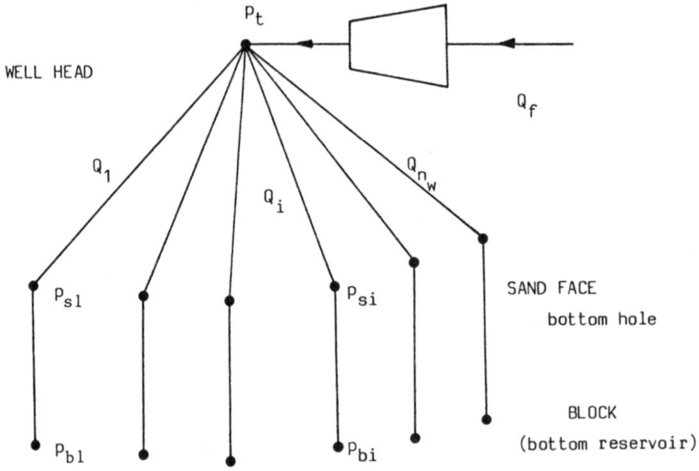

Figure 9-7. Schematic of individual well connections to the surface and compression.

Figure 9-7 represents the system in injection mode where Q_f std MMCF/D is being delivered to compressor station inlet. The individual well flowrates are designated Q_1, Q_2, ...Q_i, and the compressor discharge pressure p_t. While in actuality the well head pressures on individual wells will be different from one another, this particular simulation ignores the pressure drop that will occur differently from each well head to a common point leading to the compressor station. If and when desired, one can use one of the many pipeline network simulators as a subroutine to the present simulation program to include specific pressure drops in the gathering lines.

The unknown are, in general, Q_i, p_t, p_{s_i}, and Q_f. p_{b_i} are block average pressures computed from the reservoir simulation through the pseudo-pressures.

Since we have four unknowns, we must have four equations for each i:

1. Sand face to block:

$$Q_i = \pm C_{1i} \left[\pm (p^2_{si} - p^2_{bi}) \right]^{ni} \tag{9-24}$$

2. From well head to sand face:

$$Q_i = \pm C_{2i} [\pm (C_{3i} p^2_t - p^2_{si})]^{1/2} \tag{9-25}$$

3. From compressor station to individual wells:

$$Q_f = \sum_{i=1}^{n_w} Q_i \tag{9-26}$$

4. From inlet to discharge of compressor station:

$$Q_f = \frac{\pm HP}{4\left[\left(\frac{p_t}{p_d}\right)^{\frac{k-1}{k}} - 1\right] + 3.5} \tag{9-27}$$

Equation 9-24 is recognized as the usual back pressure equation determined from isochronal test data. Equation 9-25 is the vertical single-phase gas flow equation through the wells. Equation 9-26 is simple material balance. Equation 9-27 is the relationship between the compressor horsepower and the total flowrate by adiabatic compression modified by the constant 3.5 HP/(MMCF/D) term to account for the efficiency.

In order to solve Equations 9-24 through 9-27, p_{si} is eliminated between Equations 9-24 and 9-25, resulting in:

$$\left(\pm \frac{Q_i}{C_{1i}}\right)^{1/n_i} + \left(\pm \frac{Q_i}{C_{2i}}\right)^2 = \pm (C_{3i} p^2_t - p^2_{bi}) \tag{9-28}$$

Next define:

$$C_6 = | C_{3i} p^2_t - p^2_{bi} | \tag{9-29}$$

which, when substituted in Equation 9-28, gives:

$$f(Q) = \left(\frac{Q}{C_{1i}}\right)^{1/n_i} + \frac{Q^2}{C^2_{2i}} - C_6 = 0 \tag{9-30}$$

The transcendental equation (9-30) is solved by the Newton Raphson technique where:

$$Q \leftarrow Q - \frac{f(Q)}{f'(Q)} \text{ by successive iterations}$$

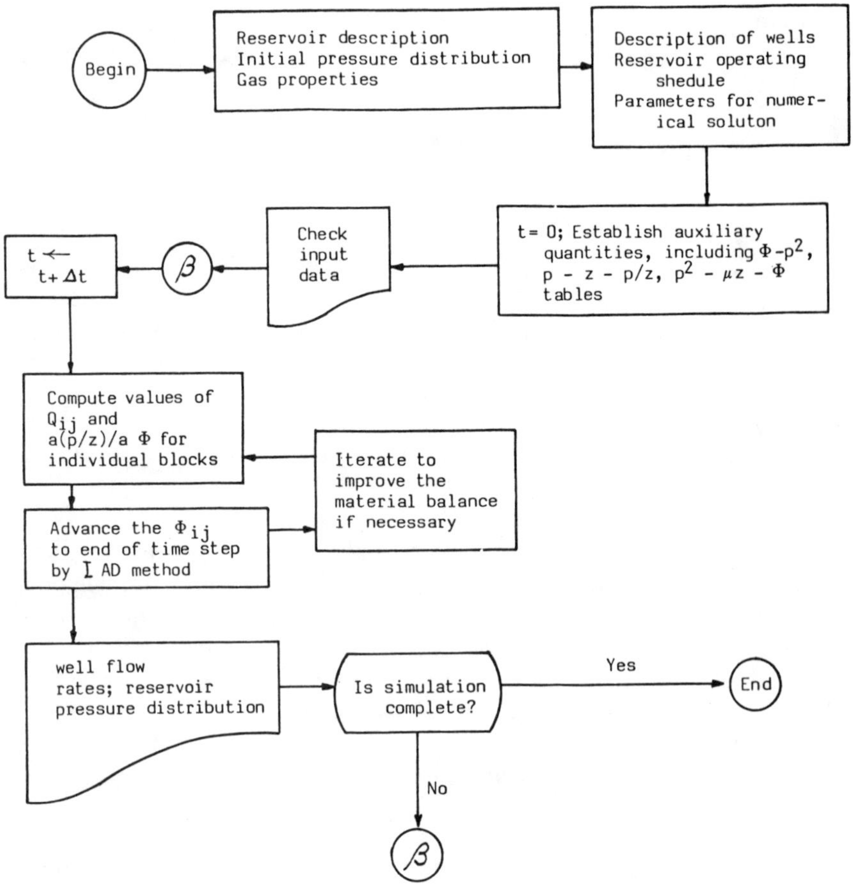

Figure 9-8. Summary flow diagram for the simulation of underground storage. Reproduced by permission of J. O. Wilkes.

Once the computer converges on a solution for Q, the simulator sets $Q_i = \pm Q$, depending upon $C_{3i}\ p^2_t >< p^2_{bi}$.

Figure 9-8 shows the summary of flow diagram for the simulation.

References

1. Aronofsky, J. S. and R. Jenkins. "Unsteady Flow of Gas through Porous Media; One-Dimensional Case," Proc., First U.S. National Congress Applied Mechanics, J. E. Edwards, Inc., Ann Arbor, Michigan, 1952, 763.

2. Jenkins, R. and J. S. Aronofsky. "Unsteady Radial Flow of Gas through Porous Media," *J. Appl. Mech.* (1953) 20, 210.
3. Bruce, G. D., D. W. Peaceman, H. H. Rachford, and J. Price. "Calculations of Unsteady State Gas Flow through Porous Media," *Trans. AIME* (1953) 198, 79.
4. Douglas, J., Jr., D. W. Peaceman, and H. H. Rachford. "Calculations of Unsteady-State Gas Flow in a Square Reservoir," *Trans. AIME* (1955) 204, 190.
5. Govier, G. W. "Interpretation of the Results of Back Pressure Testing of Gas Wells," *Trans. AIME* (1961) 511–514.
6. Smith, R. V. "Unsteady Gas Flow into Gas Wells," *J. Petro. Tech.* (1961) 13, 1151–1159.
7. Tek, M. R., K. H. Coats, and D. L. Katz. "The Effect of Turbulence on Flow of Natural Gas through Porous Reservoirs," *J. Petro. Tech.* (July 1962) 799–806.
8. Swift, G. W. and O. G. Kiel. "The Prediction of Gas Well Performance Including the Effect of Non-Darcy Flow," *J. Petro. Tech.* (1962) 14, 791–798.
9. Carter, R. D., S. C. Miller, Jr., and H. G. Riley. "Determination of Stabilized Gas Well Performance from Short Flow Tests," *J. Petro. Tech.* (1963) 15, 651–658.
10. Carnahan, B., H. A. Luther, and J. O. Wilkes. *Applied Numerical Methods*, New York: Wiley, 1969.
11. Katz, D. L., M. R. Tek, K. H. Coats, M. L. Katz, S. C. Jones, and M. C. Miller. "Movement of Underground Water in Contact with Natural Gas," Monograph, American Gas Association, New York, 1963.
12. Tek, M. R., J. O. Wilkes, D. L. Katz, D. A. Saville, B. D. Bhalla, and L. G. Sherman. "New Concepts in Underground Storage of Natural Gas," Monograph, American Gas Association, New York, 1966.
13. Ibrahim, M. A., M. R. Tek, and D. L. Katz. "Threshold Pressure in Gas Storage," Monograph, American Gas Association, Arlington, Virginia, 1970.

10

SPECIAL TOPICS IN UNDERGROUND STORAGE

As underground storage of natural gas has matured and evolved, much has been learned from experiences resulting from day-to-day operation of reservoirs. The purpose of this chapter is to highlight and share with the reader several topics where recent developments have resulted in new and helpful storage technology.

ISOTOPIC IDENTIFICATION OF SUSPECTED LEAKAGE FROM STORAGE RESERVOIRS

It is well known that natural gas may originate from two distinctly different processes.[1] One is related to thermal decomposition of organic matter under the thermodynamic and kinetic conditions resulting from sedimentation and deep burial in the crust of the earth. The timetable involved in this process is measured in geologic scales of millions of years. The natural gas so formed corresponds to a vast majority of producing gas reservoirs and is referred to as *petrogenic*.

The other process involves bacterial reduction of cellulose into methane. The cellulose comes from decomposing organic matter near the surface. Some refer to the chemical reaction as *methanation*. The bacteria causing it are called *methanomas*. This gas is said to be of *bacteriogenic* (or *biogenic*) origin. Swamp gas or marsh gas is an example of bacteriogenic natural gas. *Drift* gas found in glacial drifts near the surface is reported to be formed by bacterial decomposition of buried peats.[2,3]

Sometimes, a mixture of local bacteriogenic and migrated petrogenic gas is found and identified in shallow formations. In areas where pipeline quality gas is stored underground, occasionally the gas migrates away from the storage horizon. The migration is sometimes across structural saddles between adjacent fields. The migrated gas in the underground tends to move toward the surface driven by buoyancy. On its way to surface, if and when the gas encounters a structure delineated by an impervious caprock, it accumulates in a zone called a *collector zone*. When such collector zones are discovered, it becomes possible and technically feasible to produce such collector zones and recycle the gas back to the reservoir whence it came. Such projects occasionally involve conflicts between the owner of the storage gas and the owner of surface leases. In order to resolve such conflicts, it becomes important to be able to:

1. Identify the storage gas in samples
2. If storage gas is found mixed with some native biogenic or petrogenic gas, determine the extent of it in the mixture

Needless to say, such problems may have enormous legal, economic, and environmental significance. In such cases, use of isotopic analysis of methane to identify storage gas has been most successful. One definite advantage of isotopic analysis over chemical analysis is that chemical composition is subject to fractionation while the gas is in transit through porous media; isotopic composition is not.

Isotopic Composition of Methane

Natural carbon we encounter on earth is made up of three distinct isotopes.

One of the isotopes* is carbon 14, usually designated C^{14}. It is called radiocarbon and known to originate in the upper atmosphere by chemical reaction between N^{14} and neutrons generated by cosmic rays. The radiocarbon becomes oxidized to CO_2 and gets mixed with the rest of the CO_2 in the atmosphere. The ratio of radiocarbon in CO_2 found in all living things (plants and animals) in equilibrium with the atmosphere is constant and equal to about 1×10^{-10} percent of C^{14} relative to all carbon in CO_2. When the living things die, that ratio ceases to be constant and continually decreases because C^{14} is unstable and slowly decays back to N^{14}. The half-life time of that decaying process is 5,730 years.[4] The time since the disruption of equilibrium can be estimated by determining the proportion of C^{14} remaining in organic waste. This is the basis of *radiocarbon dating*. Radiocarbon dating, however, is only good for things that have been dead for 40,000 to 50,000 years. Due to analytical limitations, C^{14} cannot be detected in samples older than about 50,000 years.

* Isotope means same atomic number, similar chemical properties, but different mass number.

Now, petrogenic gas is millions of years old. It contains no detectable C^{14}. Biogenic gas, on the other hand, may be relatively young. If one has a mixture of petrogenic gas and young biogenic gas, C^{14} analysis can be used to determine how much there is of each. If the biogenic gas, however, is older than 50,000 years, radiocarbon dating becomes impractical.

In contrast to C^{14}, the other two isotopes of carbon (C^{12} and C^{13}) are both stable.

The carbon we normally encounter has about 98.9% C^{12} and only 1.1% C^{13}. The isotopic determination of the C^{13}/C^{12} ratio has been shown by Coleman[1] to be a practical and effective discriminator in identifying natural gases.

In the process developed by Coleman, the C^{13}/C^{12} ratio is expressed as per mill (‰) and evaluated in terms of a percent difference between the C^{13}/C^{12} ratio and that of a standard called $\delta\ C^{13}$ (delta carbon 13). The quantity $\delta\ C^{13}$ is defined by:

$$\delta\ C^{13}_{sample} = \frac{(C^{13}/C^{12})_{sample} - (C^{13}/C^{12})_{standard}}{(C^{13}/C^{12})_{standard}} \times 1{,}000 \qquad (10\text{-}1)$$

For example, a $\delta\ C^{13}$ value of -48 ‰ means that the particular sample is 48 parts per mill lighter than the standard (i.e., in C^{13}/C^{12} ratio).

Figure 10-1 shows the range of δ carbon 13 values encountered in various carbonaceous materials.[5]

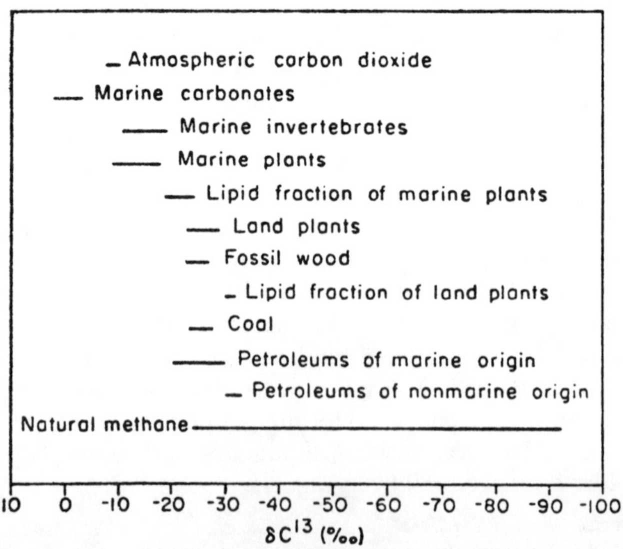

Figure 10-1. Ranges of $\delta\ C^{13}$ encountered in various carbonaceous materials.[5]

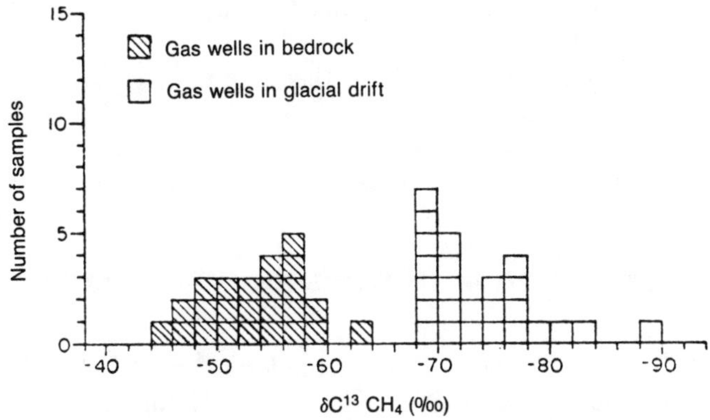

Figure 10-2. Histogram of δ C^{13} from gas wells in Illinois.[2]

Figure 10-2 shows a typical histogram comparing δ C^{13} from methane samples from petrogenic and biogenic gas in Illinois.[2]

Use of Tracers in Underground Storage

Both chemical and radioactive tracers have been used to detect and document migration of natural gas from storage reservoirs. Ethylene as chemical, tritium, krypton, and sulfur hexafluoride, as well as tritiated methane as radioactive tracers, have been used with good success in the past.[9]

SOLUBILITY AND DIFFUSION IN STORAGE

Solubility of storage gas in underground waters becomes of interest from time to time in ascertaining possible sources for gas unaccounted for in storage operations. Sometimes detection and identification of gas migration from storage projects through the indirect route of contact with aquifers require calculation of both solution and diffusion into the brines. Solubility of methane in water is given in the literature by Culberson and McKetta.[6] Solubility of natural gas in water is given in the literature by McKetta and Katz.[7] The presence of dissolved solids in liquid phase tends to decrease the solubility of natural gas in brine.

The solubility of various hydrocarbons in water may be calculated using Henry's Law for dilute solutions. The solubility is expressed in st cu ft per barrel of water. At a given reservoir temperature, it increases with pressure. It is of the order of 10 SCF of gas per barrel of water at 1,000 psia. The rate of dissolution of gas in brine is a diffusional process, controlled by the diffusion coefficient in porous media. The effective diffusion coefficients for the methane nitrogen system in dry and water-saturated porous media are given by Chen, Katz, and Tek.[8]

As shown in Figures 10-3 and 10-4, the amount of gas which will ultimately go into solution is determined from thermodynamic consideration on solubility. The rate of dissolution, on the other hand, is a transport process which can be calculated from laboratory- or field-determined diffusion coefficients.

CORROSION IN UNDERGROUND STORAGE

Corrosion has been defined as the destruction of metal due to chemical or electrochemical reaction with the environment. By comparison, it may be

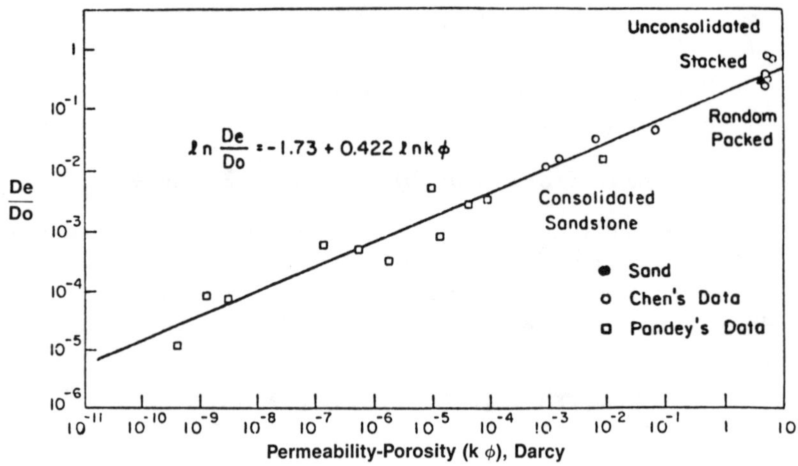

Figure 10-3. Diffusion factor vs. permeability porosity product for dry porous solids.

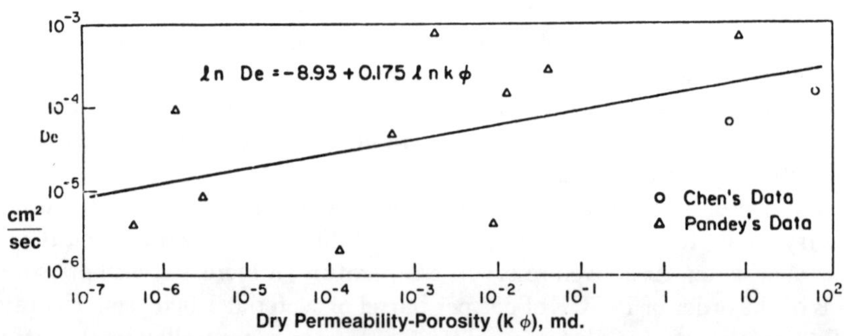

Figure 10-4. Effective diffusion coefficient vs. dry permeability porosity product for water-saturated porous solids.

noted that deterioration due to direct physical contact is not corrosion. It may be referred to as *erosion* or *wear-and-tear*. Sometimes, corrosion accompanies the physical wear-and-tear. The combined process is referred to as corrosion-erosion or corrosive wear. Deterioration of non-metals, such as cracking of paint or plastic and wearing out of seals, is not corrosion in the sense of chemical attack. Rust is a specific form of corrosion related to iron or iron-based alloys through chemical reaction leading to ferrous or ferric oxides. Non-ferrous metals do not rust but they may still corrode. Principles of chemistry and electrochemistry are important in understanding corrosion along with fundamentals of physical metallurgy. Much as biology and chemistry relate to medical science, metallurgy and chemistry relate to corrosion science.

In underground storage, monitoring and protection against corrosion has traditionally relied on past experience and rules of safety rather than precise and scientific solutions. One reason for the widespread use of empirical techniques in corrosion in underground storage is the complex nature of the problem. The many factors which influence corrosion in underground systems include soil properties, structural-compositional properties of wells and pipes, stray electrical currents, and cathodic protection systems. Underground corrosion is in general the result of several distinctly different mechanisms, such as stress corrosion and sulfur corrosion due to bacterial ingestion of methane. The primary mechanism, however, is due to galvanic or electrochemical effects. The corroding system may therefore be considered a complex galvanic cell in which three key components can be identified:

1. The well casing, or buried gathering pipe itself, represents the electrode. The casing or pipe may also be viewed as an electrical conductor connecting regions or depths of varying electrochemical potential. When sacrificial anodes are employed in cathodic protection, these may also be considered added electrodes along the pipe.
2. The second element is the soil surrounding the pipe and/or the various strata in the environment around the wellbore. They correspond to the electrolyte in the galvanic cell.
3. The third and final element of the corroding system involves impressed currents in cathodic protection and stray currents in local environment. The rate of corrosion at any given point can be shown to be directly proportional to the sum of all electrical currents at that point. These currents can be *galvanic* due to electrochemical potentials, or *electrolytic* due to external sources.

Modeling of underground pipeline or well corrosion is sometimes viewed as a current or potential distribution problem. Such modeling helps design optimized cathodic protection systems.

Monitoring is an essential part of corrosion protection systems[9]. Various logging devices are available or are being developed for wells (vertilogs) or buried pipe (line-a-logs).

In recent years, research efforts have been directed to severe problems caused by high concentrations of H_2S and CO_2 at temperatures encountered in storage operations. It has been stated on occasion that, with properly dehydrated natural gas with 1% or less CO_2 and low concentrations of H_2S and O_2, little or no significant corrosion will occur inside steel pipe. As recently as the early eighties, some storage fields were found to have corrosion problems resulting from allowable amounts of CO_2 present in pipeline gas supply. Because corrosion can proceed only when there is no protective film on the metal, corrosion inhibitors are used to provide such a protective film intermittently or, sometimes, continuously.

Occasional introduction of air into the pipelines via natural gas to dilute the gas sold without lowering the contractual heating value below a specified threshold does not appear to have resulted in severe problems of corrosion.

Corrosion literature is voluminous. Only two references will be given here to direct the reader by way of applied literature through the forums of operating section proceedings of the AGA.[10,11]

UNDERGROUND STORAGE IN SALT

Storage of LPG (liquified petroleum gas) in salt cavities was extensively practiced during the fifties. About a decade later, during the early sixties, Southeastern Michigan Gas Company initiated natural gas storage operations in a salt cavern in Marysville, Michigan near Port Huron. During the late sixties, Transco developed two solution-mined salt caverns for storage of natural gas in Eminence, Mississippi. The last two decades saw development of salt cavities for storage in France, UK, and Germany, as well. In an area of fast developing technology, storage in solution-mined void cavities, each having a volume in the range of 5 to 10 million ft^3, provides high deliverability for extreme peak shaving service.[12,13] Detailed description and full treatment of salt mining and storage technology are beyond the intent of this chapter. Those interested in theory and practice in salt storage are directed to excellent literature compiled on the subject by the Solution Mining Research Institute, a cooperative research organization with world-wide membership and reputation. A survey of salt cavern storage of natural gas was published by CER Corporation during 1980.[12]

Underground storage in salt, though still a minor practice compared to storage in depleted fields or aquifers, is an important and growing sector of the energy industry.[14] Particularly in areas such as the Gulf Coast, abounding in salt domes and in close proximity to petrochemicals, solution mining of salt not only provides an ecologically sound and safe environment for underground storage of a host of commodities, but also provides a by-product chlorine-caustic business, which blends well with ethylene-based plastics manufacture.

References

1. Coleman, D. D., W. F. Meents, C. L. Liu, and R. A. Keogh. "Isotopic Identification of Leakage Gas from Underground Storage Reservoirs—A Progress Report," Illinois Petroleum (1).
2. Coleman, D. D. "Isotopic Characterization of Illinois Natural Gas," Ph.D. Dissertation, University of Illinois, 1976.
3. Coleman, D. D. "The Origin of Drift Gas Deposits as Determined by Radiocarbon Dating of Methane," *in* R. Berger and H. E. Suess., eds., Proceedings of the Ninth International Radiocarbon Conference, Los Angeles and San Diego, June 20–26, 1976.
4. Godwin, H. "Half Life of Radiocarbon," *Nature* (1962) V. 195, p. 984.
5. Silverman, S. R. "Investigations of Petroleum Origin and Evolution Mechanisms by Carbon Isotope Studies, in Isotopic and Cosmic Chemistry," H. Craig, S. L. Miller, and G. J. Wasserburg, eds., Amsterdam: North Holland Press, 1964, pp. 92–102.
6. Culberson, O. L. and J. J. McKetta, Jr. "Solubility of Methane in Water at Pressures to 10,000 psia," *Trans. AIME* (1951) 192; 223.
7. McKetta, J. J., Jr., and D. L. Katz. "Phase Relationships of Hydrocarbon-Water Systems," *Trans. AIME* (1947) 170, 34.
8. Chen, L., D. L. Katz, and M. R. Tek. "Binary Gas Diffusion of Methane-Nitrogen through Porous Solids," *AIChE Journal* (1977) 23, No. 3, p. 336.
9. Araktingi, R. E., M. E. Bennefield, Z. Bessenyei, K. H. Coats, and M. R. Tek. "Leroy Storage Facility, Uinta County, Wyoming: A Case History of Attempted Gas Migration Control," *J. of Petro. Tech.* (Jan. 1984).
10. Marsau, D. "Detection and Inhibition of Internal Gas Pipeline Corrosion," AGA Transmission Conference, Operating Section Proceedings 1974.
11. Bush, H. E. and A. E. Beasley. "Corrosion Control in Gas Storage Wells and Lines," Paper by Baroid Division, N. L. Industries, Inc., Houston, Texas.
12. "Survey of Salt Cavern Storage of Natural Gas in United States and Canada 1980," CER Corporation, Las Vegas, Nevada, 1980.
13. Hardy, R. H., Jr. "Development of Design Criteria for Salt Cavity Storage of Natural Gas," 5th International Symposium on Salt, Hamburg, Germany, 1978.
14. Allen, Kermit. "Eminence—Natural Gas Storage in Salt Comes of Age," SPE Paper number 3433, Fall Meeting, New Orleans, 1971.

11

CASE HISTORIES IN UNDERGROUND STORAGE

Two case histories are featured in this concluding chapter. One, the case of Bistineau Gas Storage Field, Bienville Parish, Louisiana, represents conversion to and successful operation of underground storage in a highly permeable depleted dry gas field. Its story, contributed by Ralph E. Laenger on behalf of United Gas Pipeline Company, is reproduced here in detail because it represents a synthesis of many factors and attributes of storage treated separately in earlier chapters of this book. In addition, certain specific aspects such as regulatory requirements, reconditioning of wells, completion, stimulation, and detailed design of storage facilities give a real-world glimpse of sizes, rates, capacities, and layouts. This paper was a feature presentation during the special course on Underground Storage at the University of Michigan, summer of 1983. It is reproduced here with permission of the United Gas Pipe Line Company of Shreveport, Louisiana.

The second case history is also unique, because it represents storage in an aquifer where migration all the way to the surface was detected, measured, studied, and controlled. The case history of Leroy Storage was released by Mountain Fuel Resources, Inc. (then Mountain Fuel Supply Co.) and presented at the Society of Petroleum Engineers' Annual Technical Conference during 1982. It was subsequently published by the SPE in the *Journal of Petroleum Technology* during January 1984. Its highlights are reproduced here by permission of the Society of Petroleum Engineers of AIME.

CASE 1: GAS RESERVOIR WITH HIGH PERMEABILITY—BISTINEAU GAS STORAGE FIELD, BIENVILLE PARISH, LOUISIANA*

Data Summary

Storage Reservoir	Pettet Limestone at 5,250′
Acres in Storage Area	12,043
Proven Acres	4,831
Acre Feet	136,970
Average Thickness, Feet	28.35
Average Porosity, Percent	19.83
Average Connate Water Saturation, Percent	15
Initial Surface Shut-In Pressure, psia	2,180
Initial Recoverable Wet Gas, MMCF	138,160
Cumulative Wet Gas Produced, MMCF	133,939
First Injection	9-1-66
First Withdrawal	12-21-66
Arkansas Louisiana Gas Company (Arkla) Becomes Joint Interest Owner	4-25-69

INTRODUCTION

Every so often nature provides the solution to man's economic needs. Such is the case of Bistineau Gas Storage Field fulfilling at its particular location the needs of United Gas Pipe Line Company (UGPL).

A detailed study of reservoir and geological facts was essential to the plan to make Bistineau Gas Field into Bistineau Gas Storage Field. Fortunately, we had access to gas production figures and well histories. We also had the cooperation of working operators in the area. The information from various sources furnished the insight to determine the size and performance of the field.

Located in the northwest part of Louisiana on the east side of Lake Bistineau in western Bienville Parish, Bistineau Gas Storage Field suits the demands of the UGPL system for a very flexible set of parameters. Projection of needed gas supplies and deliverability for the Tyler, Shreveport, and Monroe Districts from 1961 to 1971 determined a very significant increase. We were left with two options:

* Contributed by Ralph E. Laenger, United Gas Pipe Line Company, Shreveport, Louisiana, January 1983.

1. Combination of the enlargement of South to North Louisiana 30″ facilities and the development of a large underground gas storage field for the Tyler, Shreveport, and Monroe Districts.
2. The enlargement of South Louisiana to North Louisiana 30″ facilities, including extension and enlargement of facilities delivering gas to such 30″ line.

After studying both options, the advantages and disadvantages were weighed, as follows:

I. **Combination of Storage Facilities and Enlargement of South to North Louisiana 30″ Facilities.**
   a. Advantages
      1. Provides additional deliverability for periods of heavy demand.
      2. Provides higher load factor for better performance of gas purchase contract obligations and system operation.
      3. Reduces expense of purchasing more gas reserves than actually otherwise required in order to merely provide adequate deliverability.
      4. Minimizes pipeline and compressor capacity required, both for connection of new reserves and transport capacity through new 30″ line to North Louisiana.
      5. Provides greater operating flexibility.
      6. Due to size and location, storage field can operate to advantage of practically all districts, and can have capacity increased relatively inexpensively.
      7. Costs less than enlargement of South to North Louisiana 30″ facilities, including extension and enlargement of facilities delivering gas to such 30″ line, to provide identical increased deliverabilities.
   b. Disadvantages
      1. Additional gas would be needed by 1964–65 season. The actions required to achieve storage—acquisition of storage rights, reworking and drilling wells, acquiring sites and rights of way, constructing facilities to inject cushion gas, etc.—did require appreciable time; however, the conversion of Bistineau Field to Storage was undertaken in the latter part of 1965.

II. **The Enlargement of South Louisiana to North Louisiana 30″ Facilities, Including Extension and Enlargement of Facilities Delivering Gas to Such 30″ Line.**
   a. Advantages
      1. Provides additional deliverability for periods of heavy demand.
      2. Provides greater operating flexibility.

b. Disadvantages
   1. Would require continuing purchase of additional reserves to provide needed additional deliverability.
   2. Would require continuing increase in pipeline and compressor facilities, for connection of new reserves and provision of additional transport capacity to North Louisiana.
   3. Would not improve System load factor.
   4. Would require a higher investment as compared to a combination of storage and enlargement of South to North Louisiana 30″ facilities.

REGULATORY REQUIREMENTS

Storage of interstate natural gas requires an application with the Federal Energy Regulatory Commission (now FERC, then the Federal Power Commission, FPC).

The FPC Docket No. CP66-136 (original application) was filed on October 29, 1965, and approval came on April 29, 1966.

Other governmental agencies (State and Federal) were contacted and permits received, and Bistineau Gas Storage was on its way to becoming a reality.

LAND AND RIGHTS OF WAY

Acquisition of gas storage rights was studied and reviewed for some time before the actual decision was made to convert Bistineau Gas Field into a storage reservoir. Contacting the land owners and mineral right owners requires much time and quite a bit of preliminary planning and preparation before proceeding to acquire these rights. Each land and mineral right owner was contacted, and an amount of equitable compensation was agreed upon and paid. Also, we had to purchase working interests and wells from various companies.

Generally, three-acre sites were purchased from land owners to drill or work over a well, and the wells were spaced on approximately 40 acres.

GEOLOGICAL AND ENGINEERING DATA

The field was discovered in March 1921 by the Lakeside Oil Company No. 1 State well which found shallow gas production in the Tokyo and Eagleford Formations of the Upper Cretaceous Age (1,950–2,000 ft, and at 2,550–2,600 ft). Later, deeper drilling discovered gas production in the Pettet Formation of the Lower Cretaceous Age (5,137–5,235 ft), the first completion in this formation being made in the DeSoto Oil and Gas Company (R. O. Roy) No. 1 Dunn and Olsen in January 1937. There was a total of 19 producing wells and five dry holes drilled to the Pettet Formation. It is the Pettet Formation which we selected as the gas storage reservoir. In addition to the Tokyo, Eagleford, and Pettet discoveries, a few attempts have been made to establish deeper production in the Travis Peak and Cotton Valley Formations. These attempts for commercial production have not been very successful.

From 1937 to 1966 the Lake Bistineau Field produced a total of 133.939 billion ft^3 of gas. At this time, the field was essentially depleted with the several remaining wells producing very small amounts of gas, with an estimated remaining gas reserve of 4.221 billion ft^3.

Initially, Bistineau Gas Storage was a dry gas reservoir having wells of high deliverability and having total recoverable liquids of five to six barrels per million ft^3. The production decline curve (see Figure 11-1) is almost a straight line, and from all engineering aspects is an almost textbook example of what a storage field should be. Natural gas is easily injected by the storage plant with 34,000 hp Clark compressors (three 6,000 hp and two 8,000 hp) and withdrawn by free flow, with the withdrawal rates as high as 1.2 billion ft^3 for a 24 hour period.

The Pettet Reservoir of Bistineau Gas Storage Field is a very unique and nearly perfect storage reservoir (see Figure 11-2). It is a faulted anticline with the major self-sealing fault running northwest and southeast 400 ft down-thrown (east part of field). The northern, western and southern portions of the structure are bounded by encroachment water and permeability barriers that serve as a seal to prevent any gas loss, and these are down-dip. At Bistineau the encroached water boundaries have moved very slowly over the producing life (27 years) of the field and invaded *only part* of the reservoir. In five months on withdrawal we accomplish that which covered a period of 27 years on production. The reservoir mechanism is a Pressure-Volume relationship: the more gas injected, the higher the pressure.

The top of the structure is covered by the Three Finger Lime (Sligo Formation) which is a very dense lime, impermeable, and approximately 50 ft thick. Each well that has penetrated this impervious mantle has been verified by a cement bond log as being cemented with the latest engineering practices available.

At field abandonment in 1965, the bottom-hole pressure was 180 psia. The Pettet Formation is composed of three separate zones of porosity which are oolitic limestones. During construction we completed all wells in all three zones (A, B, and C). The Northwest-Southeast fault delineates the eastern limits of the field. The minor East-West adjustment faults do not separate the reservoirs but bisect the various zones of porosity in juxtaposition to each other, thus interconnecting the reservoirs across the various fault

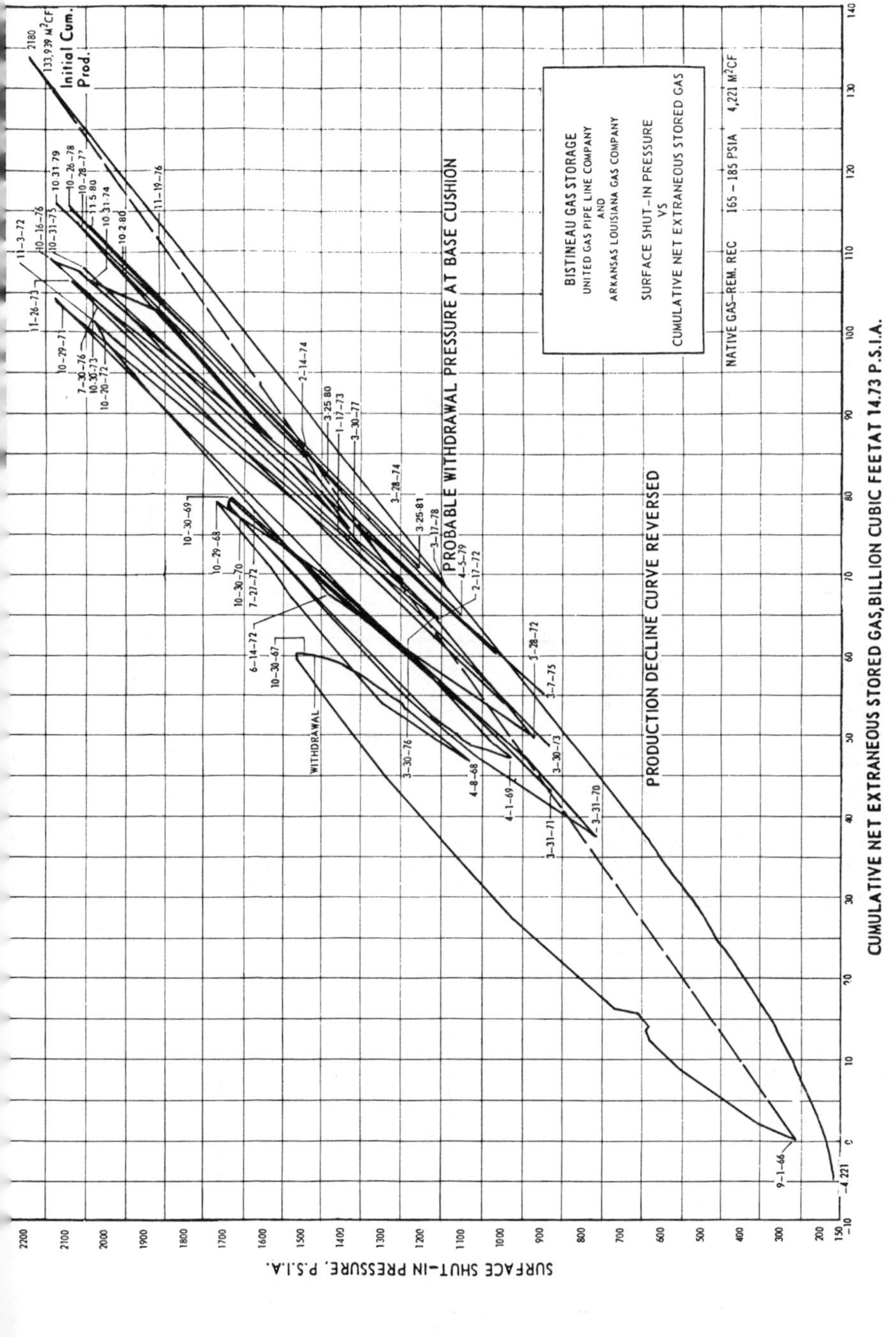

Figure 11-1. Bistineau gas storage production decline curve and subsequent storage hysteresis.

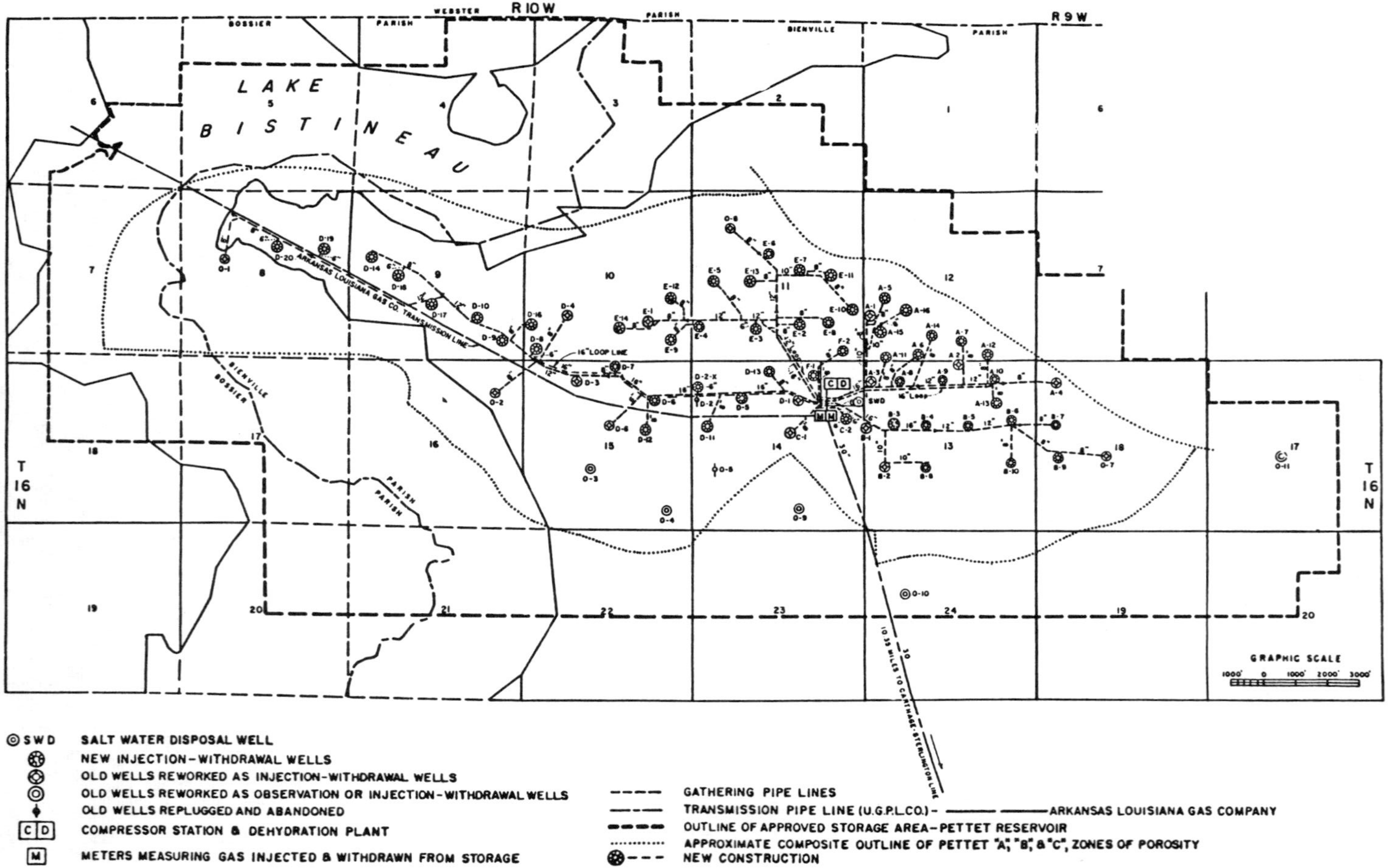

Figure 11-2. Map of Bistineau Underground Gas Storage Field.

planes. Approximately 4,831 acres are proved, with the storage rights to 12,043 acres being the total storage area.

There are 68 injection-withdrawal wells and five observation wells, and many wells are characterized by an MER of 50 MMCF/day. Some water (0.25 bbl/MMCF) is produced whenever volumes in storage reach the certificated cushion at 52.2 BCF (45.2 BCF United and 7 BCF Arkla).

As of June 8, 1979, Bistineau Gas Storage was certificated for 121 BCF, subject to maximum surface shut-in pressure of 2,175 psia. Arkansas Louisiana Gas Company (joint owner) was allowed 15 BCF, and United Gas Pipe Line Company was allowed 116 BCF injected gas in storage.

Designed deliverability capacity on March 31 with 52.2 BCF in storage (45.2 BCF United and 7 BCF Arkla) was a total of 880 MMCF/day (780 MMCF for United and 100 MMCF for Arkla).

RECONDITIONING OF OLD WELLS (FIGURES 11-3 AND 11-4)

Following is a step-by-step account of how we reconditioned the old wells in the field:

On wells that were plugged and abandoned

1. Casing strings were tied back into surface.
2. Cement Bond Log, Gamma Ray-Neutron, and Casing Inspection Logs were run to establish and verify integrity of cement bond and casing.
3. Used oil base mud.
4. Fishing operations: four workover rigs operational 24 hours a day. Each rig had drilling supervisor and tool pusher.

Operations

1. Went in well and recovered as much tubing as possible.
2. Ran Cement Bond Log to verify good bonding in caprock at approximately 5,000 feet. Ran Gamma Ray-Neutron and Casing Inspection Logs. Squeezed if bonding not adequate.
3. Set cast iron bridge plug on inside of 5½ in. casing and dumped cement on top of plug (opposite caprock—Three Finger Lime). Reason: In case hole was lost, (junked) well would be plugged.
4. After being sure well was properly plugged on bottom, inside and outside of casing, started work on top of hole.
5. Ran 7 in. wash pipe with tungsten carbide shoe, washing over 5½ in. casing till shoe made no footage.

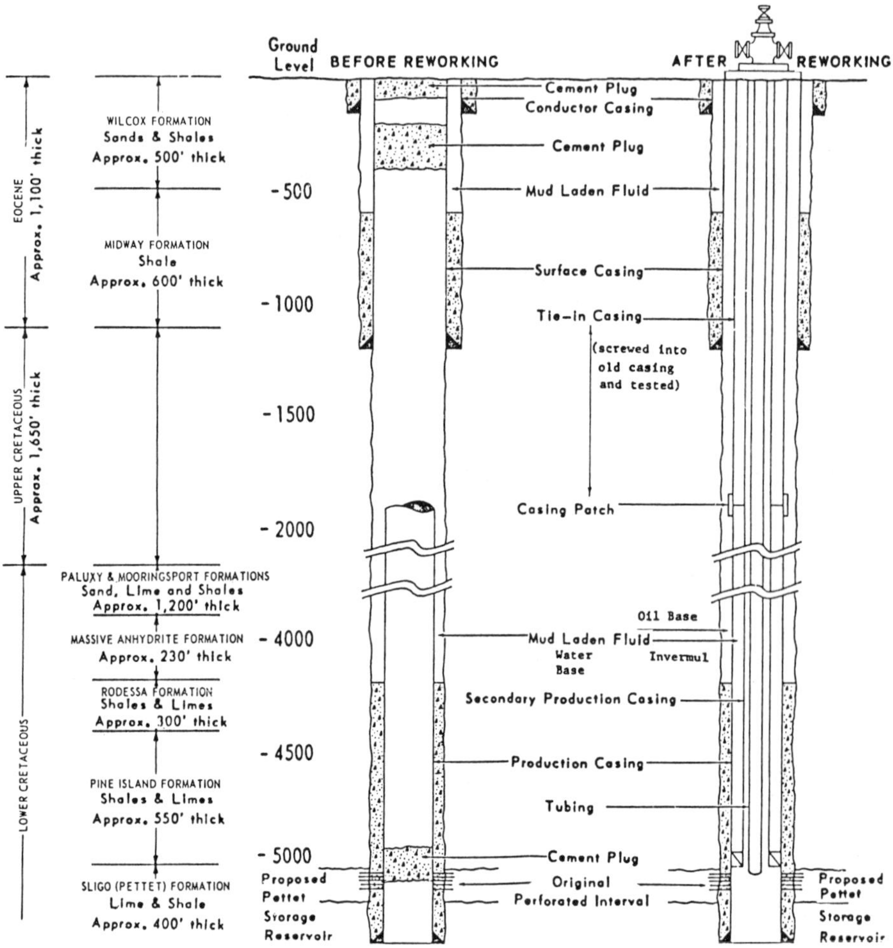

Figure 11-3. Well that was plugged and abandoned.

6. Ran small ($2^7/_8$ in.) drill pipe inside $5^1/_2$ in. casing and set on solid footing inside $5^1/_2$ in. casing, running this to surface to be used as line-up string.
7. Pulled full string of wash pipe out of hole, replaced tungsten carbide shoe.
8. Went back in hole with 7 in. wash pipe and tungsten carbide shoe to depth previously washed over.
9. Pulled $2^7/_8$ in. line-up string out of hole.
10. Continued to wash over $5^1/_2$ in. casing till we reached the point established by Cement Bond Log that there was good bond, or good casing point.

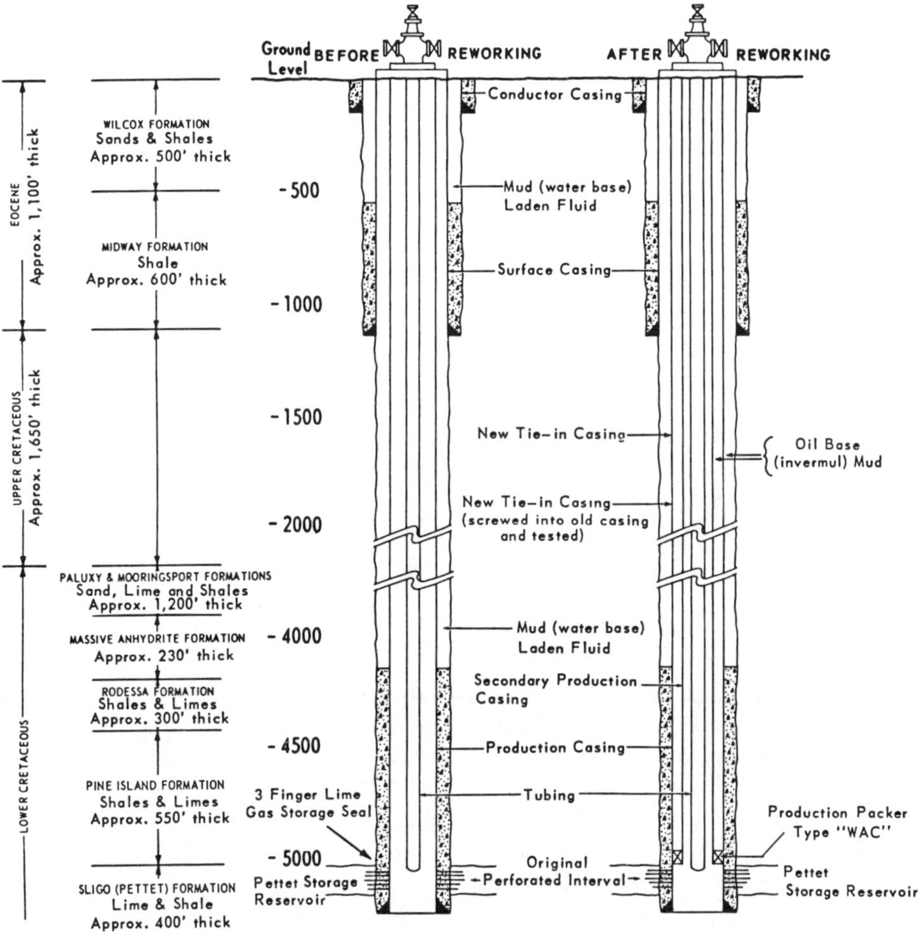

Figure 11-4. Well that was or is productive and has casing remaining in the well.

11. Picked up 7 in. wash pipe to point that exposed two or three collars of $5^1/_2$ in. casing.
12. Went in hole with electric line and string shot and shot selected collar.
13. Went in hole with $2^7/_8$ in. line-up string with back-off tool, set it inside $5^1/_2$ in. casing to a point above the collar that string shot was run in.
14. Lowered 7 in. casing down to previously washed over point.
15. Backed off $5^1/_2$ in. casing at string shot point.
16. Pulled out of hole with back-off tool and $2^7/_8$ in. drill pipe.
17. Went in hole with $2^7/_8$ in drill pipe and spear, caught top of $5^1/_2$ in. casing and pulled fish and recovered $5^1/_2$ in. casing out of hole.

18. Went in hole with 5½ in. new casing and screwed into old casing where we had backed off.
19. Tested casing with 1,500 pounds; pulled 40,000 pounds above weight of pipe.
20. Conditioned oil base mud around bottom of wash pipe at wash-over point.
21. Pulled worn shoe and wash pipe out of hole.
22. Hung 5½ in. casing in bradenhead.
23. Nippled up.
24. Went in hole with 2⅞ in. drill pipe and rock bit and displaced oil base mud with water.
25. Drilled inside 5½ in. casing, cement, and cast iron bridge plug.
26. Ran wireline packer with Position No. 1 nipple below packer and converter on top of packer (see Figure 11-3).
27. Went in hole with 3½ in. tubing and completed well with overshot receptacle on bottom of 3½ in. tubing.

Notes

1. This is a typical washover and fishing job.
2. Ran as much as 3,945 ft of 7 in. wash pipe at one point.
3. No observation wells per se were drilled.
4. Reworked 16 old wells into injection-withdrawal wells, five old wells reworked as observation wells, and two wells plugged and abandoned (23 old wells total). Of the two wells plugged and abandoned, only one well was junked, plugged, and abandoned. This one well was plugged properly before fishing operations commenced.
5. Old wells reworked as observation wells because of their locations.
6. Nine months from first workover to last workover.

DRILLING NEW WELLS—DEFINING FIELD LIMITS (FIGURE 11-5)

Following are general notes and procedures used in the drilling of new injection-withdrawal wells:

1. Wells were generally on 40 acre spacing, with surface locations taken into consideration.
2. The main concept in drilling was to use large casing to permit high deliverability rates.
3. First gas injection was September 1, 1966.

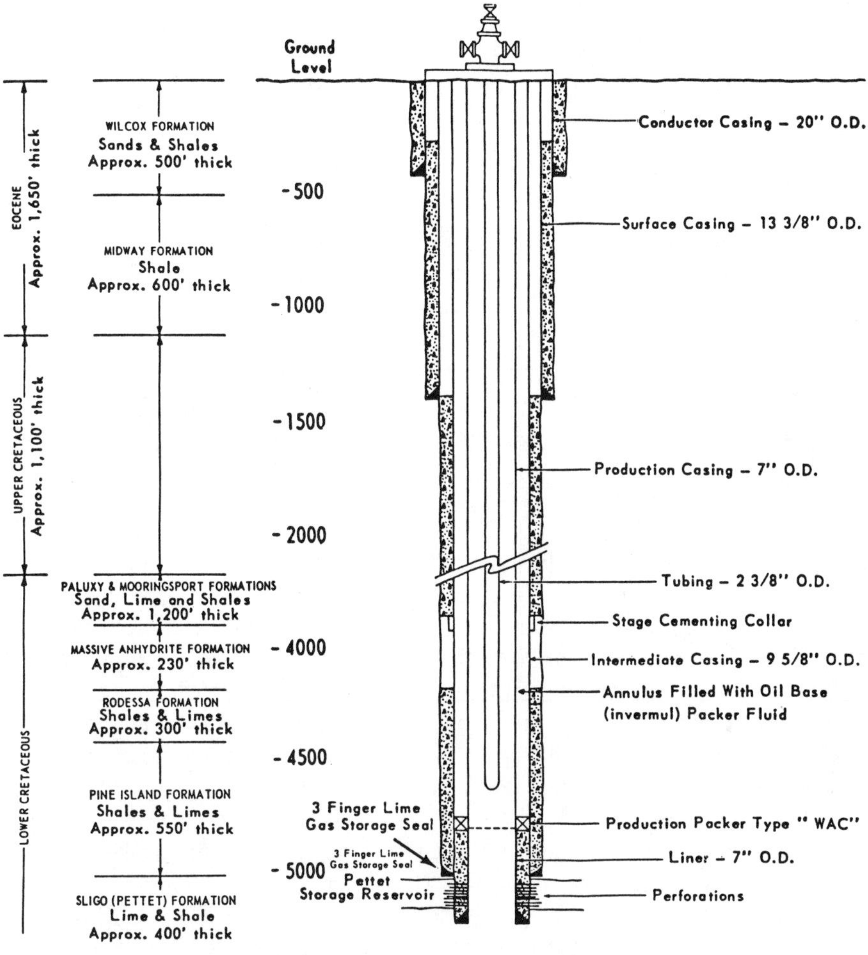

Figure 11-5. New injection-withdrawal well.

4. Drilled 20 new injection-withdrawal wells.
5. Used water base mud.

Operations

1. Drilled 24 in. hole to 400 ft.
2. String of 20 in. casing set at 400 ft and cemented to surface, with common cement plus 16% gel, plus 5% salt, and tailed with common cement plus 2% $CaCl_2$—100% excess.

3. Drilled 17½ in. hole to 1,400 ft.
4. String of 13⅜ in. casing set at 1,400 ft and cemented to surface with common cement plus 16% gel and 5% salt, 50–50 Pozmix A cement, and common cement plus 2% $CaCl_2$—100% excess.
5. Drilled 12¼ in. hole to 5,000 ft, approximately, or to middle of third finger of the Three Finger Lime of the Sligo Formation. Ran electric log surveys to 5,000 ± ft.
6. String of 9⅝ in. casing set at 5,000 ± ft and cement up into surface casing at 1,400 ft, using stage tool at approximately 3,600 ft. First-stage cement: 50–50 Pozmix A cement plus 6% gel and 1% CFR-2, common cement plus 18% salt and 1% CFR-2—25% excess. Second stage: 75–25 Pozmix A cement plus 6% gel with 1% CFR-2, 50–50 Pozmix A cement plug 4% gel with 10 lb per sack and tailed with common cement plus 18% salt and 0.75% CFR-2—25% excess.
7. Verified bonding in caprock (Three Finger Lime) with Cement Bond Log, and ran Gamma Ray-Neutron Log to surface.
8. Displaced mud with water. Drilled out a float shoe and collar shoe in 9⅝ in. and drilled 8¾ in. hole to 5,400 ft., ran electric log surveys to 5,000 ± ft, set 7 in. liner with 150 ft overlap up inside 9⅝ in. casing, bringing top of liner to approximately 4,850 ft, and cemented with 50–50 Pozmix A cement plus 12% salt and 1% CFR-2—15% excess.
9. Tested top of liner at 4,850 ft with 2,500 pounds.
10. Cleaned inside of 7 in. liner down to five ft above float shoe.
11. Ran Cement Bond Log.
12. Perforated 7 in. casing in reservoir Zones A, B, and C sections of Pettet Limestone reservoir rock.
13. Set special type wireline packer in 9⅝ in. casing, with packer tail pipe and profile nipple in top of 7 in. liner.
14. Had Position No. 1 X-nipple attached below wireline packer with converter on top to receive 7 in. overshot type receptacle (flotation seals to allow for expansion and contraction).
15. Ran 7 in. casing to surface, with 7 in. overshot type receptacle over converter of packer (seals in overshot). Just before landing 7 in. string, spotted packer fluid (Invermul F-100) between 7 in. and 9⅝ in.
16. Hung off 7 in. casing inside 7 in. tubinghead.
17. Ran 2⅜ in. kill string with Position No. 1 nipple one joint off bottom; hung this off in 2⅜ in. tubinghead.
18. Ran deliverability test after flowing well for clean-up.
19. Acidized with 20,000 gallons 28% HCl.
20. Flowed well for clean-up and ran deliverability test. Average increase of two and one-half times.
21. Well now completed for use as injection-withdrawal well.

WELL HEAD STRUCTURES AND ANNULI

Starting at the 13³/8 in. bradenhead, the first annulus is between 9⁵/8 in. and 13³/8 in. This annulus extends down to 1,400 ft to the bottom of the 13³/8 in. casing or top of cement in the 9⁵/8 in. × 13³/8 in annulus. The next annulus is between the 7 in. and 9⁵/8 in. and it extends down to the top of the packer at approximately 5,000 ft. The special packer is set in the 9⁵/8 in. casing near the top of the 7 in. liner with the 5¹/2 in. O.D. tail pipe X-nipple and nipple spacer set down in the 7 in. liner. This provides a method of setting a plug in the Position No. 1 nipple below the packer, working on just about any mechanical tool failure above the packer, then pulling the plug and bringing the well in without disturbing or in any way taking a chance of decreasing the deliverability of the well due to fluid loss. The 7 in. casing and the 2³/8 in. tubing are both gas strings. The gas mostly flows through the 7 in. casing. The 2³/8 in. tubing is used for a kill string and to lubricate chemical corrosion inhibitor into the well. This type of well head structure (see Figure 11-5) permits maximum flow of a well without any encumbrances of unnecessary equipment.

At the surface, there are two 4 in., 3,000 pound working pressure valves, one on each side of the 7 in. annulus, connected to expansion loops, then to a tee going into the 6 in. meter run. Shown are two 2 in., 3,000 pound working pressure valves that control the 2³/8 in. tubing.

There are no separation facilities at the well installation; all separation takes place at the plant.

RESERVOIR PERFORMANCE AND DELIVERY CAPACITY

All of the design and development at Bistineau Gas Storage has been based on the relation of required delivery capacity (fixed), by required working gas (fixed), by compressor and dehydration plant (fixed), by number of wells (variable), by cushion gas (variable). We have always recognized that the number of wells and cushion gas were variables that were subject to adjustment to meet the other requirements, which were fixed (as to a specific design and time). In reviewing the past history of development of this storage field, it becomes evident that we have now made optimum use of this reservoir. We drilled additional wells and adjusted the cushion gas to meet the changing demands for deliverability and working gas for our system. This one storage field on a cold day can and does supply up to one third of United's gas send-out requirements.

The requirements for our system for the 1968–69, 1969–70, and 1970–71 winters resulted in a maximum volume in storage of about 80 billion ft^3 at the end of the injection cycle, and we withdrew down to total volumes of 47, 37, and 43 billion ft^3, respectively. These operations resulted in a partial equilibrium and a reasonably stabilized reservoir performance for these conditions. We were able to obtain very high delivery capacities with reasonably low volumes of cushion gas because we were not using the total capacity of all of the reservoir, and our storage gas tended to remain near the top of the structure where most of our wells are located. These reasonably stabilized operating conditions were used for estimating long-range future operating parameters as to number of injection-withdrawal wells, requirements for cushion gas, working gas, etc. to meet the needs of our transmission system. The estimates which were made in 1969 and approved by the FPC on May 5, 1970 were used as the basis for construction and expansion in 1973.

Table 11-1 gives the working and cushion inventories, the designed deliverability, and the number of wells for 11 withdrawal seasons, starting in 1971.

Table 11-1
Operating Parameters in Bistineau Storage Unit

Year	Maximum BCF in Storage	Minimum in Storage	MMCF Cushion	Designed Deliverability	No. of Wells
1971–72	104.29	48.81	61,000	900	58
1972–73	101.34	48.49	61,000	1,000	60
1973–74	106.52	69.24	61,000	1,000	60
1974–75	106.84	53.09	61,000	1,000	60
1975–76	107.92	64.45	61,000	880	60
1976–77	109.34	66.06	52,200	880	68
1977–78	114.02	66.12	52,200	880	68
1978–79	117.07	63.47	52,200	880	68
1979–80	117.12	78.96	52,200	880	68
1980–81	114.23	68.74	52,200	880	68
1981–82	118.85	86.04	52,200	880	68

Utilization of the total reservoir at maximum capacity will result in more gas in the thinner, less permeable areas on the periphery of the reservoir where we have very few operating injection-withdrawal wells. During the 1973–74 winter we operated the storage field (UGPL portion) at high delivery capacity for "peaking" purposes. Gas was injected and withdrawn in November and December 1973 and, consequently, UGPL and Arkla had a total volume in storage of 101.47 billion ft^3 on January 1, 1974. This maintained a high pressure on the top of the structure for a long period of time and resulted in pushing about three or four billion ft^3 of gas further out on the periphery of the structure where it was not readily accessible to the injection-withdrawal wells. During the next three months of that withdrawal cycle we pulled down only to about 70 billion ft^3. For a period of several months we pushed gas further out on the structure and did not create any differential or time for much, if any, of this gas to move back to the top portion of the structure. This type of operation permits injection of more gas into the storage reservoir for a larger volume of working gas by allowing the pressure on the top of the structure to slowly decrease. The lower pressure, on the other hand, results in decreased deliverability. The operator must decide whether to place the major emphasis on high deliverability with slightly lower working gas, or slightly lower deliverability and a somewhat larger inventory of working gas.

STIMULATION

In completing a new well, the procedure taken in construction would be as follows, after setting the packer and installing the Xmas tree:

1. Flow the well to surface and clean up.
2. Determine flow capacity.
3. Acidize with 1,000 gallons HCl (15%).
4. Flow well to surface for clean up.
5. Determine flow capacity.
6. Acidize with 20,000 gallons HCl (28%).
7. Flow well to surface and determine flow capacity.
8. Well is now ready for injection or withdrawal.

This type of completion procedure was followed consistently. After the large acid treatment, the increase of a well's deliverability was usually two and one-half times.

GATHERING SYSTEM

The gathering system is composed of six lateral lines (A, B, C, D, E, and F), varying in size from 6 in., 8 in., 10 in., 12 in., and 16 in. with field pressure coming all the way to the plant. Three lateral lines (A, E, and D) are looped, and there are ten valves spotted strategically in the lateral lines in case of a line failure. Three field regulators at the plant (two 12 in. and one 6 in.) decrease maximum pressure of 2,175 psia (field pressure) down to 1,050 psia (plant pressure). This system can handle an MER* of 1.2 BCF/day with 63 BCF in storage (net extraneous gas).

Table 11-2 shows the design load carried by each lateral on peak day design deliverability.

This is a rather large gathering system. In examining Table 11-2 one can see that the "A" Lateral has the best wells. These wells are drilled in the east central part of the field.

Table 11-2
Peak Day Deliverability of Laterals

Lateral	No. of Wells	Percentage of Contribution	Approx. Volume Rate MMCF/Day Contribution
A	16	38	456
B	11	13	156
C	2	2	24
D	22	16	192
E	15	27	324
F	2	4	48
	68	100%	1,200 MMCF/day MER*

* Maximum efficiency rate

COMPRESSION

From 12,000 hp compression in December 1966, we expanded our compression to 18,000 hp in September 1969, then 26,000 hp in April 1978, and finally to 34,000 hp in May 1978. Figure 11-6 shows the major gas piping isometric of Bistineau Compressor Station. With a certificated cushion of 52.2 BCF (45.2 BCF United and 7 BCF Arkla) in storage, this storage field on withdrawal is capable of delivering to the pipe lines 880 MMCF per day with compression (731 psia well head shut-in pressure and 493 psia suction pressure).

On injection, with 74 BCF in storage, 1,330 psia shut-in pressure, 600 psia suction pressure, five engines, and 34,000 hp, we can inject 665 MMCF per day. With 700 psia suction pressure, we can inject 803 MMCF per day.

DEHYDRATION PLANT

Dehydration facilities are composed of two 525 MMCF per day dry bed type units and three 200 MMCF per day glycol units. All of these facilities are rated at 1,200 psia, 100°F for continuous operation. This amounts to 1.65 billion ft^3 capacity.

Generally, saltwater production starts when there is a long continuous sequence of high withdrawal rates. Of course, we have a Maximum Efficiency Rate on the whole field of 1.2 BCF, so we have extra capacity on dehydration, which assures day gas of seven pounds of water per MMCF, or less.

The field averages out from 0.25 barrels of salt water per MMCF to about 0.3 barrels per MMCF. This is a very small amount of salt water compared

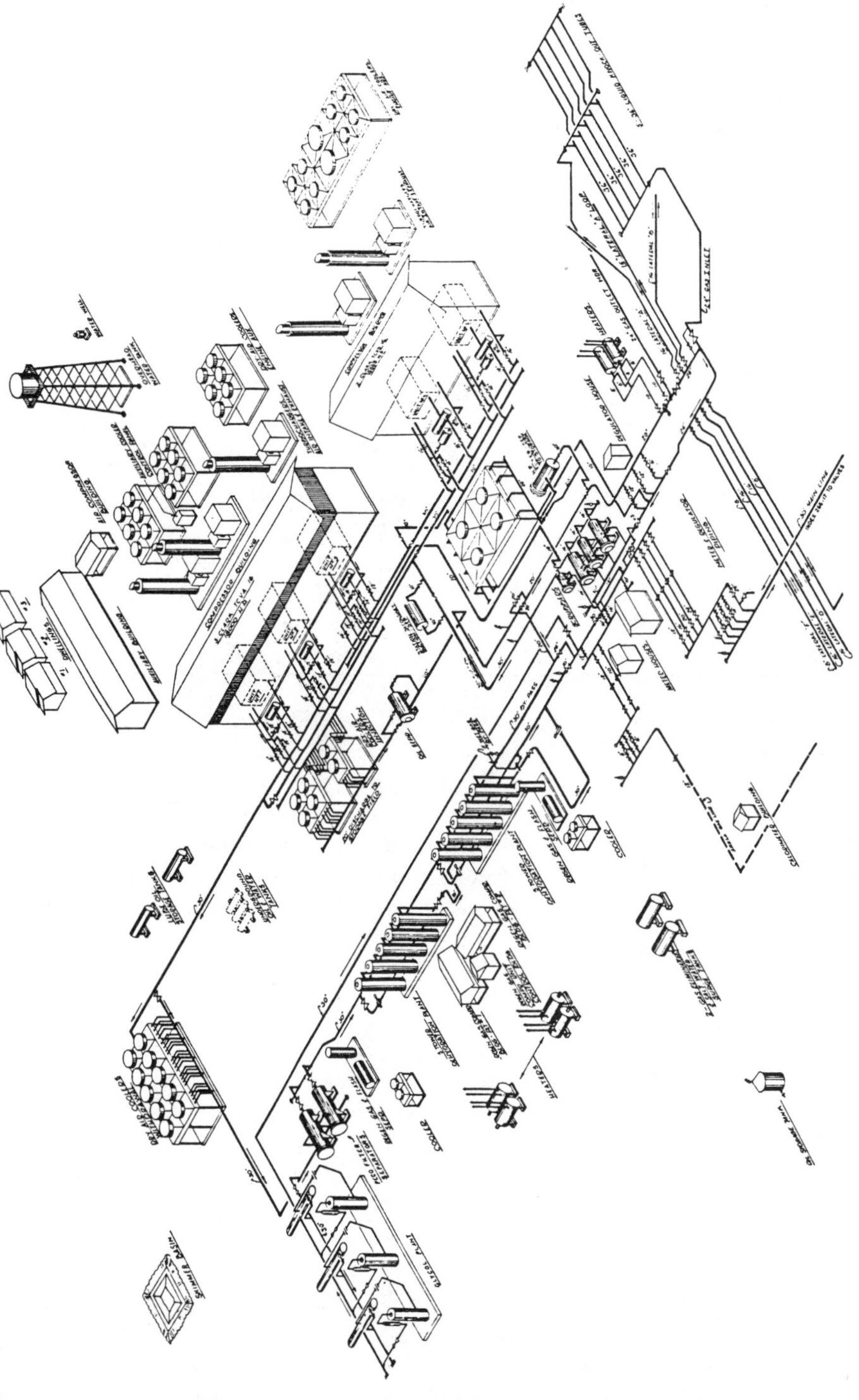

Figure 11-6. Bistineau compression station and gas piping isometric.

to a producing field, where the amount of salt water produced is usually well over this mark.

In addition to salt water entrained from the formation, a certain amount of condensed water will result from temperature and pressure conditions prevailing in the wellbores and gathering lines. It is this fresh water which is likely to cause hydrate problems in the system if not removed. This small amount of salt water seems negligible, but during withdrawals of 800 MMCF per day from storage or more, the salt water can become a problem. Formerly we had to haul our salt water production to our disposal facilities at Koran, a distance of 15 miles, but in 1979 we drilled a saltwater disposal well at the gas storage unit.

DAILY OPERATIONS AND MAINTENANCE

There are basically two operating seasons at Bistineau—injection and withdrawal—which make one cycle. The injection season usually begins around the first of April and ends around the first of November. As mentioned before, the sole purpose of gas storage is to stockpile gas in the summer months when demand is low, for use in the winter months when demand is high, to keep the pipeline load factors high. On April 1, when temperatures have warmed and existing pipelines can meet customer demands, no gas is being withdrawn from storage, and injection depends on the availability of gas to be injected. The storage unit is readied for injection by taking individual well head shut-in pressures (usually four-hour shut-ins) throughout the entire field to check our inventory. During the first months the rates of injection are not too great, but the inventory steadily increases toward its projected value. To guarantee the proper operation of plant and field equipment so that uninterrupted supply of gas can be delivered to our customers during the withdrawal season, the following maintenance takes place:

1. Install, inspect, test, adjust, maintain, operate, repair and calibrate the following equipment:
    a. Measuring equipment including: orifice meters and plates, meter tubes, flanges, fittings and piping, recording thermometers, pressure recorders, dew point recorders, and water analyzers.
    b. Control and regulation equipment including: controllers, control valves, field regulators, valve operators, positioners, pressure relief valves, and dummy valves.
    c. Miscellaneous equipment including: heaters, surface and subsurface safety valves, and related control equipment.
2. Lubricate Xmas tree valves and lateral line block valves, check all well equipment for proper operation.

3. Maintain field roads and well locations.
4. While working in the field, observe any unusual activities, such as heavy equipment digging, etc.

As the field and plant maintenance draws to an end and the storage zone is filled to capacity, all equipment is now readied for the approaching withdrawal season. As November approaches and existing transmission lines can no longer deliver enough gas to meet market demands, gas must be sent out from storage. It is turnaround time, just before (or as) withdrawal begins, that we again shut-in the field for approximately four hours and measure (by dead weight) the well head shut-in pressure at each individual well for inventory verification. As withdrawal begins, storage technicians are changed from day work to shift work. This is to assure that at any time of day or night competent and qualified personnel are at positions in the plant and to ensure that an uninterrupted supply of gas is delivered as needed. These deliveries range in volume from 0 to 1,200 MMCF/day. Also during the winter months, as gas is delivered from storage a corrosion prevention program is implemented in the wells and all lateral flow lines to the compressor station. As winter comes to a close, supplies that are needed from storage diminish, and the warmer months are ahead, our injection-withdrawal cycle is complete. There are a few details of operation which have been omitted because they are carried out year-round and not periodically, such as follows:

1. A 24-hour log of each day's events is kept on such details as volumetric rates, pressures, temperatures, wells on, and work done. This record may be referred to at a later date should the need arise, and may also be used to monitor reservoir performance by storage engineers.
2. Daily review and changing of all flow and pressure charts at the plant.
3. Weekly review of casing pressure on all wells in the field to determine the possibility of packer failure or tubing or casing leaks.
4. Maintenance and upkeep of all testing equipment such as manometers, dead weight testers, and thermometers used.
5. Upkeep of water filtering system and saltwater disposal plant and well. This ensures environmental compliance with all state and federal regulations.

Finally, a vital part of any operation is its people. Work in the storage field is accomplished by a field storage supervisor, who coordinates and oversees all jobs, and six storage technicians, the supervisor's means of completing work. Competent personnel in the right positions ensure a fast, smooth, efficient, and safe operation.

CASE 2: ATTEMPTED GAS MIGRATION CONTROL—LEROY STORAGE, UINTA COUNTY, WYOMING*

ABSTRACT

Leroy Gas Storage Facility, an aquifer storage development in Uinta County, Wyoming, operated by Mountain Fuel Supply Company**, is presented as a case history. This field represents a complex problem in aquifer storage due to the uncontrolled migration of gas to the surface. Considerable effort in reservoir engineering and planning has resulted in apparent and probable arrest of uncontrolled gas migration, although more time will be needed to evaluate and monitor the results of recent efforts.

Incorporated in the evaluation of the leakage problem have been updated geological information, locating and correcting possible well problems, computer simulation, extensive logging, tracer surveys, surface monitoring, and related engineering evaluation.

A computer program developed to simulate a unique history match, including the effect of a time- and pressure-dependent leak and a comprehensive analysis of data leading to the possible control of the leak by proper operating storage pressures, are described.

The paper documents the history and presents the status of the storage field, relating efforts expended to isolate and control the leakage. The geography, geology, historical background, reservoir description, reservoir performance, and related engineering studies relevant to the case history are included. The computer simulation of history match including the leak, analysis of the results, subsequent reservoir engineering work, discussion of results, and conclusions are presented.

INTRODUCTION

The Leroy Gas Storage Field is located in Uinta County, Wyoming in Township 16 North, Range 117 West. The field lies 80 miles west of Rock Springs, Wyoming and approximately 100 miles northeast of Salt Lake City, Utah. This particular location was of interest to Mountain Fuel Supply Company due to its favorable location relative to the pipeline supplying Salt Lake City. Interest in this area for prospective storage facilities dates back to

* Original Paper: "Leroy Storage Facility, Uinta County, Wyoming: A Case History of Attempted Gas Migration Control," by R. E. Araktingi, M. E. Benefield, Z. Bessenyei, K. H. Coats, and M. R. Tek, *Journal of Petroleum Technology*, January 1984. Abridged summary highlights presented here with permission of the Society of Petroleum Engineers of AIME.

** Now Mountain Fuel Resources, Inc.

1969. Evaluation of data obtained from exploratory Leroy Well No. 3, originally drilled by the Shell Oil Company in the summer of 1951, suggested two potential storage formations, namely the Nugget and the Thaynes. The Nugget Sandstone was rejected due to the questionable integrity of the caprock, the faulted structure, and the abnormally high pressure gradient.

Testing of the Thaynes Formation in Leroy Well No. 3 began in October, 1970 following re-entry and deepening to 3,135 ft. An extended flow test was run, producing 4,000 to 8,700 barrels of water per day with an initial static gradient of .508 psi/ft. Initial test evaluation indicated excellent transmissibility, and an expanded drilling/re-entry program followed. Leroy Wells Nos. 4, 5, and 6 were drilled and completed by Mountain Fuel Supply Company in 1971. Leroy Wells Nos. 1 and 2 were also re-entered and completed as pressure observation wells. Interference and caprock integrity tests were conducted concurrently with the re-entry and drilling programs.

During initial development, lost circulation problems occurred while drilling across the Nugget Sandstone using a water-based mud. This problem was alleviated by allowing the Nugget water to flow to the surface while drilling, and an intermediate string of casing was run upon encountering the underlying Ankareh Formation.

Two potential horizons were initially considered for storage in the Thaynes, designated the T-10 (upper horizon) and the T-20 (lower horizon). Preliminary testing with gas injection into the T-20 zone showed direct communication with the T-10 zone. Subsequent injection into the T-10 zone suggested favorable conditions for development.

Wells 7, 8, and 9 were completed by August, 1972 and 2.0 BCF of gas had been injected. Upon completion of the surface facilities, application was made to the Federal Power Commission to commence storage operations. The application was approved November 17, 1972. Following FPC approval, storage operations proceeded for the 1972–73 heating season.

The following year, inventory has been increased to 3.5 BCF and Wells Nos. 9 and 10 were completed as injection/withdrawal wells. Soon after completion of Leroy Well No. 10 in September, 1973, with an inventory of 3.667 BCF (BHP was 1,740 psia), gas began blowing out around the surface casing of Well No. 3. Preliminary Temperature surveys confirmed the blowout to be coming from the nearby Well No. 4. The failure had occurred in the Twin Creek Formation at approximately 1,360 ft with subsequent gas migration to the No. 3 well. Further investigations with noise logs supported the temperature survey findings. Repair of the damaged casing was attempted in November, 1974.

A survey of both the 9⅝ in. O.D. intermediate casing and the 7 in. O.D. production casing (both failed) indicated that the corrosion was primarily caused by sulfide ions generated by anaerobic, sulfate reducing bacteria present in the annular fluids at the time of completion or from the invasion of formation waters containing them.

Leroy Wells Nos. 11 and 12 were drilled during 1975. Repairs were attempted on Well No. 4 until May, 1977 when it was plugged and abandoned. Leroy Well No. 4A was drilled during July, 1977 as a replacement for No. 4, followed by Leroy Well No. 14 in August, 1977 as a step-out well. Well No. 13 was drilled during June of the following year as a deep aquifer observation well.

Gas migration was first confirmed on the surface as bubbling in the adjacent creek and pond during the latter part of November, 1978.

Leroy Well No. 15 was drilled during July, 1979 to locate a possible collector zone above the reservoir and was completed as a Twin Creek/Nugget observation well (no significant gas shows were apparent).

A location map is given in Figure 11-7.

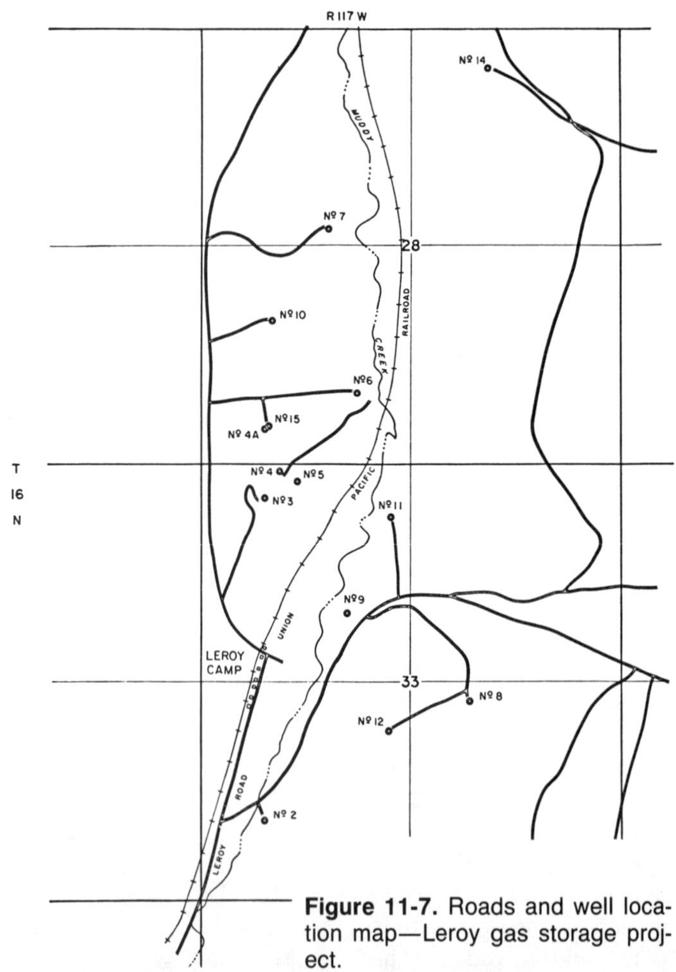

Figure 11-7. Roads and well location map—Leroy gas storage project.

GEOLOGICAL DESCRIPTION

Early geological work in the Leroy reservoir area was conducted by Shell Oil and Union Pacific Railroad during the late 1940s.

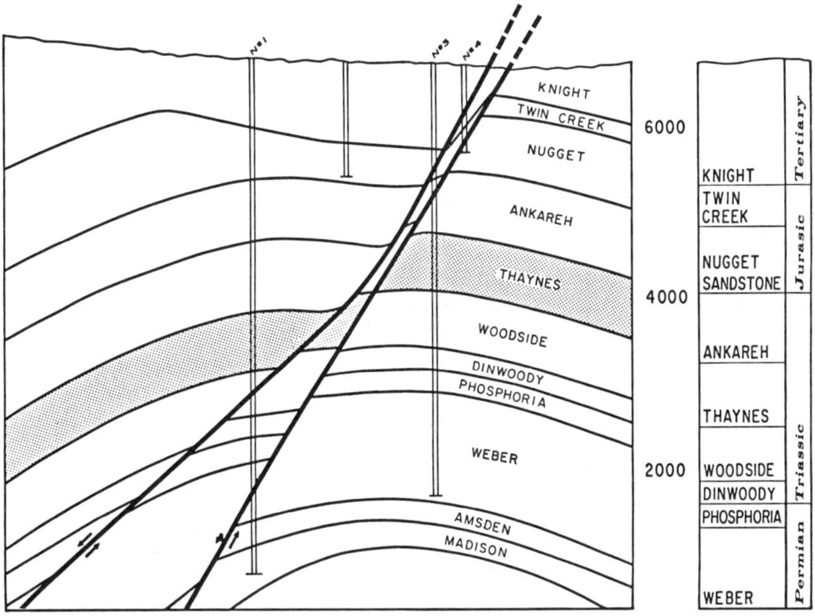

Figure 11-8. Lithologic cross section and stratigraphic sequence—Leroy gas storage project.

Figure 11-8 represents an updated geological cross section of the field. A brief summary of the stratigraphic sequence is given below:

Wasatch (Knight) Formation: Tertiary

Unconsolidated; fluvial; silts, sands and shales with limestone stringers.

Twin Creek Limestone: Jurassic

Unconformably underlies the Wasatch, grey limey, oolitic finely crystalline, some calcareous shales; dense.

Nugget Sandstone: Jurassic

Upper: Porous, coarse-grained, arkosic, very permeable.

Lower: Variable clastic beds, poorly sorted, finer-grained, less porosity/ permeability.

Ankareh Formation: Triassic

Typical red bed sequence with predominately red shales and siltstones alternating with thin sandstones and green shales.

Thaynes Formation: Triassic

Dolomite with thin limestone beds.

Upper: Grey, finely crystalline with thin-bedded anhydrite and very fine-grained sandstone beds.

Middle: Predominately red shales and siltstones with thin anhydrite and dolomite beds.

Lower: (Contains storage zones)

Top 5–8 feet: Dolomotized sandstone, very coarse grained, porous and permeable.

130–150 feet: Red shale, very calcareous, followed by 100 feet of dolomite with streaks of vugular porosity.

Woodside Formation: Triassic

Triassic Red bed sequence with predominately dense, silty shales.

Figure 11-9 represents the Thaynes structure map derived from well control and seismic surveys (based upon the top of the formation, not the top of the reservoir interval). The Thaynes represents a doubly plunging, faulted anticline. The downthrow of the major fault is to the west and divides the structure, effectively separating the reservoir. The storage reservoir was developed east of the fault with an assumed western limit at the fault. The initial reservoir pressure as of September 8, 1970, taken at 2,953 ft from ground level or + 3,809 SL, was 1,500 psig. The formation temperature was 82°F.

RESERVOIR PERFORMANCE

After the initial start-up during 1973, the following year was spent in evaluation of reservoir performance and installation of facilities.

The inventory during 1974 ranged from 3.7 to 3.8 BCF. The gas bubble pressure was close to original aquifer at 1,500 psig. Figure 11-10 represents the growth of storage inventory between 1971 and 1982. It was noted and reported that the gas bubble pressure during 1975 and 1979 had been increased to 1,830 psig, some 330 psi above the original aquifer.

During November, 1978, a survey of the project area indicated gas bubbling to the surface at two areas. One was in the vicinity of Well No. 11 and the other near Well No. 7. Two surface-gas detection surveys were conducted on April 9, 1979 and November 9, 1979, both confirming some gas migration to the surface near Wells Nos. 7 and 11.

On June 17, 1980, a Tritium tracer was injected into the T-10 zone through Well No. 4A, while all the other wells were shut-in. The Tritium tracer was identified in the very first surface sample taken near Leroy Well No. 11, nine days after injection.

Krypton-85 and Sulfur Hexafluoride tracers were injected on October 16, 1980 into Wells Nos. 3 and 10, respectively. Within 9 days, Krypton-85 was

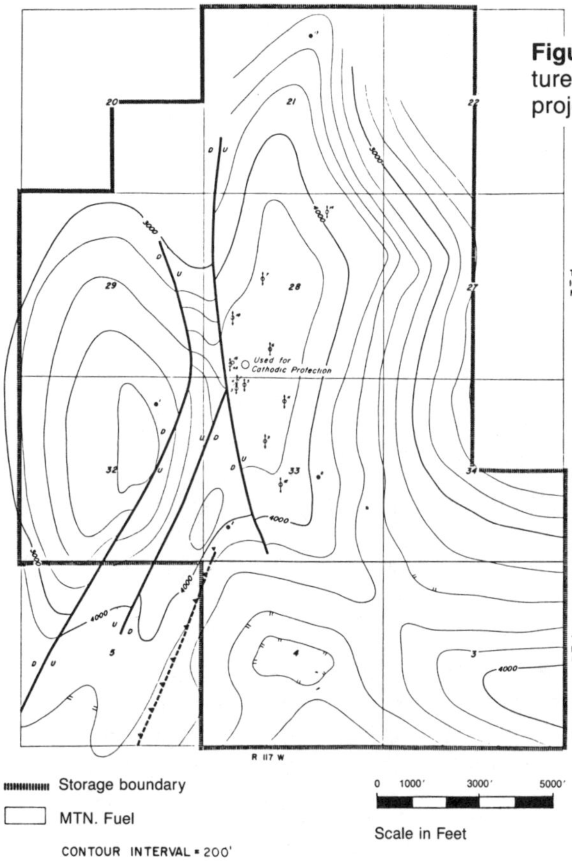

Figure 11-9. Thaynes structure map—Leroy gas storage project.

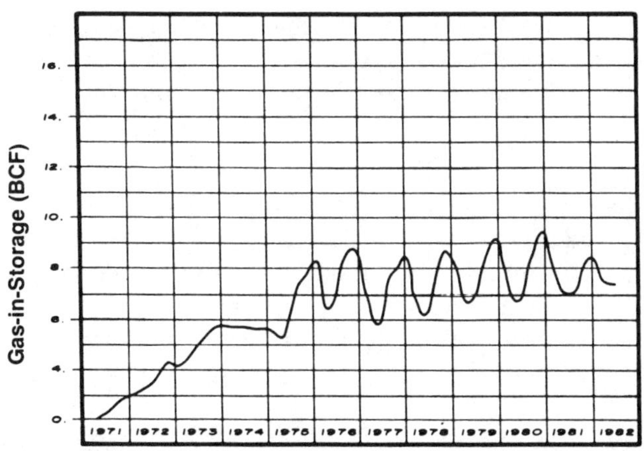

Figure 11-10. Growth of storage inventory—Leroy gas storage project.

detected in surface samples while the Sulfur Hexafluoride was confirmed at the surface 71 days after introduction.

On September 3, 1981, Tritiated Methane was injected into the reservoir through Well No. 3. Injection was continued into the other wells. The tracer appeared in surface gas samples 32 days after injection.

Due to noise log indications of gas migration behind the pipe in Leroy Well No. 4A, a workover program, designed to squeeze off any existing communication between the reservoir and near surface formations in the overburden was initiated. Subsequently, another tracer, Tritiated Ethane, was introduced on July 17, 1982, and had not appeared at the surface by the end of the summer. Previous tracer surveys indicated direct communication between the storage reservoir and the surface and it was hoped that at least part of the gas migration would be shut off by the workovers.

Subsequently, the tracer showed up 163 days after injection, indicating perhaps a longer path and reduced extent of migration.

Existing data suggest the leak event commenced sometime during 1975 and 1976.

ENGINEERING STUDIES

History Match and Simulation of Leak

During the summer of 1981, a review of storage performance in Leroy indicated the need for a computer simulated history match, including the effects of three crucial factors:

1. Water drive in the unsteady-state due to the surrounding Thaynes aquifer
2. Gas migration away from the storage horizon as evidenced by bubbling and sampling at the surface
3. Inventory-pressure data relating to the overall performance of the storage bubble.

During recent work, a simple unsteady-state aquifer model was selected and simulated on a computer. The program specifically developed for the study was based upon four parameters to include the effects of storage capacity, aquifer transmissibility, a leak-rate coefficient, and aquifer size. Using the data available from geology, reservoir behavior, and past production-pressure performance, the four parameters were bracketed through a realistic range of numerical values until a satisfactory history match was obtained.

The reservoir parameters used in history match runs relate to aquifer capacity (ϕh), aquifer transmissibility (kh), and aquifer size (r_e/r_b), and an equation simulating the leak.

Aquifer Parameter 1

Aquifer capacity parameter 1 is defined as:

$$Q_A = h\phi c r_b^2 y,\ ft^3/psi$$

where h = aquifer formation thickness, ft
ϕ = porosity (Thaynes aquifer), fraction
c = effective compressibility, volume/(volume × psi)
r_b = radius of gas bubble, ft
y = coefficient for pie shape of aquifers bounded by faults (in our case, 0.5 was used for 50% circular radial flow on aquifer and bounded by a north-south treading fault)

The value used for the Q_A parameter was (0.5 × 962.50), ft^3/psi

Aquifer Parameter 2

The aquifer transmissibility parameter 2 is defined as:

$$T = kh/\mu$$

where k = permeability (Thaynes aquifer), md
h = thickness, ft
μ = viscosity of formation water, cp

The value for T established in the course of history match runs was 75,000 md × ft/cp

The Aquifer Size Parameter

After several trial and error runs, (r_e/r_b), the ratio of exterior aquifer boundary to equivalent gas bubble radius, was determined to be approximately 20.

Simulation of Leak from Storage Horizon

In determining a suitable coefficient to quantify the leak rate in terms of storage pressures, several models were considered. These included wellbore leaks occurring at variable depths to collector zones in the overburden, seepage and bubbling from collector zones to the surface, and reservoir leaks from the storage bubble through imperfections in the caprock. It was also recognized that the time dependency of the leak rate must not only be a function of maximum storage pressures, but also the time duration when storage pressures remained above a critical value.

The Leroy gas storage hysteresis curve for the period 1975–1981 provided the early clues as well as the basic data for the development of the model which best described the leak in terms of gas storage pressures.

Figure 11-11 represents a replot of the inventory-pressure hysteresis data for the three consecutive injection seasons of 1978, 1979, and 1980. While the data reflect much "noise": effect of water movement, local gradients, etc., they show that the slope of p/z vs. inventory became consistently flatter whenever p/z exceeded 1,800 psia. Although this is probably partially due to the decreasing injection rate for pressure maintenance, it suggested to the writers a strong likelihood of a "check-valve" effect; that a pressure threshold existed, below which the aquifer storage grew with a proper hydraulic seal and above which the loss and continued migration of gas occurred. The difference between the nearly parallel upper slopes indicated that the annual leak was at a maximum between 1978 and 1979 when storage pressures remained high for an extended period of time.

Based on this and other theoretical and empirical considerations, a mathematical model was developed which fit the data and substantially improved the history match.

Figure 11-12 represents a schematic view of leakage from the storage area to the overburden, possibly from several sources (wellbores, caprocks, fractures, faults, etc.). The seepage from collector zones to the surface is also shown.

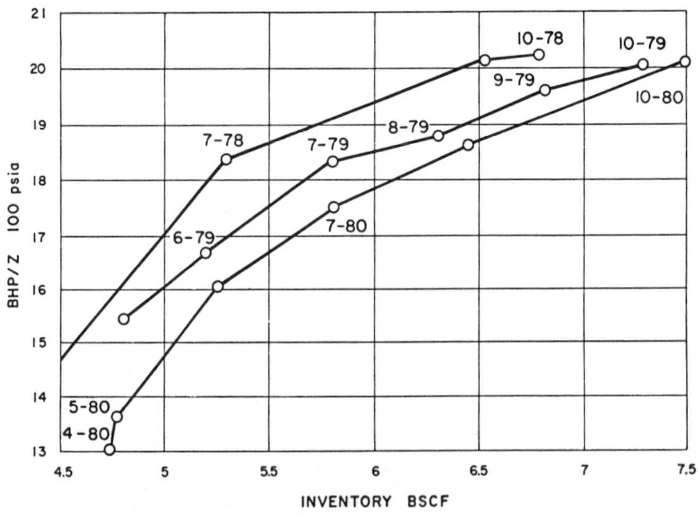

Figure 11-11. Leroy hysteresis data for 1978, 1979, and 1980.

Using pressure-inventory data and reasonable assumptions, the leak equation that best fit the data turned out to be:

$q_l = 3.74 \times 10^{-7}(p_G{}^2 - 1{,}600^2)^n$, where q_l is the daily leak rate in MMCF/D and p_G is the maximum gas bubble pressure in psia. Exponent n was assumed to be equal to 1.0.

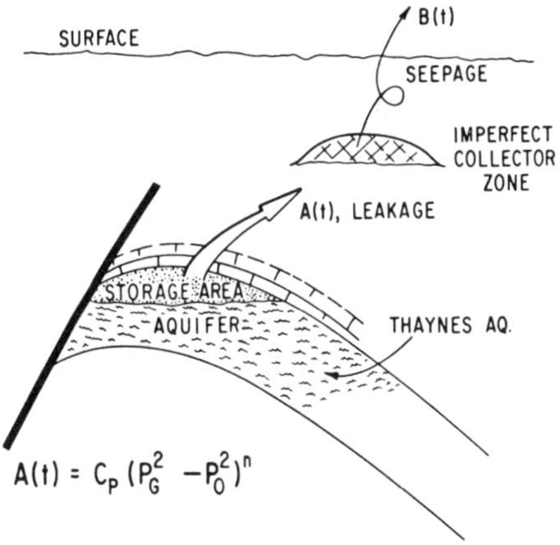

Figure 11-12. Schematic representation of migration of storage gas.

Analysis of Computer Results

The simulation of storage operations in Leroy involved a computer model developed to reflect unsteady-state water influx/efflux, reservoir volume changes due to prescribed pressure or inventory changes and, at the same time, a pressure-dependent leak occurring whenever a pressure threshold was exceeded.

Mathematical Basis for Calculation of Leak. In modeling the performance of the storage bubble developed on the aquifer, an unsteady-state material balance was invoked involving the storage gas occupying its reservoir pore volume, either inventory or pressure change due to cyclic storage operations, and water influx/efflux into the gas bubble in response to injection or withdrawal of storage gas. In addition, the resident inventory in the reservoir was corrected during each time step by the amount that would leak out whenever the pressure exceeded a threshold limit.

Accordingly, the calculations at any current time step involved the prevailing formation pressure, pore volume, and implicit calculation of water influx/efflux per Van Everdingen Hurst solutions with superpositions. These calculations also took into account the reservoir volume change due to metered inventory change and pressure-dependent leaks.

The model developed for the computer simulation of the storage reservoir consisted of a semicircular horizontal gas bubble surrounded by a porous and permeable radial flow aquifer, having lumped (ϕh) and (kh) parameters as determined by history match. The calculation procedure involved an input of either inventory or pressure schedule. When inventory (i.e., injection/withdrawal quantities) was specified, the model developed calculated the resulting reservoir pressure at each time step. The procedure also included a *leak-model* which determined the unmetered inventory loss as a function of the prevailing storage reservoir and a prescribed leak threshold pressure. When pressures were specified for each time step, the computer model calculated both the leak rate and the inventory change, updating both during each time step.

The Results. Figure 11-13 shows a sample of a history-match between observed and calculated pressures. It may be noted that agreement is better during injection while the pressure is high. The wide discrepancy during the end of each withdrawal season is due to the inaccurate pressure data. The reservoir pressures were calculated from recorded well head pressures, assuming that no liquid was present in the hole.

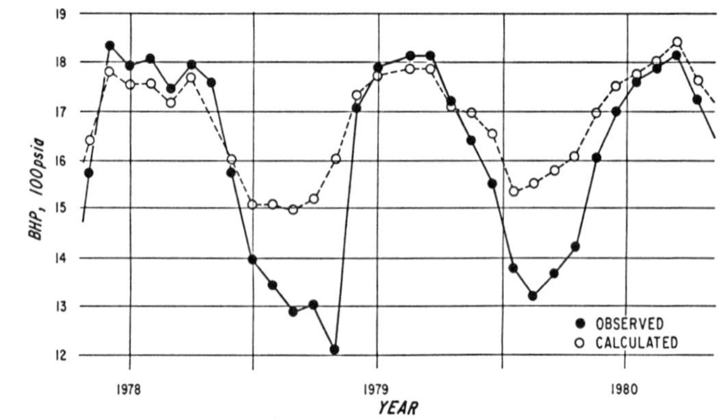

Figure 11-13. Sample of history-match results.

Figure 11-14 shows the annual leak rate obtained from the history matched computer runs. The leak process appears to have started sometime during maximum storage pressures. In the plot, the computer calculated leak rates are shown as blank dots occurring at discrete points representing the years 1976 and 1977, 1977 and 1978, etc. Also, the projected leak rates for three simulated cases are depicted in Figure 11-14. Case I represents the lowest operating pressure, while Case III represents the highest pressure. Case II corresponds to a reservoir pressure of 1,650 psig, which was the actual operating pressure during the 1981–1982 season. The cumulative leak quantities determined by the computer program have been annualized by difference then transformed to an instantaneous and continuous curve by a technique called "differencing the data." The procedure involved first obtaining the annual leak rate and plotting it as a bar-graph, as shown in Figure 11-15.

Each of the bars in Figure 11-14 corresponds in height to the cumulative leak rate for each year obtained by differencing the computer calculated cu-

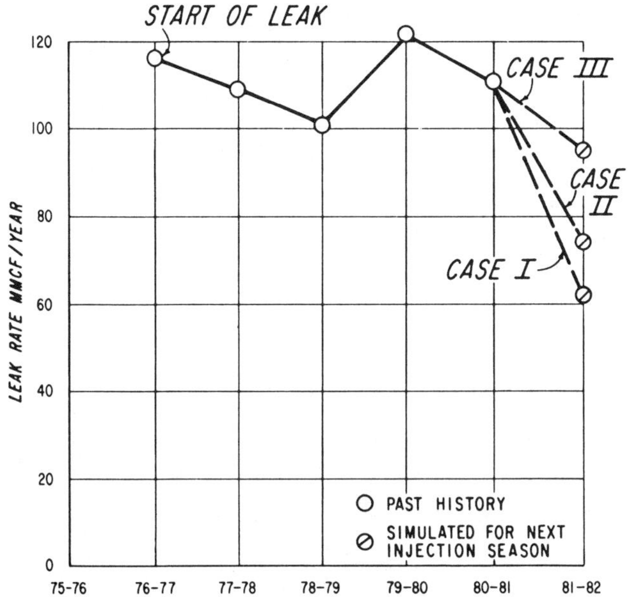

Figure 11-14. Past history of leak from pressure inventory data and computer runs.

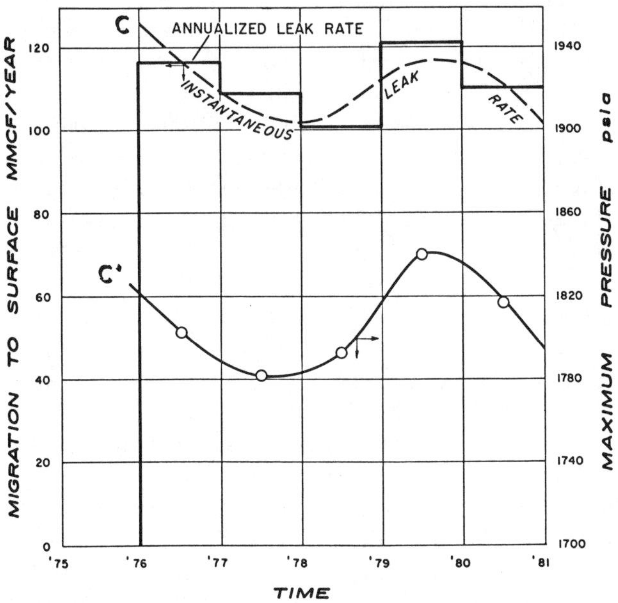

Figure 11-15. Bar-graph plot showing the relationship between the migration and maximum storage pressure in Leroy, 1976 through 1981.

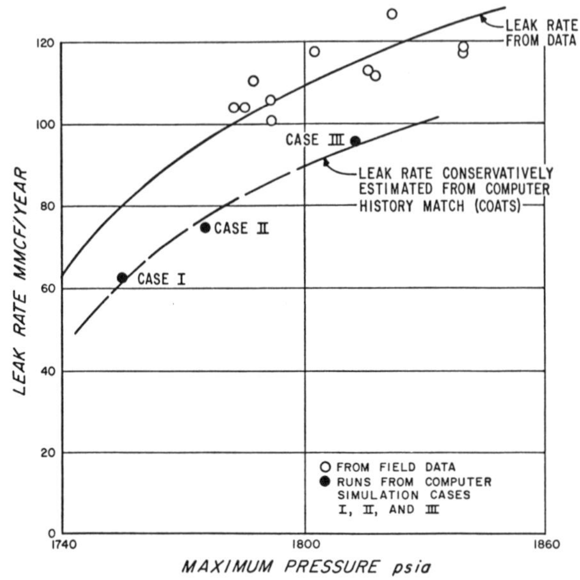

Figure 11-16. Correlation of instantaneous leak rate with maximum storage pressure.

mulative leak. The width of each bar covers, horizontally, the distance of the particular year during which the cumulative leak has occurred.

Once the five bars representing the annual gas leak are plotted side by side, a curve C was drawn in such a manner that the area underneath was equal to the sum total of the areas under the bars.

In order to provide a measure of correlation, the maximum storage pressures which occurred during the five years in question were plotted at the midpoint of each year spanning the particular time period involved. It can be seen that the curves C and C′ follow the same trend, being nearly parallel, and can be correlated against one another. This correlation is shown in Figure 11-16.

The leak rates reported by K. H. Coats, conservatively estimated from his computer runs, are shown to follow a similar trend as the approximate leak rates calculated by differencing the data.

Subsequent Reservoir Engineering Work

Continuing studies suggest that the following areas are of future importance, and they will undoubtedly be pursued:

Noise Logs: Noise logs were the earliest indication of possible gas movement near the wellbores. Logging programs have since identified mechanical problems and fluid movement behind the pipe. These logs have figured prominently at times in designing workover programs.

Temperature Surveys: Temperature surveys have been essential for location of leakage into/from the wellbores and possible migration paths behind the pipe. These surveys, coupled with noise logs, have been used concurrently with workover operations in Wells Nos. 4A and 11 during the summer of 1982.

Corrosion Analysis: Several corrosion analysis logging programs have been conducted in most of the wells and should be continued on a regular basis. Recent corrosion analysis logs have already prompted the running of new production casing strings in Well No. 11.

Tracer Programs: Injection of tracers and monitoring thereof have been an ongoing project for approximately two years and continue as positive measures of gas movement and migration to the surface.

Monitoring of Surface Bubbling: Gas bubbling has been monitored and quantitatively measured utilizing inverted barrels equipped with rotameters which are set over bubbling areas and sealed at the base. Although the data obtained in this way are inherently "noisy," they do show the following trends:

1. Some bubbling is dependent upon storage operations (indicating that migration originates from the Thaynes horizon). Bubbling at certain stations ceased altogether while most of the wells and their drainage areas were flooded during the summer. Later, after the injection operations restored dry conditions in the matrices, the bubbling resumed.
2. Some bubbling appeared invariant with storage operations (indicating that the gas migration was originating from shallow collector zones).

These trends indicate direct reservoir-to-surface leakage and also possible leakage from an intermediate collector zone.

Water Sampling: Since the Nugget formation has a high pressure gradient, it would seem logical that the water would flow to the surface. Repeated sampling of the water from the Muddy Creek could not discern apparent Nugget water following the gas leakage to the surface. This water should be easily identified due to its high salinity, but apparently cannot be detected due to mixing with large surface waters. A possible explanation would also be the bypassing of this formation by the gas via the wellbores and/or through fractures at, or near major faults.

Figure 11-17 shows the interrelationship of several sub-tasks coordinated during the Leroy Storage engineering study.

DISCUSSION OF RESULTS

On the basis of earlier history match runs, which included the effects of water movement, leak rate, and storage performance, it was possible to es-

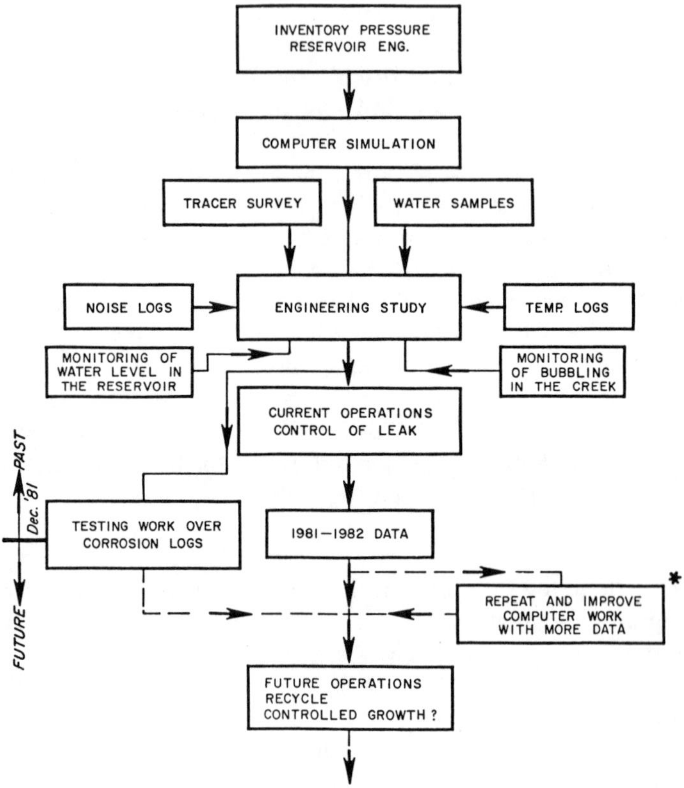

Figure 11-17. Coordination of various subtasks on Leroy storage engineering study.

tablish that the migration from the storage horizon started sometime during 1975–1976. The total cumulative amount of migration over 130 months of storage history appeared to be about 600 MMcf. As shown in Figure 11-18, the rate of leak appeared to depend upon maximum storage pressure. The inventory verification plots in the form of p/z vs. Inventory, over the many years of storage operations, also tended to support the correlation between the rate of migration and storage pressures.

Analysis of the data on hand and related engineering studies suggested that the continuing leak was controllable and did not detract substantially from the capabilities of the storage reservoir. Based on the above, the leak appears to involve pressure-triggered migration to a collector zone from which continuing seepage occurs on the surface.

As a result, after an extended shut-in period during early 1981, injection was resumed with a limited scope and extent during the summer of 1981.

* At the time of this writing this task was completed confirming earlier results.

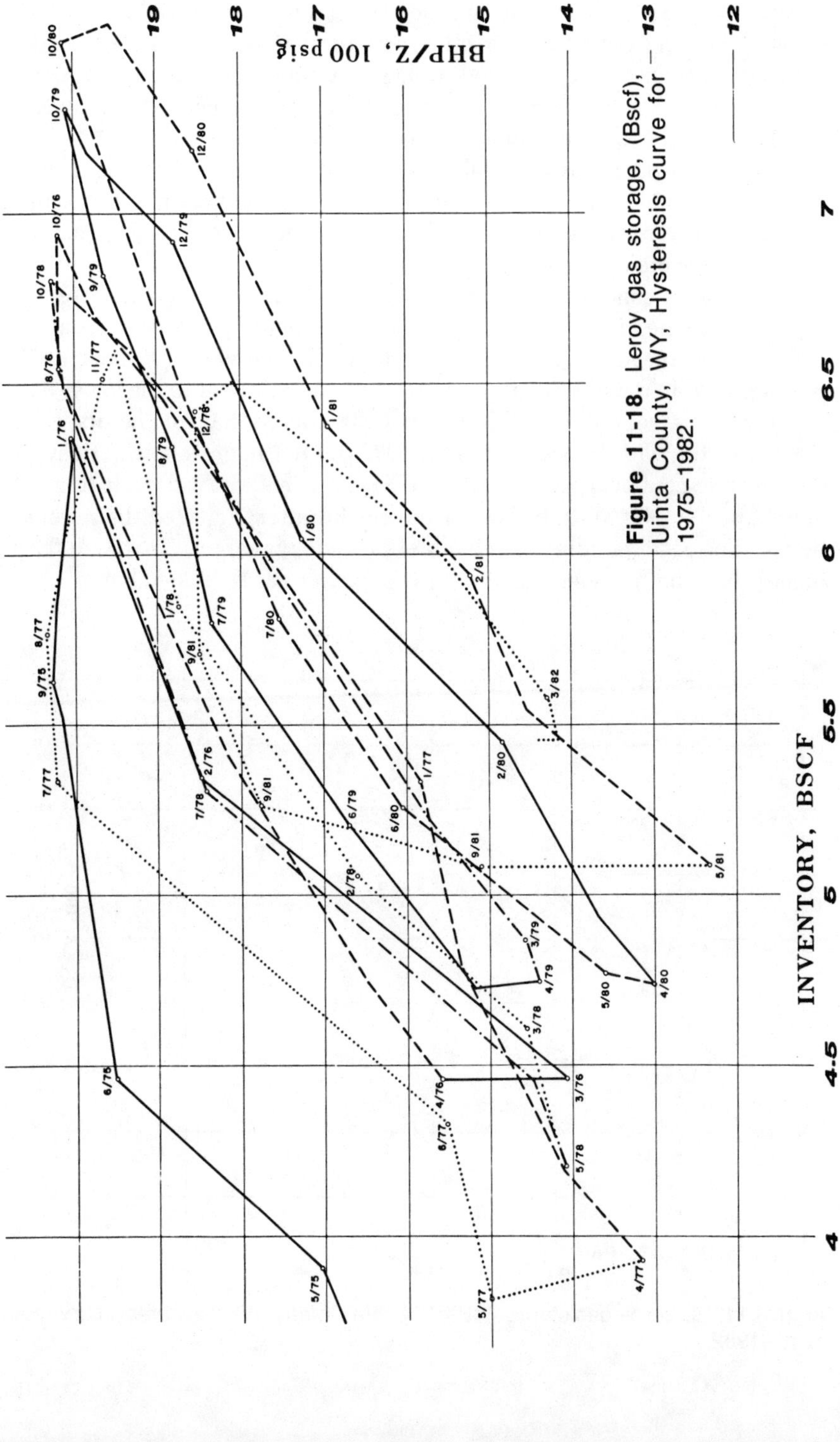

Figure 11-18. Leroy gas storage, (Bscf), Uinta County, WY, Hysteresis curve for 1975–1982.

Because the possibility existed that the leak may be due to wellbore imperfections, existing faults, and/or other structural imperfections, it was decided to continue the coordinated studies in all questionable areas. A comprehensive workover program was developed for spring, 1982. Leroy No. 4A was worked over to eliminate gas movement behind the production casing. Corrosion problems were taken care of at Leroy No. 11 by squeezing several permeable horizons and running a new production pipe. Withdrawal data were collected during the 81–82 winter, and various monitoring schemes were continued.

The pressure-content data plotted in Figure 11-19 show the latest results and compare the most recent performance with that of a year ago. It can be seen that the injection during 1981, where the BHP/z was kept under 1,850 psia, resulted in a withdrawal path that is nearly identical to the lower half of the withdrawal curve for 1980. It was significant that for the first time since 1973 the withdrawal pattern of 1982 did not migrate to the right on the pressure-content quadrant. The 1981–1982 withdrawal path stopped approximately 300 MMCF short of the gas inventory injected the summer before. The corresponding "gas bubble" pressure was approximately 100 psi higher than the previous year indicating smaller gas bubble volume.

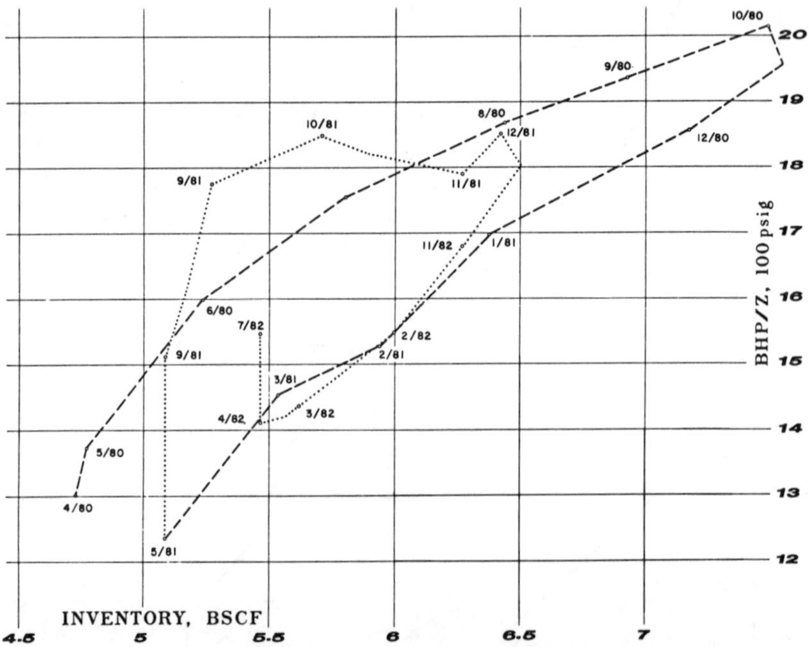

Figure 11-19. Leroy gas storage, (Bscf), Uinta County, WY, Hysteresis curve for 1980–1982.

The 300 MMCF which appeared short on top gas and added to cushion inventory was due to high water levels of the 1982 withdrawal season entrapping the gas beyond the reach of perforated intervals. Once the proper saturation and proper water levels are re-established, it was believed that the gas would revert back from cushion to top gas inventory.

Migration from the storage horizon appears to be well controlled by limiting the maximum pressure during injection. The latest gas surface detector survey indicates a receding area compared to the previous one a year ago.

CONCLUSIONS

The case history of Leroy Storage was unique because it involved definite migration away from the originally intended storage horizon, as evidenced by visual observation of seepage of gas to the surface.

Several techniques were coordinated in an effort to understand, monitor, control, and reduce the leak. These included various logging, surveying, sampling, and testing techniques; tracer work; computer simulation; and engineering analysis.

Among these, the computer simulation proved practical and effective in establishing a correlation between the leak rate and the extent of "over pressure" in the reservoir.

The final elimination of all appearance of leak to the surface is not yet on hand. There remains some future work to be done before a collector zone(s) is found, and there are other measures to be taken such as recompletion, recycling, venting, pressure control, or whatever may be indicated for the future.

Meanwhile, the case history should be significant in documenting what has been learned and how relevant computer simulations, hand-in-hand with overall engineering studies, can help in planning safe and economic gas storage operation.*

NOMENCLATURE

c_p = performance coefficient for leak, (MMCF/D)/psia^{2n}
q_l = rate of leak, MMCF/D
Q_A = aquifer parameter 1 = $\phi hcr_b{}^2y$, ft^3/psi
T = aquifer parameter 2, transmissibility = kh/μ, mdcy × ft/cp
c = effective compressibility, vol/vol × psi
h = aquifer formation thickness, ft

* As of the time of this publication, the data and all indications from Leroy storage continued to support the conclusions of this paper.

k = permeability, md
p = gas bubble pressure (stabilized), psia
p_G = maximum storage pressure, psia
r_b = radius of gas bubble (interior radial boundary for the aquifer), ft
r_e = exterior boundary of the aquifer, ft
y = coefficient for pie-shaped aquifers bounded by faults (fraction of full circular flow), dimensionless
t = time, days
z = compressibility factor, dimensionless
ϕ = fractional porosity, dimensionless
μ = viscosity of brine, cp

References

1. Interoffice Communication, Mountain Fuel Supply Company, 1970–1982.
2. Federal Power Commission, Docket #CP74-43, 1972.
3. Fincher Engineering Company, Houston, Texas, September 1974.
4. Reese, D. L. "Geologic Report - Leroy Gas Storage Field, Uinta County, Wyoming," September 1978.
5. Rocky Mountain Petroleum Consultants' Report, Salt Lake City, Utah, January 1981.
6. Tek, M. R. "Underground Storage, Theory and Practice," Notes for Intensive Summer Course, U. of Michigan, Ann Arbor, 1984.

APPENDIX A

Predicting Joule-Thomson Coefficients for Natural Gas*

JOULE-THOMSON†

The Joule-Thomson (J-T) coefficient μ is defined as the change in temperature with a very small change in pressure at constant enthalpy H.

$$\mu = \left(\frac{\partial T}{\partial P}\right)_H$$

The Joule-Thomson coefficient is useful in predicting temperature distributions in storage wells.[1] It characterizes the temperature behavior of gases in a throttling experiment. It is positive for cooling and negative for heating. The J-T coefficient is a very severe test for any equation of state.

There are not many J-T data available, and the few that are reported are only for binary, ternary, and quinary mixtures at most.[2] There have of course been a few articles in the past concerning the inversion behavior (when $\mu = 0$) of a few gases.[3,4]

The BWR equation of state has been used to predict the J-T coefficients of natural gas. This particular equation of state has been chosen for the purpose since the observed P-V-T data for hydrocarbons up to densities of twice the critical density are best represented by an eight-parameter equation.[5] The following is the order of a few popular equations of state with increas-

* This development is the result of a Term Paper Research, previously unpublished, by B. R. Pasumarti and M. R. Tek at The University of Michigan, 1982.

† As the original experiments were carried out by Joule and Kelvin both, some books refer to the effect as the *Joule-Kelvin* effect.

ing deviation from experimental data[2]: BWR equation; Beattie-Bridgeman equation; virial equation.

THERMODYNAMIC DEVELOPMENT

Joule-Thomson coefficient μ can be derived as follows:

$$\mu = \left(\frac{\partial T}{\partial p}\right)_H \tag{A-1}$$

Recall Maxwell's Equations:

$$dH = Tds + Vdp \tag{A-2}$$

$$Tds = c_p\, dT - T\left(\frac{\partial V}{\partial T}\right)_p dp \tag{A-3}$$

Substitute Tds from A-3 into A-2:

$$dH = c_p\, dT - \left[T\left(\frac{\partial V}{\partial T}\right)_p - V\right] dp \tag{A-4}$$

which gives:

$$dT = \frac{1}{c_p}\, dH + \frac{1}{c_p}\left[T\left(\frac{\partial V}{\partial T}\right)_p - V\right] dp \tag{A-5}$$

But since T = T(p, H):

$$dT = \left(\frac{\partial T}{\partial p}\right)_H dp + \left(\frac{\partial T}{\partial H}\right)_p dH$$

So it follows that:

$$\mu = \frac{1}{c_p}\left[T\left(\frac{\partial V}{\partial T}\right)_p - V\right] \tag{A-6}$$

DEVELOPMENT OF EQUATIONS FOR PREDICTING THE J-T COEFFICIENT FROM EQUATIONS OF STATE

Let us view H as a function of T and p:

$$dH = \left(\frac{\partial H}{\partial p}\right)_T dp + \left(\frac{\partial H}{\partial T}\right)_p dT$$

For an isenthalpic process dH = 0 and:

$$\left(\frac{\partial H}{\partial p}\right)_T = -\left(\frac{\partial H}{\partial T}\right)_p \left(\frac{\partial T}{\partial p}\right)_H = -c_p\mu$$

Now differentiate again with respect to T @ constant p:

$$\frac{\partial}{\partial T}\left(\left(\frac{\partial H}{\partial p}\right)_T\right)_p = -\frac{\partial}{\partial T}(c_p\mu)_p \tag{A-7}$$

but:

$$\frac{\partial}{\partial T}\left[\left(\frac{\partial H}{\partial p}\right)_T\right]_p = \frac{\partial}{\partial p}\left[\left(\frac{\partial H}{\partial T}\right)_p\right]_T \tag{A-8}$$

and:

$$\frac{\partial}{\partial p}\left[\left(\frac{\partial H}{\partial T}\right)_p\right]_T = \frac{\partial}{\partial p}(c_p)_T \tag{A-9}$$

Now, from Equations A-7 and A-9:

$$\left(\frac{\partial c_p}{\partial p}\right)_T = -\frac{\partial}{\partial T}(c_p\mu)_p \tag{A-10}$$

BWR equation of state[6]:

$$p = \frac{RT}{V} + (B_0RT - A_0 - C_0/T^2)\frac{1}{V^2} + (bRT - a)\frac{1}{V^3}$$

$$+ \frac{a\alpha}{V^6} + \frac{c}{V^3T^2}\left[\left(1 + \frac{\gamma}{V^2}\right)\exp\left(-\frac{\gamma}{V^2}\right)\right] \tag{A-11}$$

Differentiating Equation A-11 with respect to T at constant p and then substituting for $(\partial V/\partial T)_p$ from Equation 11-6, we get:

$$B_0RT - 2A_0 - \frac{4C_0}{T^2} + \frac{2bRT}{V} - \frac{3a}{V} + \frac{6a\alpha}{V^4}$$

$$C_p\mu = \frac{+\left\{\frac{5c}{VT^2} + \frac{5c\gamma}{V^3T^2} - \frac{2c\gamma^2}{V^5T^2}\right\} \exp\left(-\frac{\gamma}{V^2}\right)}{-RT - \frac{2B_0RT}{V} + \frac{2A_0}{V} + \frac{2C_0}{VT^2} - \frac{3bRT}{V^2} + \frac{3a}{V^2} - \frac{6a\alpha}{V^5}}$$

$$\overline{-\left\{\frac{3c}{V^2T^2} + \frac{3c\gamma}{V^4T^2} - \frac{2c\gamma^2}{V^6T^2}\right\} \exp\left(-\frac{\gamma}{V^2}\right)} \tag{A-12}$$

Substituting for $c_p\mu$ from A-12 in A-10 and simplifying:

$$c_p = c_p{}^\circ - R + \frac{6C_0}{VT^3} - \frac{6c}{\gamma T^3} + \left\{\frac{6c}{\gamma T^3} + \frac{3c}{V^2T^3}\right\} \exp\left(-\frac{\gamma}{V^2}\right)$$

$$+ \frac{\left[\frac{R}{V} + \frac{B_0R}{V^2} + \frac{bR}{V^3} + \frac{2C_0}{V^2T^3} - \left\{\frac{2c}{V^3T^3} + \frac{2c\gamma}{V^5T^3}\right\} \exp\left(-\frac{\gamma}{V^2}\right)\right]^2}{\frac{R}{V^2} + \frac{2B_0R}{V^3} - \frac{2A_0}{TV^3} - \frac{2C_0}{V^3T^3} + \frac{3bR}{V^4} - \frac{3a}{TV^4} + \frac{6a\alpha}{TV^7} + \left\{\frac{3c}{V^4T^3} + \frac{3c\gamma}{V^6T^3} - \frac{2c\gamma^2}{V^8T^3}\right\} \exp\left(-\frac{\gamma}{V^2}\right)} \tag{A-13}$$

Note: The specific volume V in the BWR equation has been obtained by the Newton-Raphson method of convergence.

Where $c_p{}^0$ is the specific heat of the gas at atmospheric pressure:

$$c_p{}^0 = a' + b' + c'T^2$$

The specific heat constants and the BWR constants are listed in Tables A-1 and A-2. The following are the mixtures rules adopted in finding a) the atmospheric pressure specific heat capacity and b) the BWR constants of the mixtures[5]:

(a) $$a' = \left[\sum_i a'_i x_i\right] \qquad b' = \left[\sum_i b'_i x_i\right] \qquad c' = \left[\sum_i C'_i x_i\right]$$

(b) $$B_o = \left[\sum_i x_i B_{oi}\right] \qquad A_o = \left[\sum_i \left(x_i A_{oi}{}^{1/2}\right)\right]^2 \qquad C_o = \left[\sum_i \left(x_i C_{oi}{}^{1/2}\right)\right]^2$$

$$b = \left[\sum_i \left(x_i b_i{}^{1/3}\right)\right]^3 \qquad a = \left[\sum_i \left(x_i a_i{}^{1/3}\right)\right]^3 \qquad c = \left[\sum_i \left(x_i c_i{}^{1/3}\right)\right]^3$$

$$\alpha = \left[\sum_i \left(x_i \alpha_i{}^{1/3}\right)\right]^3 \qquad \gamma = \left[\sum_i \left(x_i \gamma_i{}^{1/2}\right)\right]^2$$

Table A-1
Specific Heat Constants for Individual Hydrocarbons*

Substance	a'	$b' \times 10$	$-c' \times 10^4$
Methane	3.422	.17845	.04165
Ethylene			
Ethane	1.375	.41852	.13827
Propylene			
Propane	.4100	.6471	.22582
i-Butane	2.296	.82407	.38792
i-Butylene			
n-Butane	.072552	.7255	.22145
i-Pentane	−.282999	.47936	.2626
n-Pentane	.100532	.100532	.35369
n-Hexane	.104936	1.04936	.32397
n-Heptane	.13629	1.3629	.4877
N_2	.000677	6.77	13.99

* $C_p = a' + b'T + C'T^2$ in cal/mol°K, and T in °Kelvin

Table A-2
Specific Heat Constants*

Substance	B_0	A_0	$C_0 \times 10^{-6}$	b	a	$c \times 10^{-6}$	$\alpha \times 10^3$	$\gamma \times 10^2$
Methane	.682401	6995.25	275.763	.867325	2984.12	498.106	511.172	153.961
Ethylene	.89198	12593.6	1602.28	220678	15645.5	4133.6	731.661	236.844
Ethane	1.00554	15670.7	2194.47	2.85393	20850.2	6413.14	1000.44	302.79
Propylene	1.3263	23049.2	5365.97	4.79997	46758.6	20083.0	1873.12	469.325
Propane	1.5584	25915.4	6209.93	5.77355	57248.0	25247.8	2495.77	564.524
i-Butane	2.2039	38587.4	10384.7	10.889	117047	55977.7	4414.96	872.447
i-Butylene	1.85858	33762.9	11329.6	8.93375	102251	5387.2	3744.17	759.401
n-Butane	1.99211	38029.6	12130.5	10.2636	113705	61925.6	4526.93	872.447
i-Pentane	2.56386	4825.36	21336.7	17.1441	226902	136025	6987.77	1188.07
n-Pentane	2.51096	45928.8	25917.2	17.1441	246148	161306	7439.92	1218.86
n-Hexane	2.84835	5443.4	40556.2	28.0032	429901	296077	11553.9	1711.15
n-Heptane	3.18782	66070.6	57984.0	38.9917	626106	483427	17905.6	2309.42
Nitrogen	.73366	4496.9453	71.9523	.5084	900.06	107.254	1198.38	192.451

* Units: P = lb/sq in. absolute; v = ft³/lb moles; T = degree Rankine (°F × 459.63);

$$R = \frac{10.7335(\text{lb})(\text{ft})^3}{(\text{in.}^2)(\text{lb mole})(°\text{R})}$$

DEVELOPMENT OF CHARTS

Typical compositions of natural gas ranging from gas gravity 0.6 to 0.975 were taken from the *Handbook of Natural Gas Engineering*[6].

Two differential charts were prepared:

1. J-T coefficient μ (in °F/kPa) vs. pressure (kPa) with temperature as a parameter for different gas gravities.

2. The gas gravities were smoothed out by plotting J-T coefficient μ vs. gas gravity with pressure as the parameter, ranging from 800 kPa (117.5 psia) to 20,000 kPa (2,940 psia) at temperatures of 40, 80, 120, 160, and 200°F (4.44, 26.66, 48.88, 71.11, and 93.33°C). Plots are given for 1,600 kPa (235 psia) pressure intervals in figures 2-14 through 2-18. The J-T coefficient values can be interpolated linearly with a maximum of $\pm 2.2\%$ error.

Example A-1

Problem. A gas storage well has a total depth of 3,050 ft from the surface to datum level at mid-perforations. The geothermal temperature near the surface is 58°F and linearly increases .435°F every 100 ft of depth. The well is flowing through 7-in. casing (ID = 6.276″) at a rate of 50 MMCF/D. The flowing bottom-hole pressure is 1,561.4 psia. The flowing bottom-hole temperature is 69.4°F. Previous design calculations indicate that the overall heat transfer coefficient k between the gas flowing inside and the formation where geothermal temperature prevails outside is 5 Btu/hr × ft² × °F. Calculate and plot the estimated steady-state temperature distribution for this well.

Solution. The governing equation for the temperature distribution[1] is given as:

$$T(h) = T'(h) + \frac{mcp}{2\pi akg_t} - \frac{m}{J}\frac{g}{g_c}\frac{1}{2\pi ak} - \left[T_0' - T_0 + \frac{mcp}{2\ akg_t} - \frac{m}{2\ aK}\frac{1}{J}\frac{g}{g_c}\right]e^{\frac{-2\ ak}{mcp}h} - \mu_j(\Delta P)$$

$$T'(h) = T'_0 - \frac{h}{g_t}$$

$$h = 3{,}000 \text{ ft} \qquad T'(h) = 58°F$$

$$g_t = \frac{100}{.435} = 229.89 \text{ ft/°F}$$

$$T_0' = 58 + \frac{3{,}000}{229.89} = 71.05°F$$

$$\overline{m} = \frac{50 \times 10^6 \times 0.0764 \times 0.612}{24} = 97{,}410 \text{ lb}_m/\text{hr}$$

$$a = \frac{6.276}{12 \times 2} = 0.2615 \text{ ft}; \; k = 5 \text{ Btu/hr} \times \text{ft}^2 \times {}^\circ\text{F}$$

$$\frac{m}{2\text{ ak}} \frac{1}{J} \frac{g}{g_c} = \frac{97{,}410 \times 32.17}{2\pi \times 0.2615 \times 5 \times 778 \times 32.17} = 15.24^\circ\text{F}$$

c_p for the pipeline gas (gas gravity = 0.525) at 1,561.4 psia and 69.4°F is

$$c_p \cong 0.737 \frac{\text{Btu}}{\text{hr ft}^2 \ {}^\circ\text{F}}$$

$$\frac{\overline{m}cp}{2\pi akg_t} = 38.2309$$

$$T_0 = 69.4$$

So now the equation for temperature distribution in the well becomes:

$$T(h) - 71.05 - \frac{h}{229.89} - 15.24 + 38.2309 - (71.05 - 69.4 + 38.2309 - 15.24)_e{}^{\frac{-2\pi \times 0.26 \times 5h}{97{,}410 \times 0.737}}$$

$$T(h)_{corr} = T(h) - \mu_j\,(\Delta P)$$

Take h = 1,000 ft

$$T(h) = 71.05 - \frac{1{,}000}{229.89} - 15.24 + 38.2309 -$$

$$[71.05 - 69.4 + 38.2309 - 15.24]e^{-0.11377}$$

$$= 67.7^\circ\text{F}$$

$$T(h)_{corr} = 67.7 - \mu_j\,(\Delta P)$$

Next, calculate the pressure drop for 1,000 ft from bottom:

Q = 50,000 MCF/D $\qquad P_{bot} = 1{,}561.4$ psia

$$P_s^2 - e^s P_W^2 = 0.00667 \text{ f } Q^2 T_a^2 z^2 \frac{(e^s - 1)}{d^5} \quad \ldots. [7]$$

$$s = 0.0375 \frac{GL}{T_a z}$$

$G = 0.612$

$$P_r = \frac{1{,}561.4}{672} = 2.32$$

$$T_r = \frac{529.4}{395} = 1.34$$

$z = 0.7$

$$s = \frac{0.0375 \times 0.612 \times 1{,}000}{529.4 \times 0.7} = 0.06$$

$e^s = 1.06$

$$R_e = \frac{50 \times 10^6 \times 0.0764 \times 0.612 \times 6.276 \times 4 \times 144}{86.400 \times 12 \times 0.000672 \times 0.0145 \times 6.276^2}$$

$$= 6{,}760{,}396$$

$\epsilon/D = 0.0003$

$f = 0.015$

$$P_{1{,}000\,ft} = \left[\frac{\overline{1{,}561.4}^2}{1.06} - \frac{0.00667 \times 0.015 \times \overline{50{,}000}^2 \times \overline{5{,}294}^2 \times \overline{0.7}^2 \times \overline{0.06}}{1.06 \times (6.\overline{275})^2}\right]^{1/2}$$

$$= 1{,}509.97 \text{ psia}$$

where $P_o = 1{,}561.4$ psia $= 10{,}624.5$ kPa
$P_{1{,}000}$ ft $= 1{,}509.97$ psia $= 10{,}274.6$ kPa
$\bar{P} = 1{,}535.685$ psia $= 10{,}449.568$ kPa

$$\bar{T} \frac{69.4 + 67.7}{2} = 68.55°F$$

μ_j (from charts) at 68.55°F and 10,450 kPa

$$= 6.0631 \times 10^{-3}\ °F/kPa$$

$= .041256\ °F/psia$

$\mu_j(\Delta P) = 0.041256 \times 51.43 = 2.1218°F$

$T(h) = 67.7 - 2.1218 = 65.5782°F$

$T_{measured} = 66.0°F$

These calculations are summarized in Table A-3.

Table A-3
Temperature Distribution, Example Problem A-1*

h ft	T(h) °F	μ_j °F/psi	$T(h)_{corrected}$ °F	$T_{measured}$ °F
0	69.4			69.4
		.04125		
1,000	67.7		65.58	66.0
		.0427		
2,000	63.93		61.8	62.25
		.0443		
3,000	60.5		58.3	57.0

* The reader may note that this example problem is the same for which a detailed engineering solution was given in Chapter 4. The version in the appendix is repeated to give an idea of the use of J-T charts.

ACCURACY

As mentioned earlier, very few J-T data are available, and the J-T data especially for ternary and quinary mixtures are even fewer. The theoretical J-T coefficients have been compared to the experimental J-T coefficients (on differentiating the experimental isenthalps) of three mixtures reported by Alhert and Wenzel[2] in the temperature range of 40°F (4.44°C) to 100°F (37.77°C) and 2,500 (367.4) to 15,000 (2204.4 psia) kPa. This comparison is illustrated in Figure A-1. A maximum deviation of $\pm 4.0\%$ was observed. Actually, the deviation could probably be even less since the experimental determination of J-T coefficients is very difficult and sometimes subject to large errors, and the method of fitting the experimental data[2] with simple polynomials can be expected to yield spurious slopes at the limits of the data.

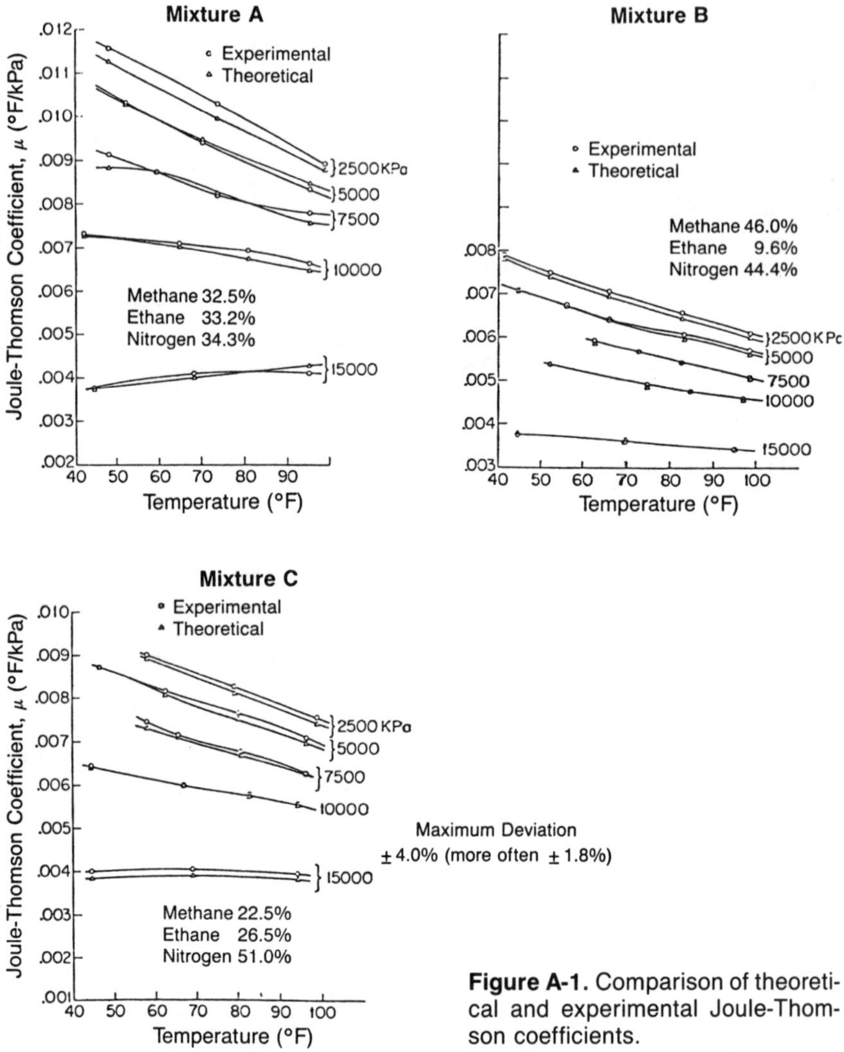

Figure A-1. Comparison of theoretical and experimental Joule-Thomson coefficients.

COMPUTER PROGRAM FOR J-T COEFFICIENTS

```
C      THIS PROGRAM CALCULATES THE JOULE THOMSON COEFFICIENT OF
C      ANY GAS MIXTURE
C      THE BENEDDICT WEBB RUBIN EQUATION OF STATE HAS BEEN USED
C      BO.AO.CO.B.A.C.ALPH.GM- BWR CONSTS FOR SINGLE COMPONENTS(IN
C      ENGLISH UNITS)
C      CP- SP.HEAT AT ANY TEMP AND PRESS  LB CUFT/SQ.IN LBMOLE DEGF
C      CPO- SP.HEAT OF MIXTURE AT ATM. PRESSURE CALS/(GM-MOLE)(DEG K)
C      CPOS- SP.HEAT AT ATM.PRESSURE  LB CUFT/SQ.IN LBMOLE DEGF
C      GRAVTY- NATURAL GAS GRAVITY
C      PMWT- MOLECULAR WEIGHT
```

```
C      PRESS- ABSOLUTE PRESS IN PSIA
C      PRSKPA-  PRESSURE IN KILOPASCALS
C      R- GAS CONSTANT (LBS)(CU.FT)/(SQ.INCH)(LB MOLE)(DEGREE RANKINE)
C      SBO,SAO,SCO,SB,SA,SC,SALPH,SGM- BWR CONSTS FOR MIXTURE
C      SPA,SPB, SPC- SPECIFIC HEAT CONSTS IN METRIC UNITS
C      TEMPF- TEMPERATURE DEGREE FARENHEIT
C      TEMPK- TEMPERATURE DEGREE KELVIN
C      TEMPR- TEMPERATURE DEGREE RANKINE
*      TJOULT- JOULE-THOMSON COEFFICIENT  DEGF/KPA
C      V- SPECIFIC VOLUME  CU.FT/LB-MOLE
C      X-MOLE FRACTIONS OF COMPONENTS IN NATURAL GAS
C
C
      DIMENSION X(10),PMWT(10),AO(10),BO(10),CO(10),A(10),B(10),C(10),
     1ALPH(10),GM(10),SPA(10),SPB(10),SPC(10)
      NAMELIST /DATA/ R,X,PMWT,AO,BO,CO,A,B,C,GM,ALPH,SPA,SPB,SPC
    1 READ(5,DATA,END=999)
      WRITE(6,DATA)
      SBOT=0.
      SAOT=0.
      SCOT=0.
      SBT=0.
      SAT=0.
      SCT=0.
      SALPHT=0.
      SGMT=0.
C
C      CALCULATIONS OF BWR CONSTS USING MIXTURE RULE
C
      DO 20 I=1,10
      SBOT=SBOT+X(I)*BO(I)
      SAOT=SAOT+X(I)*AO(I)**0.5
      SCOT=SCOT+X(I)*CO(I)**0.5
      SBT=SBT+X(I)*B(I)**0.333
      SAT=SAT+X(I)*A(I)**0.333
      SCT=SCT+X(I)*C(I)**0.333
      SALPHT=SALPHT+X(I)*ALPH(I)**0.333
      SGMT=SGMT+X(I)*GM(I)**0.5
   20 CONTINUE
      SBO=SBOT
      SAO=SAOT**2
      SCO=SCOT**2
      SB=SBT**3
      SA=SAT**3
      SC=SCT**3
      SALPH=SALPHT**3
      SGM=SGMT**2
C
C      CALCULATION OF GAS GRAVITY
      GRAVTY=0.
      DO 30 I=1,10
      GRAVTY=GRAVTY+X(I)*PMWT(I)/29.0
   30 CONTINUE
C
C      CALCULATION OF SPECIFIC VOL USING NEWTON RAPHSON CONVERGENCE
C
C
      WRITE(6,200)
      WRITE(6,800)X(1),X(2),X(3),X(4),X(5),X(6),X(7),X(8),X(9),X(10)
      TEMPI=20.0
      DO 70 K=1,9
      TEMPF=TEMPI+K*20.
      TEMPK=(TEMPF-32.0)*0.5555+273.0
      TEMPR=TEMPF+460.0
      CPO=0.
      DO 80 I=1,10
      CPO=CPO+(SPA(I)+SPB(I)*TEMPK+SPC(I)*TEMPK**2)*X(I)
   80 CONTINUE
      CPOS=CPO*5.402
      WRITE(6,400)GRAVTY,TEMPF
      WRITE(6,500)
      PRESSI=0.
      DO 50 J=1,25
      PRSKPA=PRESSI+J*800
      PRESS=PRSKPA/6.804
      V=R*TEMPR/PRESS
```

```
    7 FV=(R*TEMPR+(SBO*R*TEMPR-SAO-SCO/TEMPR**2)/V+(SB*R*TEMPR-SA)/V**2+
     1SA*SALPH/V**5+(1.0+SGM/V**2)*SC*EXP(-SGM/V**2)/V**2/TEMPR**2)/
     1PRESS-V
      FPV=(SAO+SCO/TEMPR**2-SBO*R*TEMPR)/PRESS/V**2-(SB*R*TEMPR-SA)*2.0/
     1PRESS/V**3-5.0*SA*SALPH/PRESS/V**6+(SGM/V**2-1.0-2.0*SGM/V**2+ SGM
     1**2./V**4)*2.0*EXP(-SGM/V**2)*SC/V**3/TEMPR**2/PRESS-1.0
      ERR=ABS(FV/V)
      V=V-FV/FPV
      IF (ERR.GT..0001)GO TO 7
C
C      CALCULATION OF SPECIFIC HEAT
C
      CP=CPOS-R+6.0*SCO/V/TEMPR**3-6.0*SC/SGM/TEMPR**3+
     1(6.0*SC/SGM/TEMPR**3+3.0*SC/V**2/TEMPR**3)*EXP(-SGM/V**2)+
     3((R/V+SBO*R/V**2+SB*R/V**3+2.0*SCO/V**2/TEMPR**3-
     4(2.0*SC/V**3/TEMPR**3+2.0*SC*SGM/V**5/TEMPR**3)*EXP(-SGM/V**
     52))**2)/(R/V**2+2.0*SBO*R/V**3-2.0*SAO/TEMPR/V**3-
     62.0*SCO/V**3/TEMPR**3+3.0*SB*R/V**4-3.0*SA/TEMPR/V**4+
     66.0*SA*SALPH/TEMPR/V**7+(3.0*SC/V**4/TEMPR**3+
     73.0*SC*SGM/V**G/TEMPR**3-
     82.0*SC*SGM**2/V**8/TEMPR**3)*EXP(-SGM/V**2))
C
C      CALCULATION OF JOULE THOMSON COEFFICIENT
C
      PROD=(SBO*R*TEMPR-2.0*SAO-4.0*SCO/(TEMPR**2)+2.0*SB*R*TEMPR/V-
     23.0*SA/V+6.0*SA*SALPH/V**4+(5.0*SC/V/TEMPR**2+5.0*SC*SGM/V**
     23/TEMPR**2-2.0*SC*SGM**2/V**5/TEMPR**2)*EXP(-SGM/V**2))/(-R*TEMPR-
     32.0*SBO*R*TEMPR/V+2.0*SAO/V+2.0*SCO/V/TEMPR**2-
     43.0*SB*R*TEMPR/V**2+3.0*SA/V**2-6.0*SA*SALPH/V**5-
     5(3.0*SC/V**2/TEMPR**2+3.0*SC*SGM/V**4/TEMPR**2-
     62.0*SC*SGM**2/V**6/TEMPR**2)*EXP(-SGM/V**2))
      TJOULT=PROD/CP/6.8045
      WRITE(6,600)PRSKPA,V,CP,TJOULT
   50 CONTINUE
   70 CONTINUE
C      FORMAT SPECIFICATIONS
  500 FORMAT(7X,'PRESSURE   ',13X,'V',18X,'SPECIFIC HEAT',31X,'JOULETHOM
     1SON COEFF.'/7X,'KPA  ',8X,'CU FT/LB MOLE',10X,'LB CUFT/SQ IN LBMOL
     2E DEGF',24X,'DEGF/KPA'//)
  600 FORMAT(7X,F8.2,6X,F10.6,18X,F9.5,35X,F12.9)
  400 FORMAT(//////20X,'GAS GRAVITY"',F6.3//20X,'TEMP=',F6.2,'DEGF'//)
  800 FORMAT(//20X,'COMPOSITION:'//32X,'METHANE:',F6.5/32X,'ETHANE:',
     1F6.5/32X,'PROPANE:',F6.5/32X,'I-BUTANE:',F6.5/32X,'I-PENTANE:',
     2F6.5/32X,'N-BUTANE:',F6.5 /32X,'N-PENTANE:',F6.5/32X,'HEXANE:',
     3F6.5/32X,'HEPTANES:',F6.5/32X,'NITROGEN:',F6.5//)
  200 FORMAT('1',18X,'PROGRAM OUTPUT: JOULE THOMSON COEFFICIENT',//)
      GO TO 1
  999 STOP
      END
```

NOMENCLATURE

a′, b′, c′ = specific heat constants of the components at atmospheric pressure, (cals)/(gm-mole) (°K), (cals)/(gm-mole) $(°K)^2$, (cals)/gm-mole) $(°K)^3$, respectively

a_i, b_i, c_i, A_{0i}, B_{0i}, C_{0i} = BWR constants of the individual components

$$\text{Units: } \frac{(\text{lb})(\text{ft}^3)^3}{(\text{in.})^2(\text{lb-mole})^3}, \ (\text{ft}^3/\text{lb-mole})^2,$$

$$\frac{(\text{lb})(\text{ft}^3)^3(°\text{R})^2}{(\text{in.})^2(\text{lb-mole})^3}, \ \frac{(\text{lb})(\text{ft}^3)^2}{(\text{in.})^2(\text{lb-mole})^2},$$

$ft^3/(lb\text{-}mole)$, $\frac{(lb)(°R)(ft^3)^2}{(in.)^2(lb\text{-}mole)^2}$, respectively

a, b, c, A_0, B_0, C_0 = BWR constants of the mixture

$c_p°$ = specific heat at atmospheric pressure, (cals)/(gm-mole)°K

c_p = specific heat at (lb)(ft^3)/(in.)^2(lb-mole)(°F), any pressure

H = enthalpy (lb)(ft^3)/(in.)^2(lb-mole)

P = pressure, psia

R = gas constant, 10.7335 (lb)(ft)^3/(in.)^2(lb-mole)(°R)

s = entropy $\frac{(lb)(ft^3)}{(lb\text{-}mole)(in.^2)(°R)}$

T = temperature, °R in the expression for μcp and cp, °K in the expression for cp°

V = specific volume, (ft^3)/(lb-mole)

X = mole fraction

μ_j = Joule-Thomson coefficient, °F/psia

α_i, γ_i = BWR constants of individual components, (ft^3/lb-mole)^3 and [(lb)(ft^3)^2]/[(in.)^2(lb-mole)^2], respectively

α, γ = BWR constants of mixture

References

1. Tek, M. R. "Design of Storage Fields," Proceedings, AGA Natural Gas Transmission Conference, Salt Lake City, Utah, May 5–7, 1980.
2. Alhert L. C. and L. A. Wenzel. *AIChE Journal* (1969) 15, 256.
3. Juris, K. and L. A. Wenzel, *AIChE Journal,* (1972) 18, 684.
4. Miller, D. G. *Ind. Eng. Chem. Fundam.* (1970) 9(4), 585.
5. Benedict, M., G. B. Webb, and L. C. Rubin. *Chem. Eng. Progress* (1951) 47.
6. Katz, D. L. *Handbook of Natural Gas Engineering*, McGraw-Hill Book Co., 1959.

APPENDIX B

Derivation of Steady-State Gas Flow from the Forchheimer Equation

Starting with the Forchheimer equation first written in Darcy units:

$$\frac{dp}{dr} = \frac{\mu}{kA} q + \beta\rho \frac{q^2}{A^2} \tag{B-1}$$

derive the same equation in practical field units then integrate it for steady-state radial turbulent flow of gas.

First convert the Forchheimer equation from Darcy to field units.

$$\text{Recall } v = \frac{k}{\mu}\frac{dp}{dr} \quad \text{(for Darcy flow)}$$

$$q = \frac{kA}{\mu}\frac{dp}{dr}$$

$$\frac{\overset{\text{atm}}{dp'}}{\underset{\text{cm}}{dr'}} = \frac{\overset{\text{cp}}{\mu'}}{\underset{\text{Darcy}}{k'}\underset{\text{cm}^2}{A'}} \overset{\text{cc/sec}}{q'} + \underset{\text{1/cm}}{\beta'}\overset{?}{\rho'} \frac{\overset{\text{cc/sec}}{q'^2}}{\underset{\text{cm}^2}{A'^2}}$$

$$\rho' = \frac{\text{atm} \times \text{sec}^2}{\text{cm}^2} = 14.7 \times 144 \, \frac{\text{lb}_f}{\text{ft}^2} \frac{\text{sec}^2}{\text{cm}^2}$$

$$= \frac{14.7 \times 144 \times 32.17}{\text{ft}^2 \times (1/30.48)^2 \, \text{ft}^2} \text{ poundals sec}^2$$

$$= \frac{14.7 \times 144 \times 32.17}{\text{ft}^2 \times \text{ft}^2} \text{ sec}^2 \, (30.48)^2 \, \frac{\text{lb}_m \text{ft}}{\text{sec}^2}$$

$$= 14.7 \times 144 \times 32.17 \times (30.48)^2 \, \frac{\text{lb}_m}{\text{ft}^3}$$

So,

$$1 \text{ atm} \, \frac{\text{sec}^2}{\text{cm}^2} = 6.26 \times 10^7 \, \frac{\text{lb}_m}{\text{ft}^3} \qquad \text{(B-2)}$$

Now, we can proceed to field units:

$$\frac{1}{14.7} \frac{dp}{dr} \frac{1}{30.48} = \frac{\mu \, 10^3}{kA(30.48)^2} \frac{q \times 1{,}000(30.48)^3}{86{,}400} + \frac{\beta}{30.48} \rho \frac{1}{6.26 \times 10^7}$$

$$\frac{q^2 \times 10^6 \times (30.48)^6}{(86{,}400)^2 \, A^2 \, (30.48)^4}$$

$$\frac{dp}{dr} = \frac{30.48 \times 14.7 \times 1{,}000 \times 10^3 \times (30.48)^3}{(30.48)^2 \, 86{,}400} \frac{\mu q}{k}$$

$$+ \frac{14.7 \times 30.48 \times 10^6 \times (30.48)^6 \times 30.48 \times q^2}{30.48 \times 6.26 \times 10^7 \times 86{,}400^2 \times (30.48)^4 \, A^2}$$

$$\frac{dp}{dr} = 158{,}171.73 \frac{\mu q}{kA} + 2.92 \times 10^{-8} \frac{\beta \rho q^2}{A^2} \qquad \text{(B-3)}$$

with p in psia
r in ft
μ in cp
q in MCF/D
k in mdcy
A in ft^2 and
q in local MCF

Now, the solution for steady-state radial flow:

$$\frac{dp}{dr} = 158{,}171 \frac{\mu q}{kA} + 2.92 \times 10^{-8} \frac{\beta \rho q^2}{A^2}$$

$$\rho = \frac{29G}{zRT} p$$

$$R = \frac{10.73 \times \text{psia} \times \text{ft}^3}{\text{lb mole } ^\circ\text{R}}; A = 2\pi rh; q = Q \frac{p_{ST}}{p} z \frac{T}{T_{ST}}$$

$$\frac{dp}{dr} = 158{,}171 \frac{\mu}{k2\pi rh} \frac{Q\,14.7zT}{p520} + 2.92 \times 10^{-8} \frac{\beta 29Gp}{z10.73T} \left(\frac{Q^2\, \overline{14.7}^2\, z^2\, T^2}{p^2 520^2 4\pi^2 r^2 h^2} \right)$$

$$= \frac{158{,}171 \times 14.7}{2\pi\ 520} \frac{\mu zTQ}{rkh} + \frac{2.92 \times 10^{-8} \times 29 \times 14.7^2}{10.73 \times 5202 \times 4 \times \pi^2} \left(\frac{\beta GTz}{r^2h^2} Q^2 \right)$$

$$= 711.64 \frac{\mu TzQ}{rkh} + 1.60\ 10^{-12} \frac{\beta GTz}{r^2h^2} Q^2$$

$$\int_{p_w}^{p_b} pdp = 711.64 \frac{\mu TzQ}{kh} \int_{r_w}^{r_b} \frac{dr}{r} + 1.60 \times 10^{-12} \frac{GTzQ^2}{h^2} \int_{r_w}^{r_b} \frac{dr}{r^2}$$

$$p_b^2 - p_w^2 = 1{,}423.28 \frac{\mu Tz}{hk} \left(\ln \frac{r_b}{r_w} \right) Q + 3.2 \times 10^{-12} \frac{\beta GTz}{h^2} Q^2 \left(\frac{1}{r_w} - \frac{1}{r_b} \right) \quad \text{(B-4)}$$

APPENDIX C

Differential Equations Describing the Flow of Real Gas

For the flow of real gas in the most general case (Darcy or laminar-inertial-turbulent regime), with both pressure- and temperature-dependent viscosity, compressibility, and z-factor, the equations may be derived as shown in the following.

In radial polar coordinates and in reference to cylindrical disk shape radial flow model, one can readily combine the constitutive equation (Darcy's law modified for turbulence effects in the manner of Forchheimer) with the equation of state and continuity:

$$-\frac{dp}{dr} = \frac{\mu}{k} v_r + \beta \rho v_r^2 \tag{C-1}$$

For radial flow in the horizontal and + r direction, the Forchheimer equation may be expressed for computational convenience in a modified form of Darcy's law:

$$v_r = -\delta_r \frac{k}{\mu} \frac{\partial p}{\partial r} \tag{C-2}$$

where δ_r is a dimensionless modifier related to turbulence coefficient β.

The equation of state and continuity are expressed as follows:

$$\rho = \frac{Mp}{zRT} \tag{C-3}$$

$$\frac{1}{r}\frac{\partial}{\partial r}(r\rho v_r) = -\phi\frac{\partial \rho}{\partial t} \tag{C-4}$$

In combining Equations C-2 and C-3 in C-4:

$$\frac{1}{r}\frac{\partial}{\partial r}\left[r\frac{pM}{zRT}\left(-\delta_r\frac{k}{\mu}\frac{\partial p}{\partial r}\right)\right] = -\frac{\phi\partial}{\partial t}\left(\frac{pM}{zRT}\right) \tag{C-5}$$

which can be simplified as:

$$\frac{1}{r}\frac{\partial}{\partial r}\left[r\,\delta_r\frac{p}{z\mu}\frac{\partial p}{\partial r}\right] = \frac{1}{k}\frac{\partial}{\partial t}\left(\phi\frac{p}{z}\right) \tag{C-6}$$

The pseudo-pressure has been defined as:

$$m(p) = 2\int_o^p \frac{p}{\mu z}\,dp \tag{C-7}$$

If we note that:

$$\frac{\partial m}{\partial r} = \frac{2p}{z\mu}\frac{\partial p}{\partial r} \tag{C-8}$$

In substituting C-8 in C-6, one obtains:

$$\frac{1}{r}\frac{\partial}{\partial r}\left[r\,\delta_r\frac{\partial m}{\partial r}\right] = \frac{2\phi}{k}\frac{\partial}{\partial t}\left(\frac{p}{z}\right) \tag{C-9}$$

Next, the term on the right-hand side may be developed as:

$$\frac{\partial}{\partial t}\left(\frac{p}{z}\right) = \frac{1}{z}\frac{\partial p}{\partial t} + p\frac{\partial}{\partial t}\left(\frac{1}{z}\right)$$

$$= \frac{1}{z}\frac{\partial p}{\partial t} + p\frac{\partial}{\partial p}\left(\frac{1}{z}\right)\frac{\partial p}{\partial t} \qquad \text{(chain rule)}$$

$$= \frac{1}{z}\frac{\partial p}{\partial t} + p\left(-\frac{1}{z^2}\frac{\partial z}{\partial p}\right)\frac{\partial p}{\partial t} \tag{C-10}$$

Recall definition of compressibility:

$$c = -\frac{1}{V}\left(\frac{\partial V}{\partial P}\right)_T \tag{C-11}$$

for a real gas:

$$c = \frac{1}{p} - \frac{1}{z}\left(\frac{\partial z}{\partial p}\right)_T \tag{C-12}$$

from Equation C-12:

$$\frac{1}{z}\frac{\partial z}{\partial p} = \frac{1}{p} \tag{C-13}$$

Substitute C-13 in C-10:

$$\frac{\partial}{\partial t}\left(\frac{P}{z}\right) = \frac{\partial p}{\partial t}\left[\frac{1}{z} - \frac{P}{z}\left(\frac{1}{p} - c\right)\right]$$

$$= c\,\frac{P}{z}\frac{\partial p}{\partial t} \tag{C-14}$$

Therefore, Equation C-9 becomes:

$$\frac{1}{r}\frac{\partial}{\partial r}\left[r\,\delta_r\,\frac{\partial m}{\partial r}\right] = \frac{2\phi}{k}\,c\,\frac{p}{z}\frac{\partial p}{\partial t} \tag{C-15}$$

But, from the definition of real gas potential:

$$\frac{\partial m(p)}{\partial t} = \frac{2p}{z\mu}\frac{\partial p}{\partial t}$$

$$\frac{\partial p}{\partial t} = \frac{z\mu}{2p}\frac{\partial m(p)}{\partial t} \tag{C-16}$$

Substitute in C-15:

$$\frac{1}{r}\frac{\partial}{\partial r}\left[r\,\delta_r\,\frac{\partial m}{\partial r}\right] = \frac{2\phi}{k}\,c\,\frac{p}{z}\frac{z\mu}{2p}\frac{\partial m}{\partial t}$$

which simplifies readily to:

$$\frac{1}{r}\frac{\partial}{\partial r}\left[r\,\delta_r\,\frac{\partial m}{\partial r}\right]=\frac{\phi\mu c}{k}\frac{\partial m}{\partial t} \tag{C-17}$$

For Darcy flow where $\delta_r = 1.0$, a more general form of C-17 expressed in terms of Laplacian may be given as:

$$\nabla^2 m(p)=\frac{\phi\mu c}{k}\frac{\partial m(p)}{\partial t} \tag{C-18}$$

APPENDIX D

Line Source Solution to Unsteady-State Flow

The constant terminal rate solution to the radial diffusivity equation is basic and is frequently used by engineers in the analysis of data from oil, water, and gas wells. The main dependent parameter is pressure in the case of oil or water (liquid flow) and pseudo-pressure, as shown in Appendix C for the case involving the flow of real gas.

A step-by-step derivation is given here for the case of liquid flow:

Consider $p = p(r,t)$ (Radial, unsteady state)

The partial differential equation describing the flow through porous media, in the unsteady state is:

$$\nabla^2 p = \frac{\phi\mu c}{k}\frac{\partial p}{\partial t} \quad \text{(D-1)}$$

which can be written, for the radial case, as:

$$\frac{\partial^2 p}{\partial r^2} + \frac{1}{r}\frac{\partial p}{\partial r} = \frac{\phi\mu c}{k}\frac{\partial p}{\partial t} \quad \text{(D-2)}$$

Equation D-2 may be written by inspection as:

$$\frac{1}{r}\frac{\partial}{\partial r}\left(r\frac{\partial p}{\partial r}\right) = \frac{\phi\mu c}{k}\frac{\partial p}{\partial t} \quad \text{(D-3)}$$

GEOMETRY OF FLOW SYSTEM

Horizontal disk shape porous media, permeable matrix:

Inner boundary $r = r_w$
Outer boundary $r = \infty$
Thickness = h

Assumptions

- Homogeneous matrix k and ϕ constant
- Isotropic $k_x = k_y = k_z = k$
- Constant and small compressibility c
- Constant viscosity
- Small pressure gradients
- Fully penetrating well geometry

Boundary Conditions

Inner boundary. Constant Terminal Rate q = constant at $r = r_w$, which is expressed as:

$$(2\pi rh)\frac{k}{\mu}\frac{\partial p}{\partial r} = q \text{ at } r = r_w \tag{D-4}$$

which may be written as:

$$r\frac{\partial p}{\partial r} = \frac{q\mu}{2\pi kh} \text{ at } r = r_w \tag{D-5}$$

Outer boundary.

$p \rightarrow p_o$ for $r \rightarrow \infty$ for all t

Initial Condition

$p = p_o$ at $t = 0$ for all r

Full Derivation of Line-Source Solution

For constant terminal rate the boundary condition D-5 will be modified as:

$$\lim_{r \to 0} r \frac{\partial p}{\partial r} = \frac{q\mu}{2\pi kh} \text{ for all t} \tag{D-6}$$

Use Boltzmann's Transformation:

$$s = \frac{\phi\mu c r^2}{4kt} \tag{D-7}$$

$$\frac{\partial s}{\partial r} = \frac{\phi\mu c r}{2kt} \tag{D-8}$$

and

$$\frac{\partial s}{\partial t} = -\frac{\phi\mu c r^2}{4kt^2} \tag{D-9}$$

Now, effecting the above transformation through the partial differential equation D-3 by invoking the chain rule:

$$\frac{1}{r}\frac{d}{ds}\left(r\frac{\partial s}{\partial r}\frac{dp}{ds}\right)\frac{\partial s}{\partial r} = \frac{\phi\mu c}{k}\frac{\partial s}{\partial t}\frac{dp}{ds} \tag{D-10}$$

Next, substitute for $\frac{\partial s}{\partial r}$ and $\frac{\partial s}{\partial t}$ from Equations D-8 and D-9:

$$\frac{1}{r}\frac{\phi\mu c r}{2kt}\frac{d}{ds}\left(r\frac{\phi\mu c r}{2kt}\frac{dp}{ds}\right) = \frac{\phi\mu c}{k}\left(-\frac{\phi\mu c r^2}{4kt^2}\right)\frac{dp}{ds} \tag{D-11}$$

$$\frac{1}{r}\left(\frac{\phi\mu c r}{2kt}\right)\frac{d}{ds}\left(2s\frac{dp}{ds}\right) = -\left(\frac{\phi\mu c r}{2kt}\right)^2\frac{dp}{ds} \tag{D-12}$$

$$\frac{d}{ds}\left(2s\frac{dp}{ds}\right) = -2s\frac{dp}{ds} \tag{D-13}$$

$$\frac{d}{ds}\left(s\frac{dp}{ds}\right) = -s\frac{dp}{ds} \tag{D-14}$$

Equation D-14, now an ordinary, total differential equation may be solved simply by the change of a variable:

$$\lambda = \frac{dp}{ds} \tag{D-15}$$

$$\frac{d}{ds}(s\lambda) = -s\lambda \tag{D-16}$$

$$s\frac{d\lambda}{ds} + \lambda = -s\lambda \tag{D-17}$$

$$\frac{d\lambda}{ds} = -\lambda\frac{(s+1)}{s} \tag{D-18}$$

$$\frac{d\lambda}{\lambda} = \left(-1 - \frac{1}{s}\right) ds \tag{D-19}$$

$$\ln \lambda = -s - \ln s + C_1 \tag{D-20}$$

$$\lambda = C_2 \frac{e^{-s}}{e^{\ln s}} \tag{D-21}$$

$$\lambda = C_2 \frac{e^{-s}}{s} \tag{D-22}$$

Now, using the inner boundary condition to solve for the integration constant C_2:

$$\lim_{r \to 0} r\frac{\partial p}{\partial r} = \frac{q\mu}{2\pi kh} \tag{D-23}$$

Transformed boundary condition:

$$\lim_{r \to 0} r\frac{dp}{ds}\frac{\partial s}{\partial r} = \frac{q\mu}{2\pi kh} \tag{D-24}$$

but:

$$\frac{\partial s}{\partial r} = \frac{\phi\mu c r}{2kt} \tag{D-25}$$

$$\lim_{r \to 0} \frac{r\phi\mu c r}{2kt}\frac{dp}{ds} = \frac{q\mu}{2\pi kh} \tag{D-26}$$

$$\lim_{s \to 0} 2s\frac{dp}{ds} = \frac{q\mu}{2\pi kh} \tag{D-27}$$

Now recall:

$$\frac{dp}{ds} = \lambda = C_2 \frac{e^{-s}}{s}$$

$$\lim_{s \to 0} 2s\, C_2 \frac{e^{-s}}{s} = \frac{q\mu}{2\pi kh} \qquad \text{(D-28)}$$

Therefore:

$$C_2 = \frac{q\mu}{4\pi kh}$$

and:

$$\frac{dp}{ds} = \frac{q\mu}{4\pi kh} \frac{e^{-s}}{s} \qquad \text{(D-29)}$$

Now, integrating Equation D-29:

$$\int_{p_i}^{p} dp = \frac{q\mu}{4\pi kh} \int_{\infty}^{s} \frac{e^{-s}}{s}\, ds \qquad \text{(D-30)}$$

The limits on the integrals of Equation D-30 are specified as follows:

- The pressure p is integrated from its initial value p_i to its current value p.
- The limits on the second integral correspond to the limits on the first; i.e., when $p = p_i$ at $t = 0$.
- $s = \infty$ by definition Equation D-7.
- When later p becomes p, s becomes equal to current s at $t = t$.

Therefore:

$$p = p_i + \frac{q\mu}{4\pi kh} \int_{\infty}^{s} \frac{e^{-s}}{s} \qquad \text{(D-31)}$$

which may be written as:

$$p = p_i - \frac{q\mu}{4\pi kh} \int_{s}^{\infty} \frac{e^{-s}}{s}\, ds \qquad \text{(D-32)}$$

The integral $\int_s^\infty \frac{e^{-s}}{s}$ ds may be recognized as a function of its lower limit s or x of integration.

$$\int_x^\infty \frac{e^{-s}}{s} ds = Ei(x) \tag{D-33}$$

It is called the exponential integral. A plot of the exponential integral Ei(x) vs. x for the values of argument x between x = 0.001 and x = 5.0 is given by L. Dake[1] and is shown in Figure D-1.

It can be seen from Figure D-1 that for $x < 0.01$ the exponential integral can be approximated by a logarithmic function:

$$Ei(x) \cong - \ln x - 0.5772 \tag{D-34}$$

where the constant 0.5772 is called Euler's Constant.

An alternative approximation for the exponential integral is sometimes given as:

$$Ei(x) \sim \ln(\gamma x) \tag{D-35}$$

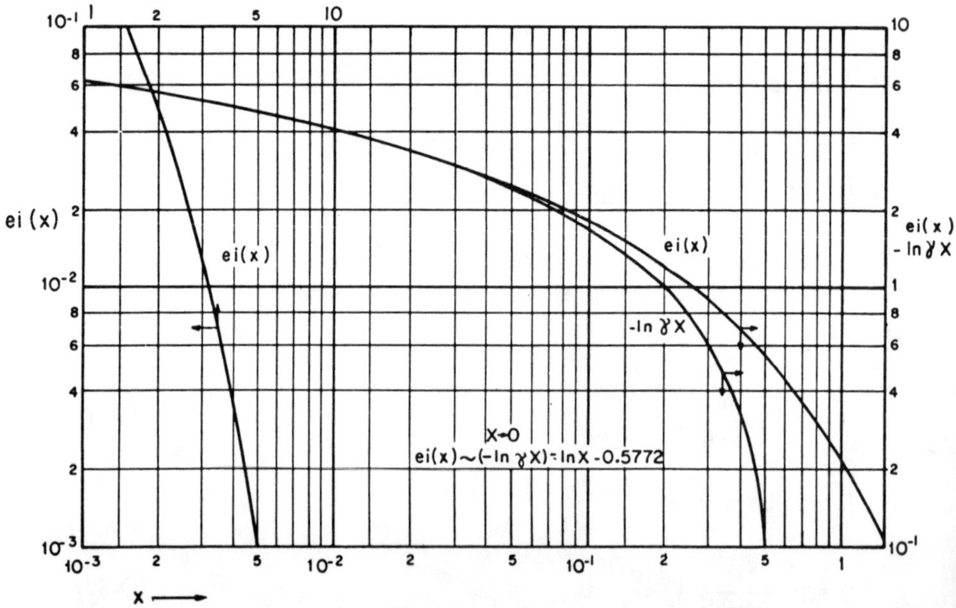

Figure D-1. The exponential integral function.

where:

$$\gamma = e^{0.5772} = 1.781$$

$$\gamma = \text{antilog}\ (0.5772)$$

Now, summarizing and substituting:

$$p_{r,t} = p_o - \frac{q\mu}{4\pi kh}\ \text{Ei}\left(\frac{\phi\mu cr^2}{4kt}\right) \tag{D-36}$$

In Equation D-36, developed above, the units are the original Darcy's law units; that is: pressure in psia, k in darcies, ϕ (fraction), c vol/vol × atm, time in seconds, μcp, $\pm q$* cc/sec, and h cm.

If the field units are to be used, then Equation D-36 becomes:

$$p = p_o + \frac{70.6(\pm q)}{kh}\ \text{Ei}\left(\frac{+\phi\mu Cr^2}{4 \times 0.00633\ kt}\right) \tag{D-37}$$

where p = pressure at r = r, psia
P_o = initial uniform pressure, psia
q = rate, barrels/day reservoir conditions (− on injection, + on production)
μ = viscosity of the liquid, cp
ϕ = porosity fraction
c = compressibility vol/vol × psi
r = radial location, ft
k = permeability, md
h = formation thickness, ft
t = time, days

If the argument of the Ei function is <0.01, then the Ei function can be represented by a natural logarithm + a well known constant under these circumstances when:

$$\frac{4kt}{\phi\mu cr^2} > \frac{1}{100} \tag{D-38}$$

$$p(r,t) = p_o - \frac{(\pm q)}{4\pi kh}\ \ln\left(\frac{\gamma\phi\mu cr^2}{4kt}\right) \tag{D-39}$$

---

* The ± before q is + for production, − for injection, in storage reservoirs.

where γ is Euler's constant = 1.78

Equation D-39 is in Darcy units. It can also be rewritten as:

$$p(r,t) = p_o - \frac{(\pm q)\mu}{4\pi kh}\left[\ln \frac{kt}{\phi\mu cr^2} + 0.80907\right] \tag{D-40}$$

If $r = r_w$ at the wellbore,

$$p_w = p_o + \frac{(\pm q)\mu}{4\pi kh} \ln\left(\frac{\gamma\phi\mu C{r_w}^2}{4kt}\right) \tag{D-41}$$

Reference

1. Dake, L. P. *Fundamentals of Reservoir Engineering*, New York: Elsevier, 1978.

APPENDIX E

Thermodynamic Analysis of Two-Phase Flow

Consider the generalized geometry of a curvilinear pipe, two phases flowing simultaneously in steady-state up the pipe. (see Figure E-1). For steady flow systems between any upstream point (1) and downstream point (2) for incompressible flow or for nearly incompressible flow:

$$U_1 + P_1V_1 + (KE)_1 + (PE)_1 + q - w =$$

$$U_2 + (KE)_2 + (PE)_2 + U_2 + P_2V_2 \tag{E-1}$$

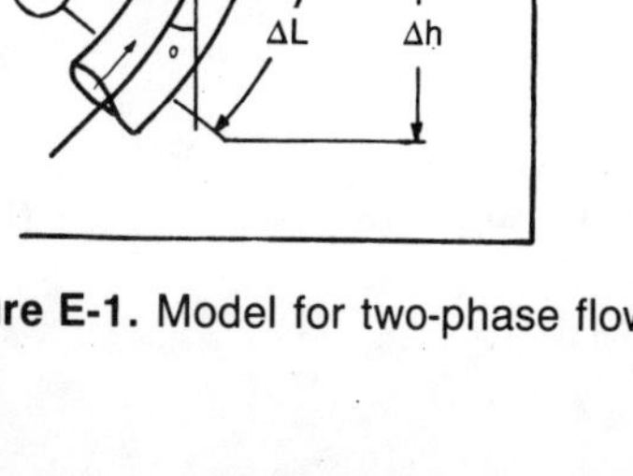

Figure E-1. Model for two-phase flow.

Equation E-1 simply states that energy entering the flow system at (1), including internal energy, the flow, work kinetic and potential energy, plus energy added between points (1) and (2) as heat, minus energy removed as shaft work, must be equal to energy remaining at point (2). Changes in internal, kinetic, and potential energies, and, again flow work, must equal the energy that had been added and removed between points (1) and (2).

$$\Delta U + \Delta\left(\frac{mv^2}{J2g_c}\right) + \Delta\left(\frac{m}{j}\frac{g\,z}{g_c}\right) + \frac{1}{J}\Delta PV = q - w \tag{E-2}$$

But,

$$\Delta(PV) = \int_1^2 PdV + \int_1^2 VdP \tag{E-3}$$

$$\Delta U = \int_1^2 Tds + \int_1^2 p(-dV) + \int_1^2 \gamma d\sigma + \int_1^2 \mu_A dm_A + \int_1^2 \mu_B dm_B + \ldots \tag{E-4}$$

Substitute everything into Equation E-2:

$$\int_1^2 Tds + \int_1^2 p(-dV) + \int_1^2 \gamma d\sigma + \int_1^2 \mu_A dm_A + \ldots. + \int_1^2 PdV + \int_1^2 VdP + \left(\frac{m}{J}\frac{v^2}{2g_c}\right) + \Delta\left(\frac{m}{J}\frac{g}{g_c}z\right) = \bar{q} - \overline{w} \tag{E-5}$$

But between states (1) and (2):

$$\int_1^2 Tds = \bar{q} + \overline{lw} \tag{E-6}$$

Substitute Equation E-6 in E-5, neglect chemical potential and surface effects, and write per lb_m basis:

$$\bar{q} + \overline{lw} + \int_1^2 \frac{dP}{\rho} + \Delta\left(\frac{1}{j}\frac{v^2}{2g_c}\right) + \Delta\left(\frac{1}{j}\frac{g}{g_c}z\right) = \bar{q} - \overline{w} \tag{E-7}$$

which in turn gives the mechanical energy balance:

$$\int_1^2 \frac{dP}{\rho} + \Delta(KE) + \Delta(PE) = -\overline{w} - \overline{lw} \tag{E-8}$$

$$\Delta\frac{p}{\rho} + \frac{\Delta v^2}{2g_c} + \Delta\frac{g}{g_c}h = -w - lw, \text{ where h designates} \tag{E-9}$$

height above datum

Equation E-9 is energy balance for flow systems, called by some mechanical energy balance for steady-state incompressible flow between any 2 points.

Differential form of Equation E-9 for the case $-\overline{w} = 0$:

$$\frac{dP}{\rho} + d\left(\frac{v^2}{2g_c}\right) + d\left(\frac{g}{g_c}h\right) = -\frac{f\,dl\,v^2}{2g_cD} \tag{E-10}$$

Let $\frac{f\,\rho\,v^2}{2g_cD} = \tau_f$ (friction gradient)

$$\frac{dp}{dl} + \rho\frac{2v\,dv}{2g_cdl} + \rho\frac{g}{g_c} = -\tau_f;\text{ since for vertical flow } dh = dl$$

$$+\frac{dp}{dl} + \frac{\rho}{g_c}v\frac{dv}{dl} + \frac{\rho g}{g_c} = -\tau_f$$

$$-\frac{dp}{dl} - \frac{\rho\,v}{g_c}\frac{dv}{dl} = \rho\frac{g}{g_c} + \tau_f \tag{E-11}$$

Now, if v is viewed as mixture velocity:

$$\frac{Q_L + Q_G}{A} = v;\text{ if } Q_L << Q_G \text{ then, } dV = \frac{dQ_G}{A}$$

$$pQ = znRT;\; Q = zn\frac{RT}{p};\; dQ = -znRT\frac{dp}{p^2}$$

$$dQ = -\frac{znRT}{p}\frac{dp}{p},\text{ but } \frac{znRT}{p} = Q$$

$$dQ = -Q\frac{dp}{p};\;\frac{dQ}{A} = -\frac{1}{A}Q\frac{dp}{p};\;\frac{\rho\,v}{g_c}\frac{dv}{dl} = -\frac{\rho}{g_c}\frac{(Q_G + Q_L)}{A}\frac{1}{A}Q\frac{dp}{p}$$

but $\rho\,(Q_G + Q_L) = W_T$ total mass flowrate

$$\frac{\rho}{g_c}v\frac{dv}{dl} = -\frac{W_T}{g_c}\frac{1}{A^2}\frac{Q}{dl}\frac{dp}{p} \tag{E-12}$$

Substitute Equation E-12 into E-11:

$$-\frac{dp}{dl} + \frac{W_TQ_G}{g_cA^2}\frac{dp}{pdl} = \rho\frac{g}{g_c} + \tau_f$$

Let the average pressure level over dp pressure decrement be $\bar{p}$ and let $\bar{\rho}$ designate the mixture density:

$$-\frac{dp}{dl}\left(1-\frac{W_T Q_G}{g_c A^2 \bar{p}}\right)=\bar{\rho}\,\frac{g}{g_c}+\tau_f$$

which in finite increment form gives:

$$-\frac{\Delta P}{\Delta L}=\frac{\bar{\rho}\, g/g_c+\tau_f}{1-\dfrac{W_T Q_G}{g_c A^2 \bar{p}}} \tag{E-13}$$

APPENDIX F

Derivation of Equations for Temperature Distribution in a Gas Well

Figure F-1 depicts a wellbore with natural gas flowing vertically upward in steady-state.

Let m be the mass flowrate in lb_m/hr; h, the distance from datum level at the bottom; T, the temperature of the gas; and T′, the geothermal (undisturbed) temperature of the earth around the wellbore.

We know that $T' = T'(h)$, and we shall assert $T = T(h)$.

Total energy balance written between point (1) upstream and point (2) downstream gives:

$$\Delta H + \Delta(KE) + \Delta(PE) = q - w \quad \text{(F-1)}$$

where H denotes the enthalpy of the gas.

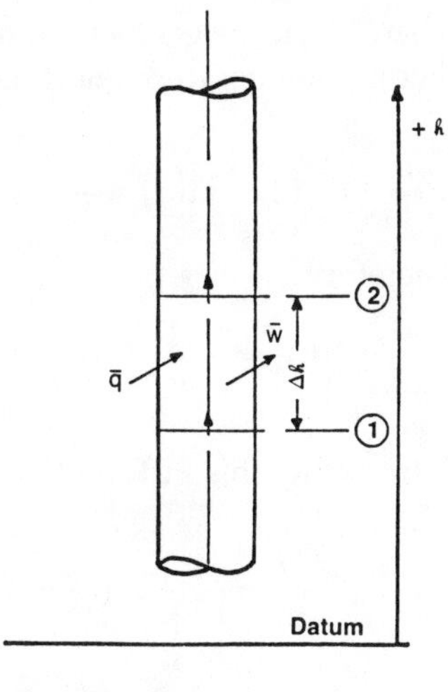

Figure F-1. Wellbore with flowing natural gas.

For all practical cases considered, it is well known that the change of kinetic energy per pound mass flowing between points (1) and (2) is negligible compared to other terms, so (KE) = 0. Furthermore, there are no pumps, compressors, turbines, etc. between (1) and (2), so w = 0.

With each term in Equation F-1 expressed in BTUs, Equation F-1 may be written for the case of an ideal gas as follows:

$$m\,C_p[T(h + \Delta h) - T(h)] + \frac{m}{J}\frac{g}{g_c}\Delta h = 2\pi a \Delta h\, k\,[T'(h) - T(h)] \qquad \text{(F-2)}$$

In Equation F-2, J is the mechanical equivalent of heat; a, the inside radius; and k, the overall heat transfer coefficient between the gas inside the well and the environment temperature (assumed constant) prevailing outside the wellbore.

Dividing both sides of Equation F-2 by Δh and allowing $\Delta h \rightarrow 0$:

$$mC_p \frac{d}{dh}[T(h)] + \frac{m}{J}\frac{g}{g_c} = 2\pi ak[T'(h) - T(h)] \qquad \text{(F-3)}$$

For the geothermal temperature prevailing outside $T'(h)$ from known previous observations, we can assume a linear relationship with h:

$$T'(h) = T'_o - \frac{h}{g_t} \qquad \text{(F-4)}$$

where T'_o is the geothermal temperature at datum (h = 0) and g_t is the geothermal temperature constant.

$$\frac{d}{dh}[T'(h) - T(h)] = \frac{-1}{g_t} - \frac{d}{dh}[T(h)] \qquad \text{(F-5)}$$

Therefore:

$$\frac{d}{dh}[T(h)] = -\frac{1}{g_t} - \frac{d}{dh}[T'(h) - T(h)] \qquad \text{(F-6)}$$

Substituting this in Equation F-3 we get:

$$mC_p - \left(\frac{1}{g_t} - \frac{d}{dh}[T'(h) - T(h)]\right) + \frac{m}{j}\frac{g}{g_c} = 2\pi ak[T'(h) - T(h)] \qquad \text{(F-7)}$$

Let us divide then by 2 ak and let:

$$\frac{mC_p}{2\pi ak} = L;\; T'(h) - T(h) = \theta(h); \qquad \text{(F-8)}$$

$$-\frac{L}{g_t} - L\frac{d\theta}{dh} = -\frac{m}{2\pi ak}\frac{1}{j}\frac{g}{g_c} + \theta \tag{F-9}$$

Furthermore let:

$$\frac{m}{2\pi ak}\frac{1}{j}\frac{g}{g_c} = \alpha, \text{ and } \frac{L}{g_t} = \beta$$

With the above substitutions:

$$-\beta - L\frac{d\theta}{dh} = -\alpha + \theta$$

$$L\frac{d\theta}{dh} + \theta = \alpha - \beta \tag{F-10}$$

which is recognized as the linear total differential equation for dependent variable θ, non-homogeneous and with constant coefficients.

Now, the solution of Equation F-1:

$$\theta = \theta_1 + \theta^*$$

where θ_1 is the general solution to the homogeneous equation corresponding to F-10 and θ^* is a particular solution to Equation F-10.

For the homogeneous part:

$$L\frac{d\theta_1}{dh} = -\theta_1$$

$$L\frac{d\theta_1}{\theta_1} = -dh$$

$$\int_C^{\theta_1}\frac{d\theta_1}{\theta_1} = \int_o^h -\frac{1}{L}\,dh$$

$$\ln\frac{\theta_1}{C} = -\frac{h}{L}$$

$$\theta_1 = Ce^{-h/L}$$

Also:

$\theta^* = \alpha - \beta$ is the particular solution.

The general solution is:

$$\theta = Ce^{-h/L} + \alpha - \beta \text{ at } h = 0 \; \theta = \theta_o; \; \theta_o + \beta - \alpha = C$$

Now, substituting back to the original variables:

$$T'(h) - T(h) = (T'_o - T_o + \beta - \alpha)\, e^{\frac{-2\pi ak}{mC_p}h} + \frac{m}{2\pi ak}\frac{1}{j}\frac{g}{g_c} - \frac{mC_p}{2\pi akg_t}$$

$$T(h) = T'(h) - \left(T'_o - T_o + \frac{mC_p}{2\pi akg_t} - \frac{m}{2\pi ak}\frac{1}{j}\frac{g}{g_c}\right) e^{\frac{-2\pi ak}{mC_p}h} - \frac{m}{2\pi ak}\frac{1}{j}\frac{g}{g_c} + \frac{mC_p}{2\pi akg_t}$$

But, $T'(h) = T'_o - \frac{h}{g_t}$; so:

$$T(h) = T'_o - \frac{h}{g_t} - \frac{m}{2\pi ak}\frac{1}{j}\frac{g}{g_c} + \frac{mC_p}{2\pi akg_t} - \left[T'_o - T_o + \frac{mC_p}{2\pi akg_t} - \frac{m}{2\pi ak}\frac{1}{j}\frac{g}{g_c}\right] e^{\frac{-2\pi ak}{mC_p}h} \qquad \text{(F-11)}$$

It can be seen from Equation F-11 that for very deep wells as $h \rightarrow \infty$. T(h) would approach an asymptote parallel to the undisturbed geothermal gradient displaced to the right by the amount: $\left(\frac{L}{g_t} - \frac{m}{2\pi ak}\frac{g}{jg_c}\right)$ indicating that the higher m or the rate of flow is, the higher the eventual temperature of the gas at the surface.

Also, the higher the heat transfer coefficient between the well and the equipment, the closer the asymptotic limit to the geothermal line.

Once the temperature of the gas at the surface is calculated using Equation F-11, the assumption of ideal gas inherent in Equation F-11 is equivalent to temperature calculation on an isobaric path. The final temperature can be corrected approximately by a further isenthalpic temperature drop from bottom-hole pressure to actual surface pressure.

For temperature distribution in the flow of real gas, Equation F-11 must be modified as:

$$T(h) = T'(h) + \frac{mC_p}{2\pi akg_t} - \frac{m}{j}\frac{g}{g_c}\frac{1}{2\pi ak}$$

$$- \left[T_o' - T_o + \frac{mC_p}{2\pi akg_t} - \frac{m}{2\pi ak}\frac{1}{j}\frac{g}{g_c}\right] e^{\frac{-2\pi ak}{mC_p}h} + \mu_j\,(\Delta p) \qquad \text{(F-12)}$$

The procedure given above in the Mollier chart is equivalent to a path from initial pressure and temperature to the final pressure and temperature of two distinct parts:

1. Isobaric sub-path from bottom to top at constant bottom-hole pressure.
2. Isenthalpic sub-path from bottom pressure and calculated top temperature to actual top temperature and actual top pressure.

APPENDIX G

Calculation of Virtual Inventory

CONSTANT VOLUME STORAGE FIELDS

Consider that ΔQ billions have been removed from the reservoir, rather quickly before any appreciable water movement takes place, dropping the p/z value from $(p/z)_i$ to $(p/z)_{i+1}$.

$$\frac{p}{z} = \frac{RT\ n}{V}$$

$$\frac{p}{z} = \frac{RT\ Q \times 10^9}{V\ 379}$$

$$(p/z)_i = \frac{RT\ 10^9\ Q_i}{V\ 379}$$

$$(p/z)_{i+1} = \frac{RT\ 10^9\ Q_{i+1}}{V\ 379}$$

$$(p/z)_i - (p/z)_{i+1} = \frac{RT\ 10^9}{V\ 379}\ Q;$$

where, as shown in Figure G-1, Q represents the amount of gas withdrawn on p/z vs. content quadrant.

$$\frac{RT\ 10^9}{V\ 379} = \frac{(p/z)_i - (p/z)_{i+1}}{\Delta Q}$$

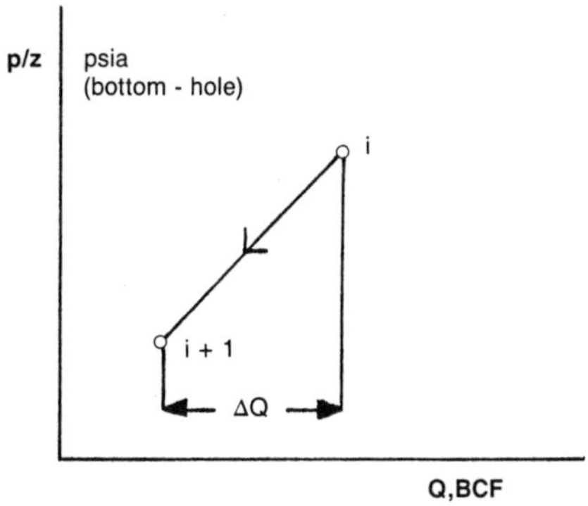

Figure G-1. Q, BCF (Procedure for calculation of "virtual inventory" at Point 1).

but

$$\frac{RT\ 10^9}{V\ 379} = \frac{(p/z)_i}{Q_i}$$

From the equations above it follows that:

$$Q_i = \frac{(p/z)_i\ \Delta Q}{(p/z)_i - (p/z)_{i+1}} \tag{G-1}$$

Equation G-1 is the one used for verification of inventory on volumetric, i.e., constant volume reservoirs.

VARIABLE VOLUME (STORAGE FIELDS SUBJECT TO WATER DRIVE)

When the reservoir volume changes from V_1 to V_2 due to water influx/efflux, when the inventory changes from Q_1 to $Q_2 = Q_1 - \Delta Q$, the following procedure permits calculation of the inventory Q_1 from data on metered ΔQ and gaged $\Delta(p/z)$.

$$Q_1 = (p/z)_1 \frac{V_1}{RT}$$

$$Q_2 = (p/z)_2 \frac{V_2}{RT}$$

$$Q_1 - Q_2 = \Delta Q$$

$$\Delta Q = \frac{1}{RT} [(p/z)_1 V_1 - (p/z)_2 V_2]$$

But $RT = (p/z)_1 \dfrac{V_1}{Q_1}$

Substitute in preceding equation:

$$\frac{1}{Q_1} \Delta Q = \frac{(p/z)_1 V_1 - (p/z)_2 V_2}{(p/z)_1 V_1}$$

$$Q_1 = \frac{\Delta Q \ (p/z)_1}{(p/z)_1 - (p/z)_2 \dfrac{V_2}{V_1}} \qquad \text{(G-2)}$$

Equation G-2 is the equation for inventory verification when the reservoir is not volumetric but subject to calculable volume change due to influx/efflux of water.

APPENDIX H

Calculation of Equalized Reservoir (p/z) for Two Adjacent Communicating Structures

Volumetric Case (no water drive)

$$Q_A = (p/z)_A \frac{V_A}{RT}$$

$$Q_B = (p/z)_B \frac{V_B}{RT}$$

$$(p/z)_C = (p/z)_A \frac{V_A}{V_A + V_B} + (p/z)_B \frac{V_B}{V_A + V_B} \qquad \text{(H-1)}$$

Appropriate units:

Q in BCF
p/z in psia
V in cubic feet pore volume

Variable Volumes (active water drive)

When both reservoirs are subject to water influx/efflux, the combined equilibrium $(p/z)_C$ for both can be calculated from individual p/z, pseudopressure, and individual BCF contents, as follows:

$$(p/z)_C = \frac{Q_A + Q_B}{\dfrac{Q_A}{(p/z)_A} + \dfrac{Q_B}{(p/z)_B}} \qquad \text{(H-2)}$$

APPENDIX I

Calculating Gas Loss from Apparent Trends in Storage Inventory

Consider Figure I-1, the total gas stored, as book (metered) inventory vs. time. It can be seen over many years that both maximum and minimum inventory curves follow some trend indicating that, while the maximum inventory grew, it had also become necessary to have more and more cushion gas to sustain storage operations.

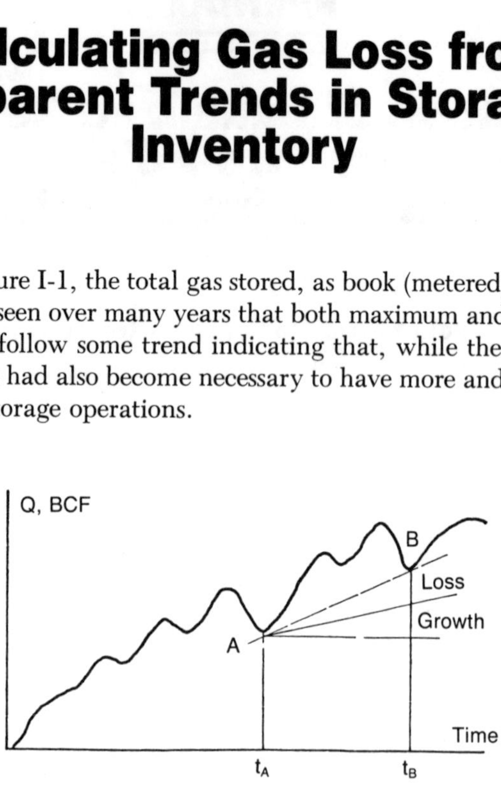

Figure I-1. Procedure for calculating gas loss from apparent trends of book inventory.

If it is possible to identify those withdrawal seasons, such as A and B, during which all of the top gas was produced from the storage field, $Q_B - Q_A$ represents the differentially larger amount of cushion gas necessary at the year t_B when compared to cushion gas required at t_A. The increase in minimum cushion gas required may be attributed to two causes:

1. The growth in actual reservoir pore volume between years t_A and t_B.
2. The gas loss that may have occurred from the storage inventory between the years t_A and t_B.

If the growth can be calculated from pressure over z-data and the inventory, and expressed in terms of percent growth, then the amount can be subtracted from the percentage of $(Q_B - Q_B)/Q_A$, the remainder indicating the gas which must have been lost during the years between t_B and t_A. The calculation of net pore volume is only approximate, as it is based on book inventory and available pressure data.

If the pore volume of the reservoir at anytime is designated as V, then V can be calculated from:

$$pV = znRt \tag{I-1}$$

or:

$$V = \frac{nRT}{(p/z)} \tag{I-2}$$

where n = No. of moles book inventory

$$n = \frac{Q \times 10^9}{379}$$

where Q = the book inventory, BCF

Accordingly:

$$V = \frac{Q \times 10^9 \ RT}{379 \ (p/z)} \tag{I-3}$$

where V = net pore volume, ft^3

$$R = \text{the gas constant} = 10.72 \ \frac{\text{psia} \times \text{ft}^3}{\text{lb-mole °R abs}}$$

T = formation temperature, °R abs

(p/z) = equilibrium (p/z) from pressure surveys bottom-hole, psia

APPENDIX J

The International System of Units (SI) and Its Relationship to U.S. Customary Units—Common Equivalents and Conversions

Approximate Common Equivalents

1 inch	= 25 millimeters
1 foot	= 0.3 meter
1 yard	= 0.9 meter
1 mile	= 1.6 kilometers
1 sq. in.	= 6.5 sq. centimeters
1 sq. foot	= 0.09 sq. meter
1 sq. yard	= 0.8 sq. meter
1 acre	= 0.4 hectare*
1 cubic inch	= 16 cu. centimeters
1 cubic foot	= 0.03 cu. meter
1 cubic yard	= 0.8 cu. meter
1 quart (lq)	= 1 liter*
1 gallon	= 0.004 cu. meter
1 ounce (avdp)	= 28 grams
1 pound (avdp)	= 0.45 kilogram
1 horsepower	= 0.75 kilowatt
1 millimeter	= 0.04 inch
1 meter	= 3.3 feet

* common term not used in SI

1 meter	= 1.1 yards
1 kilometer	= 0.6 mile
1 sq. centimeter	= 0.16 sq. inch
1 sq. meter	= 11 sq. feet
1 sq. meter	= 1.2 sq. yards
1 hectare*	= 2.5 acres
1 cu. centimeter	= 0.06 cu. inch
1 cu. meter	= 35 cubic feet
1 cu. meter	= 1.3 cubic yards
1 liter*	= 1 qt. (lq)
1 cubic meter	= 250 gallons
1 gram	= 0.035 ounces (avdp)
1 kilogram	= 2.2 pounds (avdp)
1 kilowatt	= 1.3 horsepower

Conversions Accurate to Parts Per Million

inches × 25.4	= millimeters
feet × 0.3048	= meters
yards × 0.9144	= meters
miles × 1.609 34	= kilometers
sq. inches × 6.4516	= sq. centimeters
sq. feet × 0.092 903 0	= sq. meters
sq. yards × 0.836 127	= sq. meters
acres × 0.404 686	= hectares
cubic inches × 16.3871	= cu. centimeters
cubic feet × 0.028 316 8	= cu. meters
cubic yards × 0.764 555	= cu. meters
quarts (lq) × 0.946 353	= liters
gallons × 0.003 785 41	= cu. meters
ounces (avdp) × 28.349 5	= grams
pounds (avdp) × 0.453 592	= kilograms
horsepower × 0.745 700	= kilowatts

millimeters × 0.039 370 1	= inches
meters × 3.280 84	= feet
meters × 1.093 61	= yards
kilometers × 0.621 371	= miles
sq. centimeters × 0.155	= sq. inches
sq. meters × 10.7639	= sq. feet
sq. meters × 1.195 99	= sq. yards
hectares × 2.471 05	= acres

* common term not used in SI

cu. centimeters × 0.061 023 7	= cu. inches
cu. meters × 35.3147	= cu. feet
cu. meters × 1.307 95	= cu. yards
liters × 1.056 69	= quarts (lq)
cubic meters × 264.172	= gallons
grams × 0.035 274	= ounces (avdp)
kilograms × 2.204 62	= pounds (avdp)
kilowatts × 1.341 02	= horsepower

APPENDIX K

Additional Literature on High Velocity, Inertial Turbulence, and Non-Darcy Flow Through Porous Media*

Abdulvagabov, A. L. *Izv. Vyss. Uch. Zaved., Nefti i Gas* (1961) No. 4, 83 (Russian).

Aris, Rutherford. *Vectors, Tensors and the Basic Equations of Fluid Mechanics*, Prentice-Hall, 1962.

Aronofsky, J. S. and R. Jenkins. "A Simplified Analysis of Unsteady Radial Gas Flow," *Trans. AIME* (1954) Vol. 201, 49.

Bakhmeteff, B. A. and N. V. Feodoroff. "Flow through Granular Media," *Trans. AIME* (1937) *59:* A 97.

Barenblatt, G. I. "On Some Non-Stationary Problems in the Flow of Liquids and Gases in Porous Media," *Prik. Mat. Mekh* (1952) *16*, 67.

Bird, R. B. *Chem. Eng. Sci.* (1957) Vol. 6, 123.

Brons, F. and V. E. Marting. "The Effect of Restricted Fluid Entry on Well Productivity," *Trans. AIME* (1961) 222, 172.

Brown and Associates. *Unit Operations*, John Wiley & Sons, Inc., 1950, p. 136.

Brownell, L. E., H. S. Drombrowski and C. A. Dickey. *Chem. Eng. Prog.* (1950) 46, 415.

Brownell, L. E. and D. L. Katz. "Flow of Fluids Through Porous Media," *Chem. Eng. Prog.* (1947) 43, 537.

* These references were compiled as a result of research on high velocity gas flow by the author during the period 1957–1977.

Bruce, G. H., D. W. Peaceman, H. H. Rachford, and J. D. Rice. "Calculations of Unsteady-State Gas Flow Through Porous Media," *Trans. AIME* (1953) *198*, p. 79.

Carlson, A. J. and J. E. Eastman. "Factors Influencing Permeability Measurements," *Petr. Tech.* (May 1940), TP 1196.

Carter, R. D., S. C. Miller, and H. G. Riley. "Determination of Stabilized Gas Well Performance from Short Flow Tests," SPE 400, Paper presented at 37th Ann. Fall Meeting of AIME, (Oct. 1962).

Casse, F. S. and H. J. Ramey. "The Effect of Temperature and Confining Pressure on Single Phase Flow in Consolidated Rocks," SPE paper 5877, 1976.

Cornell, D. "Flow of Gases Through Consolidated Porous Media," Ph.D. thesis, U. of Michigan, Ann Arbor, 1952.

Cornell, D. and D. L. Katz. "Flow of Gases through Porous Media," *Ind. Eng. Chem.* (1953), *45*, 2145.

Cornell, D. and D. L. Katz. "Pressure Gradients in Natural Gas Reservoirs," *Trans. AIME* (1953) 198, 61.

Cullender, M. H. "The Isochronal Performance Method of Determining the Flow Characteristics of Gas Wells," *Trans. AIME* (1955) Vol. 204, 137.

Darcy, Henry. *Les Fontaines Publiques de la Ville de Dijon*, Paris: Victor Dalmont, 1856.

Darin, S. R. and J. L. Huitt. "Effect of a Partial Monolayer of Propping Agent on Fracture Flow Capacity," *Trans. AIME* (1960) *219*, 31.

Douglas, J., Jr. and D. W. Peaceman. "Numerical Solution of Two-Dimensional Heat Flow Problems," *AIChE Journal* (Dec. 1955) p. 505.

Douglas, J., Jr., D. W. Peaceman, and H. H. Rachford, Jr. "Calculation of Unsteady-State Gas Flow Within a Square Drainage Area," *Trans. AIME* (1955) *204*, 190.

Douglas, J., Jr., D. W. Peaceman, and H. H. Rachford, Jr. "A Method for Calculating Multi-Dimensional Immiscible Displacement," *Trans. AIME* (1959) *216*, 297.

Dranchuk, P. M. and L. J. Kolada. "Interpretation of Steady Linear Visco-Inertial Gas Flow Data," *J. Cdn. Pet. Tech.* (Jan.–March 1968) *36*.

Dupuy, M. "Models Mathematiques A Penetration Partielle et a Permeabilities Variées Realises Pour Le Bureau Reservoir de La C.F.P.A.," *Revue de L'Institut Francais du Petrole* (Oct. 1961) *XVI*, No. 10, 1067.

Elenbaas, J. R. and D. L. Katz. "A Radial Turbulent Flow Formula," *Trans. AIME* (1948) Vol. 174, 25.

Ericksen, J. L. *Handbuch der Physik* III/1, edited by S. Flugge, Springer-Verlag, Berlin, 1960.

Fancher, G. H. and J. A. Lewis. "Flow of Simple Fluids through Porous Media," *Ind. Eng. Chem.* (1933) Vol. 25, 1139.

Fancher, G. H., J. A. Lewis, and K. B. Barnes. "Some Physical Characteristics of Oil Sands," *Bull. 12*, Pennsylvania State College Mineral Industries Experiment Station (1933) 65; *Ind. Eng. Chem.* (1933) 25, 1139.

Ferrandon, Jean. *Genie Civil* (1948) Vol. 125, 24.

Forchheimer, P. "Wasserbewegung durch Boden," *Zeitz, ver deutsch, Ing.* (1901) *45*, 1731.

Forchheimer, P. *Zeits, ver Deutsch, Ing.* (1901) Vol. 45, 1782.

Geertsma, J. "Estimating the Coefficient of Inertial Resistance in Fluid Flow Through Porous Media," *Soc. Petro. Eng. J.* (Oct. 1974) 445.

Gewers, C. W. W. and L. R. Nichol. "Gas Turbulence Factor in a Microvugular Carbonate," *J. Cdn. Petro. Tech.* (April–June 1969) *51*.

Green, L. and P. Duwez. "Fluid Flow Through Porous Metals," *J. Appl. Mech.* (1951) 18, 39.

Hartsock, J. H. and J. E. Warren. "The Effect of Horizontal Hydraulic Fracturing on Well Performance," *Trans. AIME* (1961) *222*, 1050.

Houpeurt, A. "An Analogue Study of the Transient Radial Flow of Gases in Porous Media," *Rev. Inst. Franc. Pet.* (April–June, 1953), *8*, p. 129.

Houpeurt, A. "Etude Analogigue de L'Ecoulement Radial Circulaire Transitoire des Gaz Dans les Milieux Poreu," *Revue de L'Institut Francais du Petrole et Annales des Combustibles Liquides* (1953), VIII, No. 4, 5, and 6.

Houpeurt, A. "On the Flow of Gases in Porous Media," *Rev. Inst. France, Pet.* (November 1959), *14*, 11, p. 1468.

Hubbert, M. King. "Darcy's Law and the Field Equations of the Flow of Underground Fluids," *Trans. AIME* (1956) 207, 222.

Hubbert, M. King. "The Theory of Ground-Water Motion," *J. of Geol.* (1940) 48, 785.

Iffly, R. "Etude de l'Ecoulement des Gaz dans les Milieux Poreu—Application a la Determination de la Morphologie des Roches," A la Faculte des Sciences de l'Universite de Paris, March 1956.

Janicek, J. D. and D. L. Katz. "Applications of Unsteady State Gas Flow Calculations," Preprint from Res. Conf., U. of Michigan, Ann Arbor, June 20, 1955.

Johnson, T. W. and D. B. Taliaferro. "Flow of Air and Natural Gas Through Porous Media," Technical Paper 592, USBM, 1938.

Jones, K. R., W. C. Pearson, and H. G. Riley. "Some Practical Aspects of Unsteady-State Gas Flow Related to Gas Well Performance," *J. Petro. Tech.* (Jan. 1963) p. 41.

Jones, L. G. "An Approximate Method for Computing Non-Steady-State Flow of Gases in Porous Media," *Soc. Pet. Eng. J.* (Dec. 1961) *1*, 4, 264.

Jones, L. G. and W. C. Skinner. "Analog Solutions to Unsteady State Radial Gas Flow Problems," Proc. Third Internat. Conf. on Analog Computation, Opatija, Yugoslavia, Sept. 1961.

Katz, M. L. and M. R. Tek. "A Theoretical Study of Pressure Distribution and Fluid Flux in Bounded, Stratified Porous Systems with Crossflow," *SPE Journal* (March 1962) *2*, No. 1, 68.

Kiel, O. G., Ph.D. Dissertation, University of Oklahama, 1963.

Kim, V. Yu. "On An Approximate Solution to the Problem of Non-Stationary Filtration of Liquids under Elastic Conditions," *Izv. Akad, Nauk, Otd. Tekh. Nauk* (1960) 2, p. 153.

Klinkenberg, L. J. "The Permeability of Porous Media to Liquids and Gases," *Drill. and Prod. Prac.*, API (1941) *230.*

Lefkovits, H. C., P. Hazebrook, E. E. Allen, and C. S. Matthews. "A Study of the Behavior of Bounded Reservoirs Composed of Stratified Layers," *Trans. AIME* (1961) *222*, II-43.

McCarthy, D. G. and E. C. Barfield. "The Use of High-Speed Computers for Predicting Flood-Out Patterns," *Trans. AIME* (1958) *213*, 139.

McConnell, A. J. *Applications of Tensor Analysis*, New York: Dover, 1957.

Muskat, M. *The Flow of Homogeneous Fluids through Porous Media*, Ann Arbor: J. W. Edwards, Inc., 1946.

Nisle, R. G. and F. H. Poettmann. "Calculation of the Flow and Storage of Natural Gas in Pipe" *Petro. Eng.* (1955) 27, D14, C36, D37.

Pendergrass, J. D. and V. J. Berry, Jr. "Pressure Transient Performance of a Multilayered Reservoir with Crossflow," *SPE Journal* (Dec. 1962) *2*, No. 4.

Poettmann, F. H. and R. E. Schilson. "Calculation of the Stabilized Performance Coefficient of Low Permeability Natural Gas Wells," *Trans. AIME* (1959) Vol. 216, 240.

Polubarinova-Kochina, P. Ya. *Theory of Ground Water Movement*, Princeton University Press, 1962.

Rawlins, E. L. and M. A. Schellhardt. *Back-Pressure Data on Natural Gas Wells and Their Application to Production Practices*, U.S. Bureau of Mines, Monograph 7, 1936.

Richtmyer, R. D. *Difference Methods for Initial Value Problems*, New York: Interscience Publishers, Inc., 1957, p. 103.

Rowan, G. and M. W. Clegg. "An Approximate Method for Transient Radial Flow," *Soc. Petro. Eng. J.* (Sept. 1962) 2, 3, 225.

Russell, D. G. and M. Pratts. "Performance of Layered Reservoirs with Crossflow-Single-Compressible-Fluid Case," *SPE Journal* (March 1962) *2*, No. 1, 53.

Scheidegger, A. E. *Handbuch der Physik* VIII/2, edited by S. Flugge, Springer-Verlag, Berlin, 1963.

Schellhardt, M. A. "Application of the Back-Pressure Method for Determining Absolute Open Flows of Large Gas Wells," *U.S. Bureau of Mines R.I.*, 3767 (August 1944).

Schneebeli, G. "Sur la Theorie des Ecoulements de Filtration," *La Houille Blanche* (1953) 186, No. Special A.

Sheldon, J. W., C. D. Harris, and D. Bavly. "A Method for General Reservoir Behavior Simulation on Digital Computers," SPE Paper number 152, presented at the Fall Meeting of AIME, 1960.

Shilov, G. E. *Theory of Linear Spaces*, Prentice-Hall, 1961.

Slattery, J. C. *J. Colloid Sci.* (1961) Vol. 16, 431.

Smith, R. V. "Unsteady-State Gas Flow into Gas Wells," Paper 1565-G, Presented before the 35th Annual Fall Meeting of SPE.

Swift, G. W. and O. G. Kiel. "The Prediction of Gas Well Performance Including the Effect of Non-Darcy Flow," *J. Petro. Tech.* (July 1962) p. 791.

Tek, M. R. "Development of Generalized Darcy Equation," *J. Petro. Tech.* (June, 1957) Vol. 9, 45.

Tek, M. R. Private Communication 102-50-54R.

Tek, M. R., K. H. Coats, and D. L. Katz. "The Effect of Turbulence on the Flow of Natural Gas through Porous Reservoirs," *J. Petro. Tech.* (July 1962) p. 799.

Tek, M. R., M. L. Grove, and F. H. Poettmann. "Method for Predicting the Back Pressure Behavior of Low Permeability Natural Gas Wells," *Trans. AIME* (1957) 210, 302.

Vacher, J. P. and V. Cazabat. "Ecoulement des Fluides Dans Les Milieux Avec Communication," *Revue de L'Institut Francais du Petrole* (Oct. 1961) *XVI*, No. 10, 1147.

Van Everdingen, A. F. "The Skin Effect and Its Influence on the Productive Capacity of a Well," *Trans. AIME* (1953) *198*, 171.

Van Everdingen, A. F. and W. Hurst. "The Application of the La Place Transformation to Flow Problems in Reservoirs," *Trans. AIME* (1949) Vol. 186, 305.

Warren, J. E. and H. S. Price. "Flow in Heterogeneous Porous Media," *Trans. AIME* (1961) *222*, II-153.

Wong, S. W. "Effect of Liquid Saturation on Turbulence Factors for Gas-Liquid Systems," *J. Cdn. Petro. Tech.* (Oct.–Dec. 1970) *274*.

Wright, D. E. "Non-Linear Flow Through Granular Media," *J. Hydraulic Div.*, Proc. ASCE (1968) *94*, 851.

Yih, C. S. *Fluid Mechanics: A Concise Introduction to the Theory*, McGraw-Hill Book Co., Inc., 1969.

INDEX

Q–R

S

T

U–Z